Biologically-Inspired Systems

Volume 9

Series editor

Stanislav N. Gorb, Department of Functional Morphology and Biomechanics, Zoological Institute, Kiel University, Kiel, Germany

Motto: Structure and function of biological systems as inspiration for technical developments.
Throughout evolution, nature has constantly been called upon to act as an engineer in solving technical problems. Organisms have evolved an immense variety of shapes and structures from macro down to the nanoscale. Zoologists and botanists have collected a huge amount of information about the structure and functions of biological materials and systems. This information can be also utilized to mimic biological solutions in further technical developments. The most important feature of the evolution of biological systems is multiple origins of similar solutions in different lineages of living organisms. These examples should be the best candidates for biomimetics. This book series will deal with topics related to structure and function in biological systems and show how knowledge from biology can be used for technical developments in engineering and materials science. It is intended to accelerate interdisciplinary research on biological functional systems and to promote technical developments. Documenting of the advances in the field will be important for fellow scientists, students, public officials, and for the public in general. Each of the books in this series is expected to provide a comprehensive, authoritative synthesis of the topic.

More information about this series at http://www.springer.com/series/8430

Lars Heepe · Longjian Xue
Stanislav N. Gorb
Editors

Bio-inspired Structured Adhesives

Biological Prototypes, Fabrication, Tribological Properties, Contact Mechanics, and Novel Concepts

Springer

Editors
Lars Heepe
Department of Functional Morphology and Biomechanics, Zoological Institute
Kiel University
Kiel
Germany

Stanislav N. Gorb
Department of Functional Morphology and Biomechanics, Zoological Institute
Kiel University
Kiel
Germany

Longjian Xue
School of Power and Mechanical Engineering
Wuhan University
Wuhan
China

ISSN 2211-0593 ISSN 2211-0607 (electronic)
Biologically-Inspired Systems
ISBN 978-3-319-59113-1 ISBN 978-3-319-59114-8 (eBook)
DOI 10.1007/978-3-319-59114-8

Library of Congress Control Number: 2017943120

Printed on acid-free paper

This Springer imprint is published by Springer Nature
The registered company is Springer International Publishing AG
The registered company address is: Gewerbestrasse 11, 6330 Cham, Switzerland

This book is dedicated to the memory of Dr. Uwe Hiller (Münster University, Germany), the pioneer of ultrastructural and experimental works on gecko adhesion

Preface

Performance of Biological and Bio-inspired Attachment Systems in Real Environment

The field of biological and bio-inspired adhesive systems underwent enormous developments during the past 15–20 years. Starting from curiosity-driven science with questions like "Why does the gecko or the fly not fall from the wall?", it developed into one of the most exciting interdisciplinary fields of research, where biologists, chemists, materials scientists, even theoretical physicists, and engineers meet together to unravel natures' mystery of temporary, strong, and reliable adhesion. Although enormous efforts, to understand attachment principles of biological prototypes, such as geckos, spiders, flies, and beetles were recently undertaken, until now, only very few biomimetic materials, inspired by the feet of these animals, are commercialized.

What could be the reasons for the shortage of such bio-inspired products? One potential reason could be that we are still limited in industrial processes capable of mass production of structures with such extreme complexity and small feature size. Another reason could be that we still have not properly understood functioning of biological prototypes. The latter reason might be due to the fact that we lack information about the way *how the animals use their attachment systems in real environment.* Also biologically inspired artificial attachment devices are usually tested on smooth flat substrates under laboratory conditions. This approach is perfectly fine as a first step of the biomimetic technical development, but not lead to the development of real-world applications.

Additionally, if we want to fix things quasi-permanently on the wall, we should not necessarily consider gecko-inspired materials. Geckos and Co are fast-running animals with an attachment system optimized for short-term and highly dynamic adhesion. Therefore, these attachment systems mainly rely on elaborate nanoscopic thin film-like spatula-shaped contact elements. Instead, we should search for other specialized biological prototypes, for example, organisms, which utilize long-term adhesion. Their strong and, most important, passive attachment systems mainly rely either on glue, a mushroom-shaped contact geometry, or on the combination of both.

Chapter 1 of this book is devoted to a meta-analysis of the available literature on biologically inspired adhesives. Niewiarowski et al. applied for the first time a bibliometric analysis of gecko adhesion to gain more insight from the past research for the future directions and transfer to technical applications. Using analysis of citations, co-citations, and patents, they confirmed that the theme 'gecko adhesion' in the broader context of 'biomimetic adhesion' is the most influential one. Moreover, the authors noticed that some capabilities of geckos (and other climbing animals), such as adhesion to wet and/or dirty surfaces, do not show up as major themes in the context of bio-inspired adhesives. The authors speculate that this may potentially be due to a lack of research focus on such environmental effects. The next three chapters by Heepe et al. and Gorb et al. present such attempts to understand environmental effects on the attachment ability in one of the biological model adhesive systems: the seven-spotted ladybird beetle *Coccinella septempunctata*.

In Chap. 2, the authors investigated the effect of ambient humidity on the beetle attachment ability. Surprisingly, beetles, which have a so-called wet hairy adhesive system, showed a similar dependency on relative humidity as spiders and geckos which exhibit a so-called dry hairy adhesive system. The obtained results indicate an optimal range of relative humidity with a maximum in attachment ability. At very low and very high humidity, the beetle attachment ability is considerably lower. Potential mechanisms, responsible for such nontrivial humidity dependence, are discussed.

In Chap. 3, the question "What is the dominant factor of the surface effecting the attachment ability of beetles: chemistry or roughness?" was asked. To answer that question, the authors systematically measured insect traction forces on various substrates with similar surface roughness but different surface chemistry and on substrates with similar surface chemistry but different surface roughness. Although surface chemistry indeed affected the attachment ability, the surface roughness was shown to be the dominant factor determining the beetle attachment ability.

In Chap. 4, the influence of the substrate stiffness on the attachment ability of female and male ladybird beetles was investigated. Natural habitats of climbing animals may provide a variety of substrate stiffness ranging from rigid rock surfaces to soft, biofilm covered ones. The authors used a custom-build centrifugal tester, to measure the attachment ability of beetles on smooth silicone elastomer substrates with stiffness ranging from about 0.3 MPa to about 2.2 MPa. Whereas in females, the attachment ability was not affected by the substrate stiffness within the range of tested stiffness, males showed decreasing attachment ability with decreasing substrate stiffness. This sexual dimorphism in attachment ability is explained by the contacts mechanics of a specialized, discoidal seta type in males, which is not present in females. In general, the overall level of attachment ability was reduced if compared to results obtained on rigid substrates.

In Chap. 5, structural effects on the adhesive properties of another biological prototype, spider silk anchorages, are thoroughly discussed and reviewed by Jonas

Wolff. Spider webs hang between, for example, tree trunks and plants. For the purpose of web attachment to such substrates, spiders produce a specific hierarchical structure called silk anchors, which are made of a two-compound secretion. Due to their hierarchical organization, discontinuous contact area, and material composition with embedded compliant fibers, these anchors allow secure fixation to vegetation, which is often characterized by anti-adhesive and anti-wetting properties. That is why, a proper understanding of the functional principles of silk anchors may potentially lead to novel strategies in bonding technology.

The hierarchical organization of biological attachment systems, both in climbing animals and silk anchors, is taken up as a general principle, in Chap. 6, to explain optimization strategies observed in these biological systems. Brely et al. perform, based on a multiple peeling theory, a computational study of different hierarchical and multiple peeling configurations including contact tapering, gradually changing material properties, and their combination. Further, a first detailed experimental proof of the multiple peeling theory is reported in Chap. 7. Using elastic adhesive tapes in the so-called double-peeling configuration, Heepe et al. found a very good agreement of experimentally obtained results with predictions of the multiple peeling theory.

Another general concept of adhesion in biological attachment systems is proposed in Chap. 8. Tan et al. discuss the success of biological adhesive systems considering the effective stiffness of their adhesive organs. Two successful examples in biological attachment systems with sufficiently low effective stiffness, responsible for making intimate contact with a substrate, are reviewed: (1) the hierarchical organization of gecko adhesive setae and (2) material gradients in insect adhesive setae. Based on these biological observations, current biomimetic structured adhesive materials are analyzed and improvements for next generation bio-inspired structured adhesives are proposed.

In Chap. 9, various aspects of adhesion in hairy attachment systems, such as the role of surface roughness, optimal elasticity, shear activation, hierarchical architecture including 3D configuration (see also Chaps. 5 and 6), and material gradients, are reviewed by Filippov et al. using rather simple, but effective numerical models of different parts of hairy attachment systems.

Chapter 10 by Borodich and Savencu provides a review of various models of engineering rough surfaces, including multilevel models, hierarchically structured models, and appropriate multi-scale models of contact interactions between rough surfaces with an emphasis on the historical development and evolution of such models. Moreover, a self-consistent terminology of rough surface models is introduced and further applied to bio-inspired structured adhesives.

A brief historical background of the development of bio-inspired structured adhesives is given in Chap. 11. Dan Sameoto highlights the development in contact geometry of individual adhesive elements and reviews milestones of both structured adhesives and manufacturing approaches. A detailed discussion and summary of current manufacturing possibilities in combination with advantages and disadvantages of different classes of materials with respect to the upscaling of production of such adhesives is given. Based on this information, potential future routes for commercialization are presented.

An analysis of the development of bio-inspired structured adhesives, as done in Chap. 11, reveals that currently the majority of such adhesives rely on a mushroom-shaped contact geometry of individual elements. It was shown that this tip shape provides highest normal adhesion if it is compared to other kind of geometries. A physical explanation for the enhanced adhesion of this particular geometry is reviewed in Chap. 12. Carbone et al. review their latest theoretical and experimental results on the contact formation, interfacial stress distribution, pull-off force, tilt angle dependence, and detachment dynamics of individual mushroom-shaped adhesive microstructures (MSAMSs). For a simplified T-shaped variant of MSAMSs, optimal design parameters for maximal pull-off forces are predicted. The optimal design of MSAMSs is also the topic of Chap. 13. According to a previous, alternative contact mechanical model of MSAMSs, based on a Dugdale-Barenblatt cohesive zone model, Gorumlu et al. present experimental results of the adhesion performance of MSAMSs with well-defined tip geometries of individual adhesive elements, including variations in tip wedge angle. For a comparison, the simple T-shaped variant of MSAMSs (see Chap. 12) has a tip wedge angle of 90°. The authors demonstrated good agreement between their obtained experimental results and their recent theoretical model.

The concept of fibrillar adhesive elements is extended in Chap. 14 to the concept of shape complementarity. It has been employed to achieve high and reversible adhesion between an elastomer and a soft gel. Paretkar et al. investigated two configurations of elastomer–gel interfaces: (1) a ridge channel and (2) a fibril-hole configuration. For the ridge-channel structure, reversible adhesion was observed, i.e., a re-closure was possible. For the fibril-hole structure, only a strong first-time adhesion was observed. Adhesion enhancement, if compared to a smooth control interface, was shown to mainly rely on frictional sliding during extraction of the fibril or ridge from its complementary hole or channel. Such shape complementarity concept for robust bonding of very soft materials may be of interest in medical applications, involving, e.g., skin adhesion.

So far, the main focus of this book was on the adhesion control by structural effects rather than by surface chemistry. In the Chap. 15, Chinh Ngo et al. focus on the functionalization of silicone with respect to bio-adhesion. Usually, silicone surfaces are functionalized by plasma treatments and/or grafting. Here, the authors present an alternative approach by the incorporation of block copolymers. Combining advanced atomic force microscopic techniques and bio-adhesion experiments with mussels with two model systems, the authors showed that these organisms attach preferably to block copolymer-filled coatings after immersion in water due concomitant molecular reorganization at the top surface of the copolymer-filled coatings. These observations provided evidence for the significant role played by the selected amphiphilic block copolymers to promote bio-adhesion on surface-treated silicone coatings.

In summary, this book presents an overview of current research activities on performance of biological and bio-inspired attachment systems in real environment.

Different chapters provide experimental and theoretical novices in the field of biomimetics of adhesive systems. The volume is intended for use by researchers who are active, or plan to become active, in the field. The appeal of this topic is expected to be broad, ranging from biology and biomechanics to tribology and surface engineering.

As editors, we would like to thank all the authors for contributing their beautiful work to this book. We are grateful to all referees for their promptly provided reports, which facilitated the high quality of individual manuscripts and finally the high quality of the complete work. Finally, we thank participants of the COST Action Ca15216 for providing their contribution and COST for the support of this book project.

Kiel, Germany Lars Heepe
Wuhan, China Longjian Xue
Kiel, Germany Stanislav N. Gorb

Contents

1 **A Bibliometric Analysis of Gecko Adhesion: A View of Its Origins and Current Directions** 1
Peter H. Niewiarowski, Alyssa Y. Stark and Ali Dhinojwala

2 **Impact of Ambient Humidity on Traction Forces in Ladybird Beetles (*Coccinella septempunctata*)** 21
Lars Heepe, Jonas O. Wolff and Stanislav N. Gorb

3 **Roughness Versus Chemistry: Effect of Different Surface Properties on Insect Adhesion** 33
Elena V. Gorb, Matt W. England, Tomoya Sato, Makoto Yagihashi, Atsushi Hozumi and Stanislav N. Gorb

4 **Effect of Substrate Stiffness on the Attachment Ability in Ladybird Beetles *Coccinella septempunctata*** 47
Lars Heepe, Dennis S. Petersen, Lisa Tölle, Jonas O. Wolff and Stanislav N. Gorb

5 **Structural Effects of Glue Application in Spiders—What Can We Learn from Silk Anchors?** 63
Jonas O. Wolff

6 **Optimal Adhesion Control via Cooperative Hierarchy, Grading, Geometries and Non-linearity of Anchorages and Adhesive Pads** 81
Lucas Brely, Daniele Liprandi, Federico Bosia and Nicola M. Pugno

7 **Double Peeling Mechanism Inspired by Biological Adhesive Systems: An Experimental Study** 95
Lars Heepe, Saverio Raguseo and Stanislav N. Gorb

8 **The Role of Effective Elastic Modulus in the Performance of Structured Adhesives** 107
Di Tan, Yingjie Zheng and Longjian Xue

9 Biological Microstructures with Enhanced Adhesion and Friction: A Numerical Approach 141
Valentin L. Popov, Alexander E. Filippov and Stanislav N. Gorb

10 Hierarchical Models of Engineering Rough Surfaces and Bio-inspired Adhesives 179
Feodor M. Borodich and Ovidiu Savencu

11 Manufacturing Approaches and Applications for Bioinspired Dry Adhesives 221
Dan Sameoto

12 Contact Mechanics of Mushroom-Shaped Adhesive Structures 245
Giuseppe Carbone and Elena Pierro

13 Bioinspired Mushroom-Like Fiber Adhesives 277
Serdar Gorumlu and Burak Aksak

14 Adhesion Enhancement of a Gel-Elastomer Interface by Shape Complementarity 291
Dadhichi Paretkar, Pooja Malhotra, C.-Y. Hui and Anand Jagota

15 On the Bioadhesive Properties of Silicone-Based Coatings by Incorporation of Block Copolymers 303
Thị Chinh Ngo, Radostina Kalinova, Rosica Mincheva, Audrey Beaussart, Elise Hennebert, Patrick Flammang, Yves Dufrêne, Philippe Dubois, Roberto Lazzaroni and Philippe Leclère

Index 345

Contributors

Burak Aksak Department of Mechanical Engineering, Texas Tech University, Lubbock, USA

Audrey Beaussart Biochemistry, Biophysics and Genetics of Microorganisms (BBGM), Université catholique de Louvain (UCL), Louvain-la-Neuve, Belgium; Laboratoire Interdisciplinaire des Environments Continentaux, UMR 7360 CNRS—Université de Lorraine, Vandoeuvre-les-Nancy, France

Feodor M. Borodich Cardiff School of Engineering, Cardiff University, Cardiff, Wales, UK

Federico Bosia Department of Physics and "Nanostructured Interfaces and Surfaces" Centre, Università di Torino, Turin, Italy

Lucas Brely Department of Physics and "Nanostructured Interfaces and Surfaces" Centre, Università di Torino, Turin, Italy

Giuseppe Carbone TriboLab, Department of Mechanical and Management Engineering, Politecnico di Bari, Bari, Italy

Ali Dhinojwala Department of Polymer Science, University of Akron, Akron, OH, USA

Philippe Dubois Laboratory of Polymeric and Composite Materials, Center of Innovation and Research in Materials and Polymers (CIRMAP), Research, Institute for Materials Science and Engineering University of Mons (UMONS), Mons, Belgium

Yves Dufrêne Biochemistry, Biophysics and Genetics of Microorganisms (BBGM), Université catholique de Louvain (UCL), Louvain-la-Neuve, Belgium

Matt W. England National Institute of Advanced Industrial Science and Technology (AIST), Nagoya, Japan

Alexander E. Filippov Donetsk Institute for Physics and Engineering, National Academy of Science, Donetsk, Ukraine

Patrick Flammang Laboratoire de Biologie Marine, Research Institute for Biosciences, University of Mons (UMONS), Mons, Belgium

Elena V. Gorb Department of Functional Morphology and Biomechanics, Kiel University, Kiel, Germany

Stanislav N. Gorb Department of Functional Morphology and Biomechanics, Zoological Institute, Kiel University, Kiel, Germany

Serdar Gorumlu Department of Mechanical Engineering, Texas Tech University, Lubbock, USA

Elise Hennebert Laboratoire de Biologie Marine, Research Institute for Biosciences, University of Mons (UMONS), Mons, Belgium

Lars Heepe Department of Functional Morphology and Biomechanics, Zoological Institute, Kiel University, Kiel, Germany; Mads Clausen Institute, University of Southern Denmark, Sønderborg, Denmark

Atsushi Hozumi National Institute of Advanced Industrial Science and Technology (AIST), Nagoya, Japan

C.-Y. Hui Department of Mechanical and Aerospace Engineering, Cornell University, Ithaca, USA

Anand Jagota Department of Chemical and Biomolecular Engineering, Lehigh University, Bethlehem, USA

Radostina Kalinova Laboratory of Polymeric and Composite Materials, Center of Innovation and Research in Materials and Polymers (CIRMAP), Research, Institute for Materials Science and Engineering University of Mons (UMONS), Mons, Belgium; Institute of Polymers, Bulgarian Academy of Science, Sofia, Bulgaria

Roberto Lazzaroni Laboratory for Chemistry of Novel Materials, Center of Innovation and Research in Materials and Polymers (CIRMAP), Research, Institute for Materials Science and Engineering, University of Mons (UMONS), Mons, Belgium

Philippe Leclère Laboratory for Chemistry of Novel Materials, Center of Innovation and Research in Materials and Polymers (CIRMAP), Research, Institute for Materials Science and Engineering, University of Mons (UMONS), Mons, Belgium

Daniele Liprandi Department of Physics and "Nanostructured Interfaces and Surfaces" Centre, Università di Torino, Turin, Italy

Pooja Malhotra Department of Chemical and Biomolecular Engineering, Lehigh University, Bethlehem, USA

Rosica Mincheva Laboratory of Polymeric and Composite Materials, Center of Innovation and Research in Materials and Polymers (CIRMAP), Research, Institute for Materials Science and Engineering University of Mons (UMONS), Mons, Belgium

Thị Chinh Ngo Laboratory for Chemistry of Novel Materials, Center of Innovation and Research in Materials and Polymers (CIRMAP), Research, Institute for Materials Science and Engineering, University of Mons (UMONS), Mons, Belgium; Institute of Research and Development, Duy Tan University, Da Nang, Vietnam

Peter H. Niewiarowski Department of Biology, Program in Integrated Bioscience, University of Akron, Akron, OH, USA

Dadhichi Paretkar Department of Chemical and Biomolecular Engineering, Lehigh University, Bethlehem, USA

Dennis S. Petersen Department of Functional Morphology and Biomechanics, Zoological Institute, Kiel University, Kiel, Germany

Elena Pierro Scuola di Ingegneria, Università degli Studi della Basilicata, Potenza, Italy

Valentin L. Popov Institut für Mechanik, FG Systemdynamik und Reibungsphysik, Technische Universität Berlin, Berlin, Germany

Nicola M. Pugno Laboratory of Bio-Inspired and Graphene Nanomechanics, Department of Civil, Environmental and Mechanical Engineering, Università di Trento, Trento, Italy; School of Engineering and Materials Science, Queen Mary University of London, London, UK; Ket Lab, Edoardo Amaldi Foundation, Italian Space Agency, Rome, Italy

Saverio Raguseo Department of Functional Morphology and Biomechanics, Zoological Institute, Kiel University, Kiel, Germany

Dan Sameoto Department of Mechanical Engineering, University of Alberta, Edmonton, Canada

Tomoya Sato National Institute of Advanced Industrial Science and Technology (AIST), Nagoya, Japan

Ovidiu Savencu Cardiff School of Engineering, Cardiff University, Cardiff, Wales, UK; Austrian Center of Competence for Tribology, Wiener Neustadt, Austria

Alyssa Y. Stark Department of Biology, University of Louisville, Louisville, KY, USA

Di Tan School of Power and Mechanical Engineering, Wuhan University, Wuhan, China

Lisa Tölle Department of Functional Morphology and Biomechanics, Zoological Institute, Kiel University, Kiel, Germany

Jonas O. Wolff Department of Functional Morphology and Biomechanics, Zoological Institute, Kiel University, Kiel, Germany; Department of Biological Sciences, Macquarie University, Sydney, NSW, Australia

Longjian Xue School of Power and Mechanical Engineering, Wuhan University, Wuhan, China

Makoto Yagihashi Nagoya Municipal Industrial Research Institute, Nagoya, Japan

Yingjie Zheng School of Power and Mechanical Engineering, Wuhan University, Wuhan, China

Chapter 1
A Bibliometric Analysis of Gecko Adhesion: A View of Its Origins and Current Directions

Peter H. Niewiarowski, Alyssa Y. Stark and Ali Dhinojwala

Abstract Although there has been long-standing interest in understanding how geckos achieve robust and repeatable adhesion across a wide range of surfaces and under many conditions, the underlying mechanisms have only come to light over the last 15 years. Biological interpretation of the significance of the system, as well as the potential for developing synthetic mimics, both rely on a more complete analysis and interpretation than currently exists of the gecko adhesion model outside of ideal circumstances. In this contribution, we explore the basic research literature on gecko adhesion and some related foundational work in the patent literature in an effort to provide a qualitative description of the major themes of basic research and the potential applications captured in associated patents. Our main analytical approach borrows from bibliometric methods and tools that are designed to reveal idea sources and development as dynamic networks among keywords, authors, and journals captured in co-citation data. Although our co-citation dataset includes over 1500 studies, themes emerging from the analysis suggest that gecko research efforts are distributed on many different fronts with unclear or weak linkage to applications as reflected in the patent literature. Our anecdotal impression that studies focusing on natural materials in the gecko adhesive system that are less well represented than those centered on other themes (e.g., structure, geometry, mechanics) is consistent with the bibliometric patterns revealed here.

P.H. Niewiarowski (✉)
Department of Biology, Program in Integrated Bioscience, University of Akron, Akron, OH 44325, USA
e-mail: phn@uakron.edu

A.Y. Stark
Department of Biology, University of Louisville, Louisville, KY 40290, USA
e-mail: alyssaystark@gmail.com

A. Dhinojwala
Department of Polymer Science, University of Akron, Akron, OH 44325, USA
e-mail: ali4@uakron.edu

L. Heepe et al. (eds.), *Bio-inspired Structured Adhesives*,
Biologically-Inspired Systems 9, DOI 10.1007/978-3-319-59114-8_1

1.1 Introduction

While interest in how geckos stick to surfaces has existed for millennia, it is easy to forget that our understanding of how their adhesive system works has only emerged over the last 15 years [1, 2]. Studies of form and function of the system, from the nanoscale to the whole organism, including fundamental surface science, ecological and evolutionary significance, and biomimetic applications, now number in the thousands. Because the gecko adhesive system has served as a powerful platform for interdisciplinary work, opportunities for and obstacles to further progress likely exist due to gaps in translation of information from field to field. At least two anecdotal observations support this observation. First, there does not currently exist a biomimetic version of the gecko adhesive system that captures all the performance capabilities of the natural system, and second, the hypothesis that the gecko adhesive system is a key innovation is still largely untested, though broadly accepted and commonly repeated. We ask whether a look at the scientific studies making up the gecko adhesion literature provides an opportunity to identify sources, directions and organization of the research that can aid in using trends and relationships within and among the various research themes to advance the field and fill obvious gaps as it continues to grow. In this contribution, we undertake such an analysis, using the tools of bibliometric methods [3] to describe such characteristics of the gecko adhesion literature. Our work is not meant to provide quantitative conclusions or specific recommendations, but rather to review and characterize the current state of this very diverse and rapidly expanding literature.

Bibliometric analysis uses citation, co-citation and related data from sources such as the Web of Knowledge™ to identify patterns of relationships among published studies. Emphasis is placed on visualization of the foci of research activity, history and trends within research foci, and connections and inter-relationships among foci. Such analyses can aid in the tracing of sources and consequences of intellectual threads comprising any research topic for which there is sufficient data. Bibliometric analyses can also provide insight into potential and emerging areas of research that can influence research opportunities and strategy. To our knowledge, bibliometric analysis has yet to be applied to the gecko adhesion literature.

In this paper we focus on a basic bibliometric analysis of gecko adhesion using two applications, CiteSpace III [3] and VOS [4]. Our approach is qualitative rather than quantitative. We seek to reveal trends rather than test hypotheses. However, bibliometric tools such as CiteSpace and VOS are based on theory and metrics that support robust and repeatable visualizations of the patterns embedded in the literature of interest [3, 5]. Both of the applications we chose are based on explicit algorithms that construct networks of studies based on citation, co-citation and other relationships. Both applications also use specific algorithms to construct clusters, which can be defined by a variety of criteria such as key words, journal of publication, country of origin, etc. The algorithms used by CiteSpace III are especially designed around theory of how citation relationships in research follow patterns of knowledge growth and development in fields of interest. For example, new research fields often build from a core set of studies ('intellectual base') that define a foundation of work (group of co-cited papers) actively cited by subsequent studies representing the advancing 'front' of knowledge ('research front') [6]. Intellectual bases and research fronts can have different properties and are identified in different ways through networks of shared keywords, shared citation and co-citation sets of bibliographic information [7]. We present and explore some of these patterns in the field of gecko adhesion research. We are especially interested in gaining insight into the following questions: (1) Where does gecko adhesion research fit within the broader context of bio-inspired adhesion research? (2) What are the major themes of gecko adhesion research and how are they related historically and with respect to their future progress? (3) What are the advancing fronts of research activity and are these related to translational applications of gecko adhesion as revealed in existing patents? All of these are exploratory rather than formal questions, suited to taking a first step in characterizing the literature. This contribution should also help determine if deeper, more focused analysis is warranted.

1.2 Materials and Methods

CiteSpace III and VOS accept multiple forms of input data, including citation lists like those generated by Web of Science™ (WOS). The main source of data for our analyses was two searches using the "Core Collection" database of WOS. The first data set ('bioinspired adhesion') resulted from a search using the keyword search term "(bio-inspired or biomimetic) and (stick or adhesive)" between the years 1994 and 2016. It returned 961 references (Fig. 1.1). Subsequently, we focused a second more specific search to return data ('gecko fibrillar adhesion') from the keyword search term "(fibrillar adhesion) or (gecko adhesion)," between 1994 and 2016. It returned 1669 records (Fig. 1.1).

CiteSpace III was used with its default settings (cited reference search, time slices of 1 year duration, network clustering based on terms in title, abstract, and keywords) to analyze the two datasets above. We also selected the option to prune the network with the pathfinder algorithm. Following a qualitative examination of the results, we used VOS for further insight into the structure of the CiteSpace III derived citation networks. Specifically, VOS helped us identify co-citation sub-networks associated with the 5 highest centrality score nodes from the CiteSpace III run (see Table 1.1). Centrality is a measure of how strongly a reference connects references belonging to 2 or more clusters. We also used VOS in a similar way to examine structure of citation networks associated with studies that had high 'burstness' in the CiteSpace III network. Burstness is a measure of rate of production of new references citing a focal reference across time slices.

Finally, we explored the relationships between gecko fibrillar adhesion research with patent activity in order to visualize how fundamental research is related to the development of gecko-inspired adhesive applications. The tools used to search

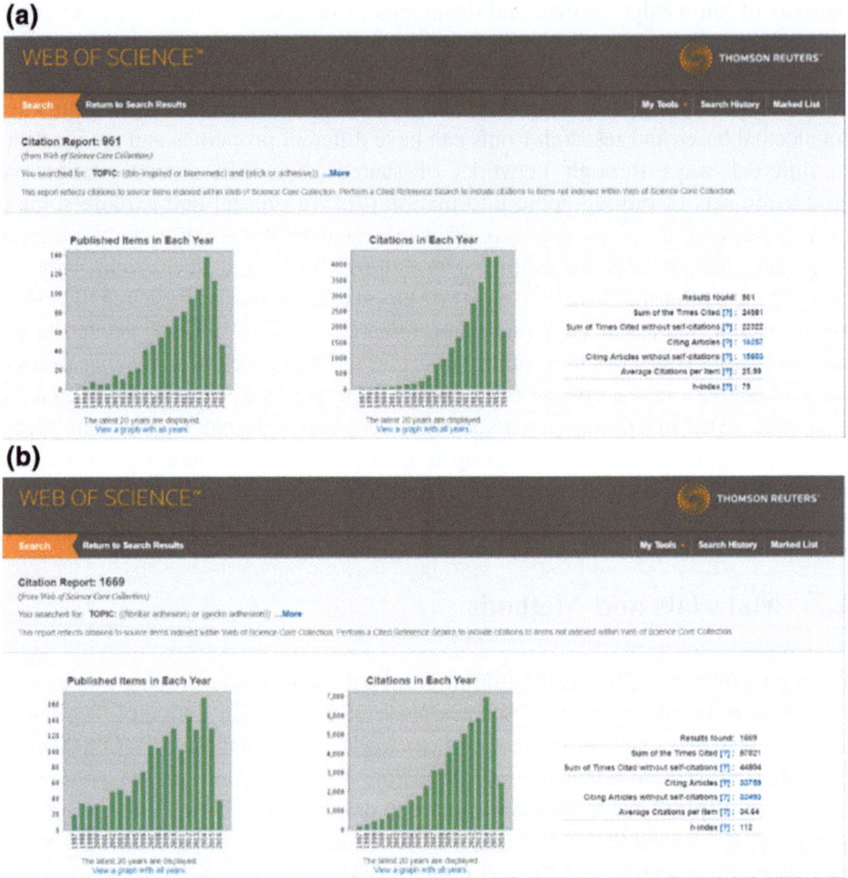

Fig. 1.1 Web of Science Core Collection data input search sources. **a** "(bio-inspired or biomimetic) and (stick or adhesive)". **b** "(fibrillaradhesion) or (gecko adhesion)"

bibliographic databases are not easily applied to patent databases so we used a two-step approach. First, using a Google patent search we identified two key initial patents that have largely launched the existing patent literature related to gecko inspired adhesives. We created landscape maps showing the patent citation networks of these two foundational patents using IPvision®. IPvision® constructs interconnection maps that are analogous to the citation maps produced by WOS when a citation analysis is requested. Subsequently, we created two datasets comprised of the non-patent citations found in these two foundational patents, and constructed bibliographic networks using VOS as described above. A combination of these two steps helps us identify connections among existing patents (granted and applications) as well as map trends in the development of the patents as related to the research literature the patents cite.

Table 1.1 Network element statistics of the top 5 references for burstness and top 5 references for centrality from the 'gecko fibrillar adhesion network'

Author, source	Year	Cluster	Sigma	Burstness	Centrality	Rank	Co-citations
Autumn et al. [8]	2000	3	240.65	**67.35**	*0.08*	1	180
Autumn et al. [14]	2002	3	1.99	**43.68**	*0.02*	2	220
Geim et al. [15]	2003	3	7.15	**26.67**	*0.08*	3	205
Arzt et al. [16]	2003	3	2.17	**22.53**	*0.04*	4	168
Sitti et al. [17]	2003	3	1.09	**22.14**	*0*	5	144
Hansen and Autumn [11]	2005	5	1	*0*	**0.12**	1	118
Lee et al. [9]	2008	0	1	*0*	**0.11**	2	70
Persson and Gorb [12]	2003	3	3.76	*13.43*	**0.1**	3	84
Jagota and Bennison [13]	2002	3	2.61	*10.21*	**0.1**	4	51
Glassmaker et al. [10]	2004	1	2.03	*7.38*	**0.1**	5	123

1.3 Results

CiteSpace III found 120 clusters among 961 references in the 'bioinspired adhesion' dataset. With the exception of 3 clusters, most clusters were small and relatively unstructured, including references that had few individual citations and low scores for centrality and burstness. Clusters 0 and 1 contained 170 references (92 and 78 respectively), had centrality scores more than twice as high as the next highest clusters (0.11 vs. 0.05) and were comprised of references that collectively accounted for a large fraction of all the co-citations in the overall network (2446 and 1363, respectively out of 5454 total). Analysis of the frequency of words used in the titles of references in Clusters 0 and 1 and the content of the studies, identified these groups of references as co-citation clusters associated with adhesion research on reversible gecko adhesion (cluster 0) and non-reversible underwater adhesion observed in biological models such as mussels (cluster 1).

Our second search, specific to adhesion as it relates to geckos, returned a larger number of references (1669) with 80 clusters identified (Fig. 1.2). Clusters 0, 1, 3, 4, 5, and 7 were comprised of 166 papers (42, 34, 31, 23, 22, and 14 respectively) totaling 8205 (86%) of a total 9538 co-citations in the network (2189, 1091, 2542, 1016, 1131, 236 respectively). The top 5 papers for centrality were distributed in 4 different clusters (0, 1, 3, and 5). Note the sigma score, a measure of innovation potential, for the Autumn et al. [8] paper (the original paper establishing that setae are the key adhesive structure in geckos and that they are directionally dependent). The top 5 papers for burstness all come from cluster 3, accounting for 917 papers in the overall network (Table 1.1) with the mean publication year of 2004, and characterized by papers including keywords such as *gecko*, *adhesive*, *friction*, *model*, and *surface* (Table 1.2).

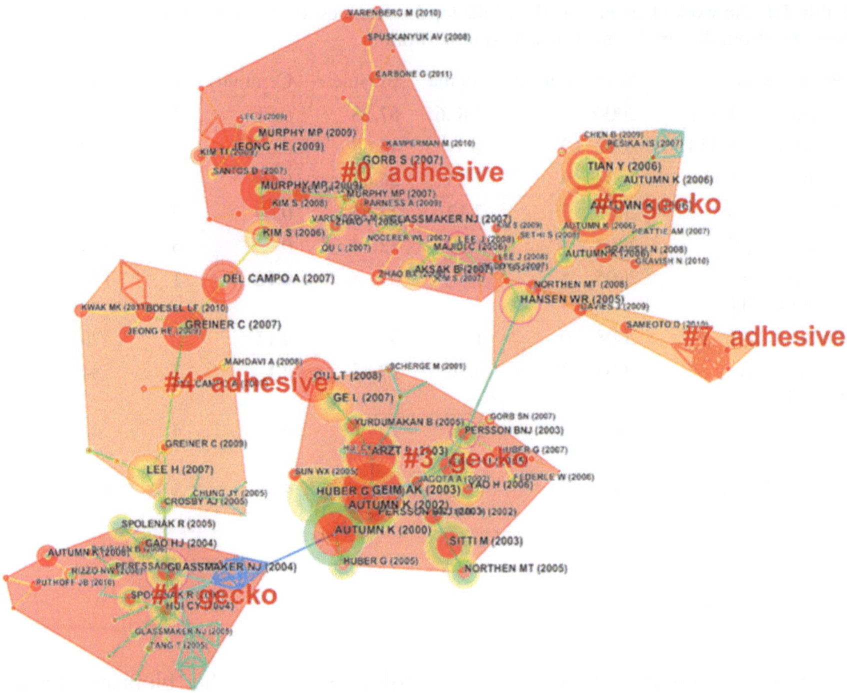

Fig. 1.2 In this figure and also Figs. 1.3, 1.4, 1.5, 1.6, 1.7 and 1.8, the network and heatmap views in VOS allow visualization of the dominant themes and their connections. Size of the node in the network view is proportional to the number of times the key word appeared in co-cited references. *Lines* connect nodes representing the top 100 pairs of shared terms in co-cited references. The heat map view reflects the density of connections reaching a node as well as the relative number of times the word appears in the title/abstract among co-cited references. *Color* of nodes and *lines* reflect cluster membership for the keywords among co-cited references. Clusters containing papers with >80% of all co-citations. Nodes with a *purple outer ring* indicate papers with high centrality (contain many co-citations between clusters), papers with large *red* sectors are highly cited papers showing bursts of co-citations

Table 1.2 Select cluster labels and mean year of publication of member papers from 'gecko fibrillar adhesion network'

Cluster	Mean year	Labels extracted from keywords of co-citation references. Ranks decrease left to right				
0	2009	Adhesive	Gecko	Surface	Contact	Dry adhesive
1	2002	Gecko	Adhesive	Surface	Contact	Model
3	2004	Gecko	Adhesive	Friction	Model	Surface
4	2007	Adhesive	Surface	Gecko	Dry adhesive	Contact
5	2005	Gecko	Adhesive	Friction	Gecko setae	Model
7	2010	Adhesive	Extreme positive allometry	Gecko	Surface	Adhesive pad

A timeline view of the top 12 clusters in the gecko fibrillar adhesion network (Fig. 1.3) shows how bursts of publication activity in gecko adhesion research can be traced to the Autumn et al. [8] paper and are concentrated in clusters 0, 1, 3, 4, and 5. The timeline view also helps us visualize that a span of about 5–8 years separates the Autumn et al. [8] paper from rapid increases in gecko inspired gecko adhesives (e.g., cluster 0-Lee et al. [9]; cluster 1-Glassmaker et al. [10]; cluster 3-Hansen and Autumn [11]). Moreover, early work in papers captured in the network in cluster 2 (matrix adhesion) has been largely independent of nearly all of the gecko or synthetic fibrillary research that followed foundational studies in the early 2000s (e.g., see clusters 8-2, Fig. 1.3).

Subsequent analysis of the 5 studies with the highest centrality scores in the 'fibrillary adhesion network' data set (Table 1.1) using VOS revealed that high centrality studies connected different aspects of adhesion research among the main clusters of the network. Hansen and Autumn [11], which ranked highest in centrality, connected mainly two out of four groups of studies described by keywords such as 'application', 'fabrication', and 'approach', on the one hand, and 'array', 'contact', 'adhesive', and 'setae', on the other. A third and fourth group of studies was only loosely connected by comparison (Fig. 1.4). Lee et al. [9] also connected four groups of studies, somewhat uniformly. The main themes of the studies were

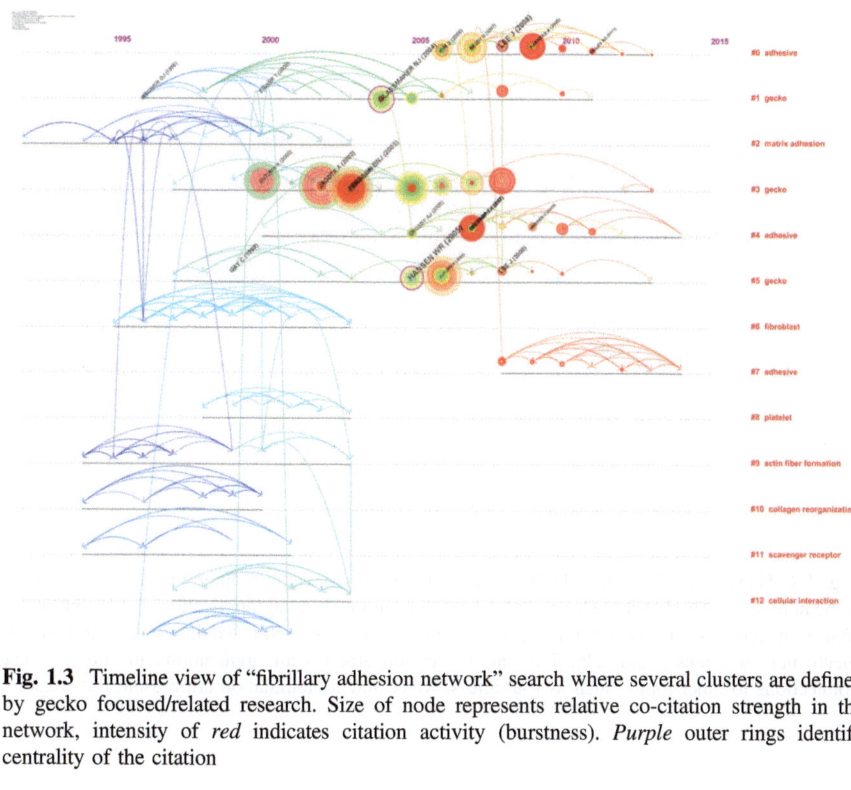

Fig. 1.3 Timeline view of "fibrillary adhesion network" search where several clusters are defined by gecko focused/related research. Size of node represents relative co-citation strength in the network, intensity of *red* indicates citation activity (burstness). *Purple* outer rings identify centrality of the citation

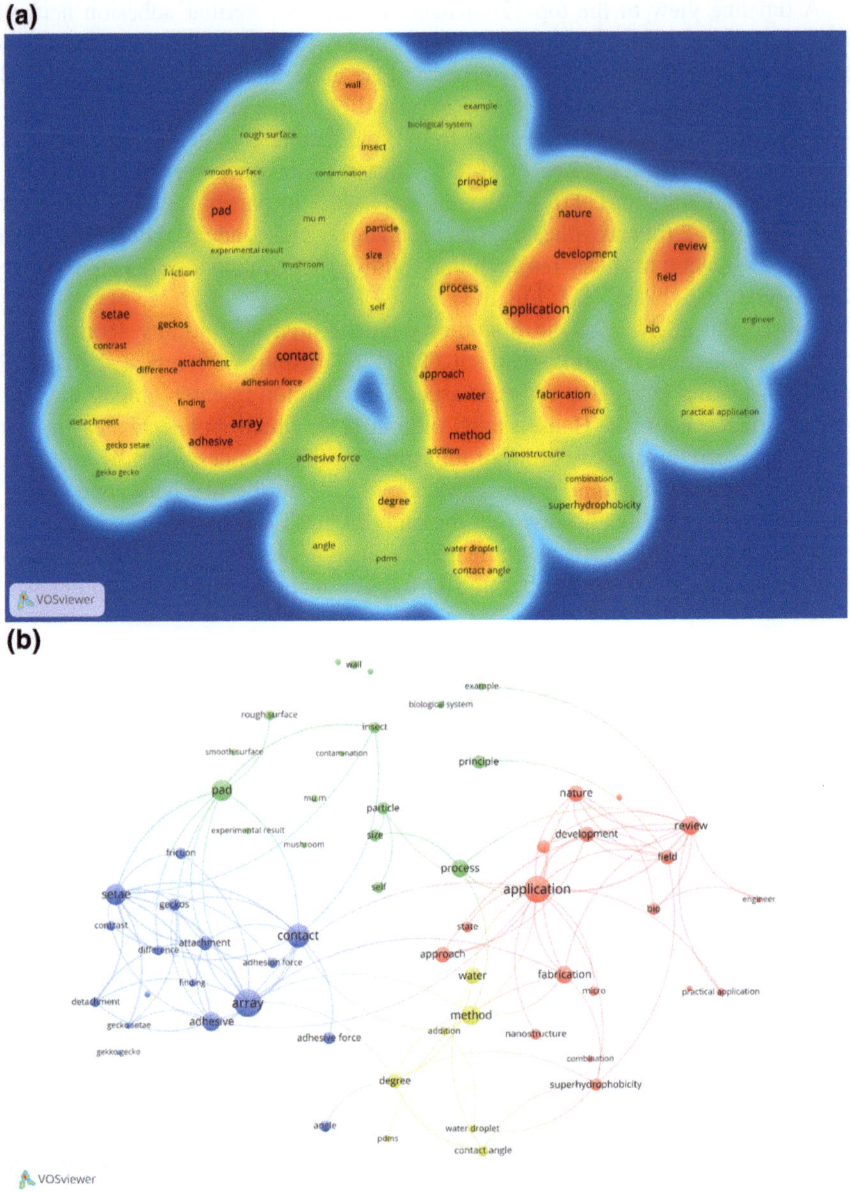

Fig. 1.4 Hansen and Autumn [11]. Note that the top 100 connections among nodes do not appear to be homogenously distributed, but instead are concentrated in the 'application' (*red*) and 'array' (*blue*) clusters. The Hansen centrality is associated with connecting studies that deal with the mechanics of contact and adhesion and those concerned with applications in mimics. The connections to other topics such as roughness, water and contamination are present but weaker

defined by terms such as 'friction', mechanism', and 'geometry', 'fiber', and 'fabrication', diameter', 'robot', and 'ability' (Fig. 1.5). Persson and Gorb [12] connected studies in three groups defined by the terms 'geometry', 'method', 'design', and 'adhesion force,' 'ability', 'spatulae', and 'rough surface', 'insect', 'setae' (Fig. 1.6). Jagota and Bennison [13] connected 5 groups of studies defined by the terms 'setae', 'principle', and 'detachment', 'direction', 'attachment', and 'feet', 'fabrication', and 'parameter', behavior', 'analysis', and 'interface', 'tip', 'pull' (Fig. 1.7). Glassmaker et al. [10] connected studies defined by the terms 'gecko', 'friction', and 'method', 'fabrication', 'process', and 'model', analysis', 'pull' (Fig. 1.8).

A Google patent search on existing granted patents and applications helped us choose two patents for further analysis: US6737160B1 [23] and US7335271B2 [14]. These two patents are related but have differences in document content, including images and the number and diversity of patent and non-patent citations included in their own references cited sections. The two patents also show distinct patent citation landscape maps and patent citation category classifications (Fig. 1.9). US6737160B1 had 8 non-patent citations, and US7335271B2 had 22. WOS Core Collection searches of the literature citing these two bibliographic data sets resulted in co-citation networks numbering 4447 and 9848 references for US6737160B1 and US7335271B2 respectively. VOS co-citation maps using keywords and journal sources help visualize the overlapping and non-overlapping literature bases associated with the two selected patents (Figs. 1.10 and 1.11). In the case of keywords, both patents are associated with literature citations tending to fall into two clusters (red and green), with sparser connections to 2 secondary clusters (blue and yellow). In the case of US6737160B1, the two primary clusters could be described as structures of the natural gecko adhesive system and associated mechanics (red), and whole-animal scale mechanics and performance associated with the adhesive system (green). For US7335271B2, the two primary clusters are associated with topics such as mechanics and performance of adhesion at various scales (green) and structure and function of synthetic mimics and the tools used in their characterization (red). Co-occurrence from the perspective of source publication similarly shows some distinctions between the two patent literature bases. US6737160B1 is primarily associated with work appearing in *Journal of Morphology, Journal of Experimental Biology, and Journal of Royal Society Interface*. US7335271B2, in contrast is mostly associated with work appearing in the journals *Langmuir, Advanced Materials, and Macromolecules*, suggesting a more application oriented set of themes. In the case of both patents, the cited literature bases include a relatively prominent node for studies in 'Journal of Morphology', but in the US6737160B1 patent [23] that node is separated from much experimental work in materials and physics/mechanics except through connections with 'Journal of Experimental Biology' and 'Journal of the Royal Society Interface'. In US7335271B2 [14], 'Journal of Morphology' clusters with 'Journal of Experimental Biology', and is connected to very prominent nodes in materials/mechanics/physics through a prominent linking node for *Langmuir* (Fig. 1.11).

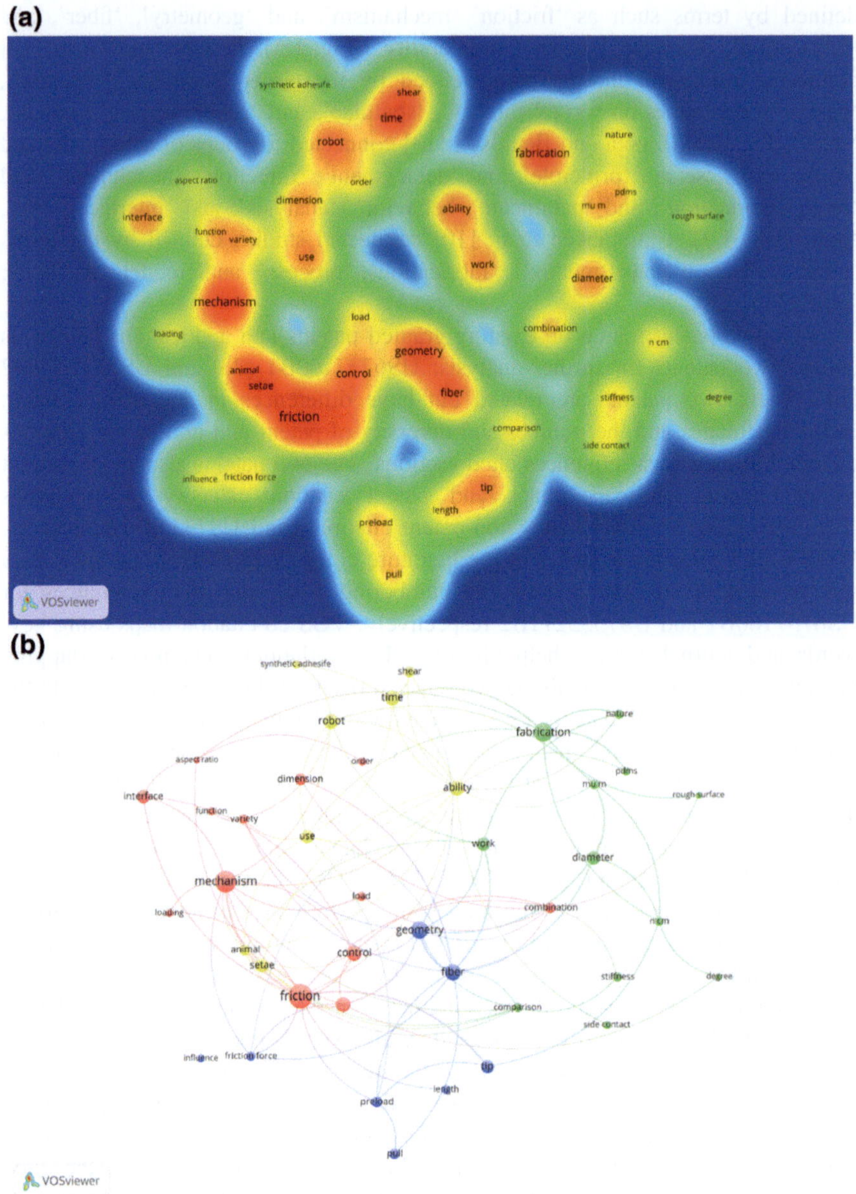

Fig. 1.5 Lee et al. [9]. Three of the four largest nodes have the majority of links in the top 100 connections. The Lee paper connects studies concerned with the mechanics and morphology of adhesion, especially as it is supported by or related to friction

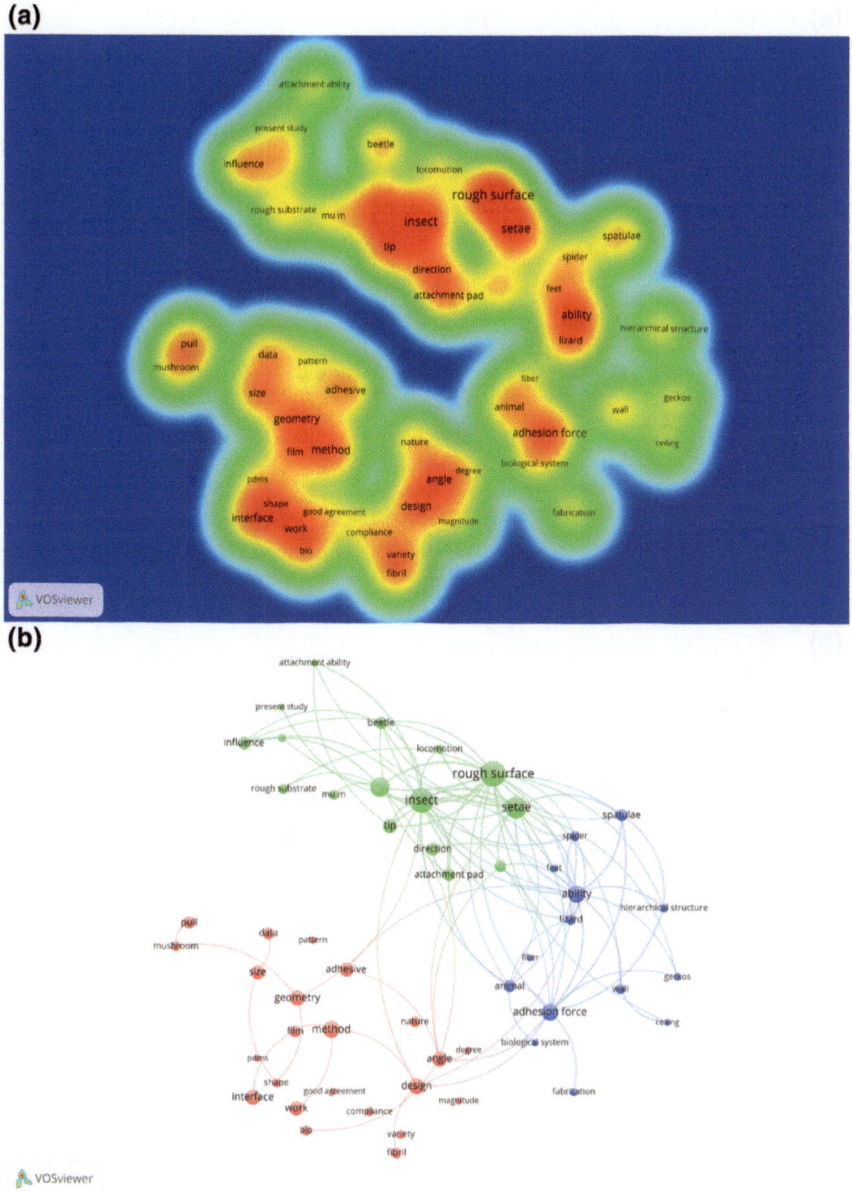

Fig. 1.6 Persson and Gorb [12]. Two of the three largest nodes (*blue* and *red*) have the majority of the top 100 connections, with the Persson paper connecting studies that combine geometry, adhesion, substrate and various animal groups with adhesive pads

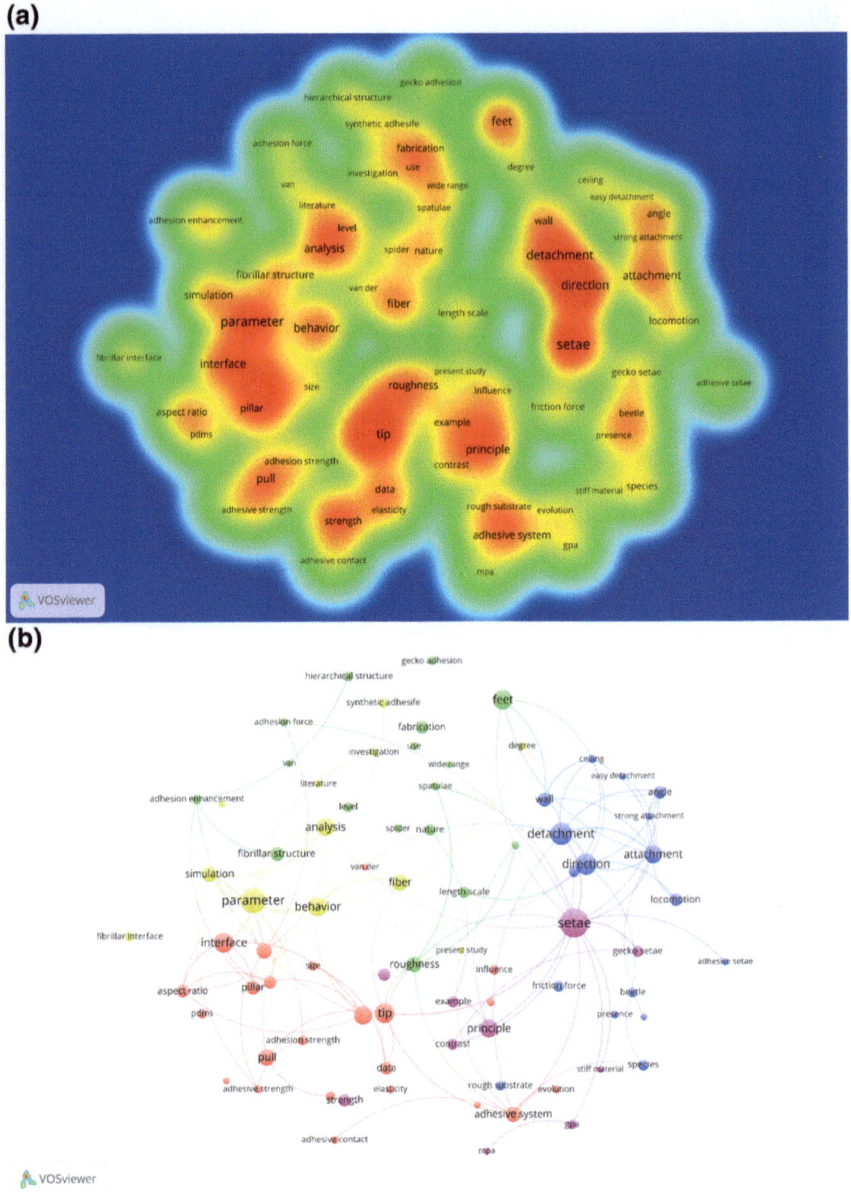

Fig. 1.7 Jagota and Bennison [13]. Three of the five clusters have a disproportionate share of the connections among a wide variety of topics apparently focused around setal morphology and mechanics in adhesion. The most weakly connected cluster (*green*) contains many topics of relatively infrequent occurrence associated with fabrication of synthetic mimics

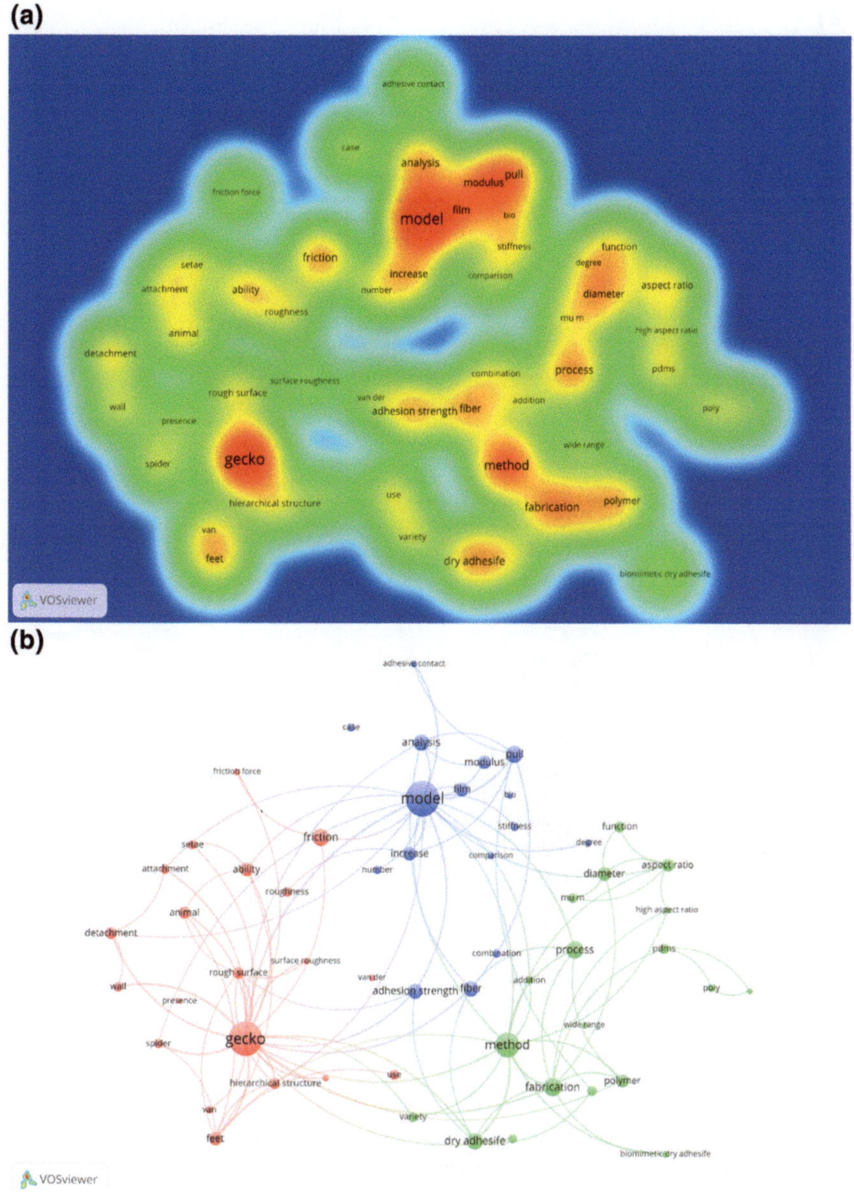

Fig. 1.8 Glassmaker et al. [10]. Note that the three clusters each have similar density of connections within and across clusters, but that 'gecko' and 'model' nodes are larger and appear as 'hotter' in the heat map suggesting that the centrality of the glassmaker study is associated with connecting studies linking gecko adhesive features with modeling and potential methods of mimicking

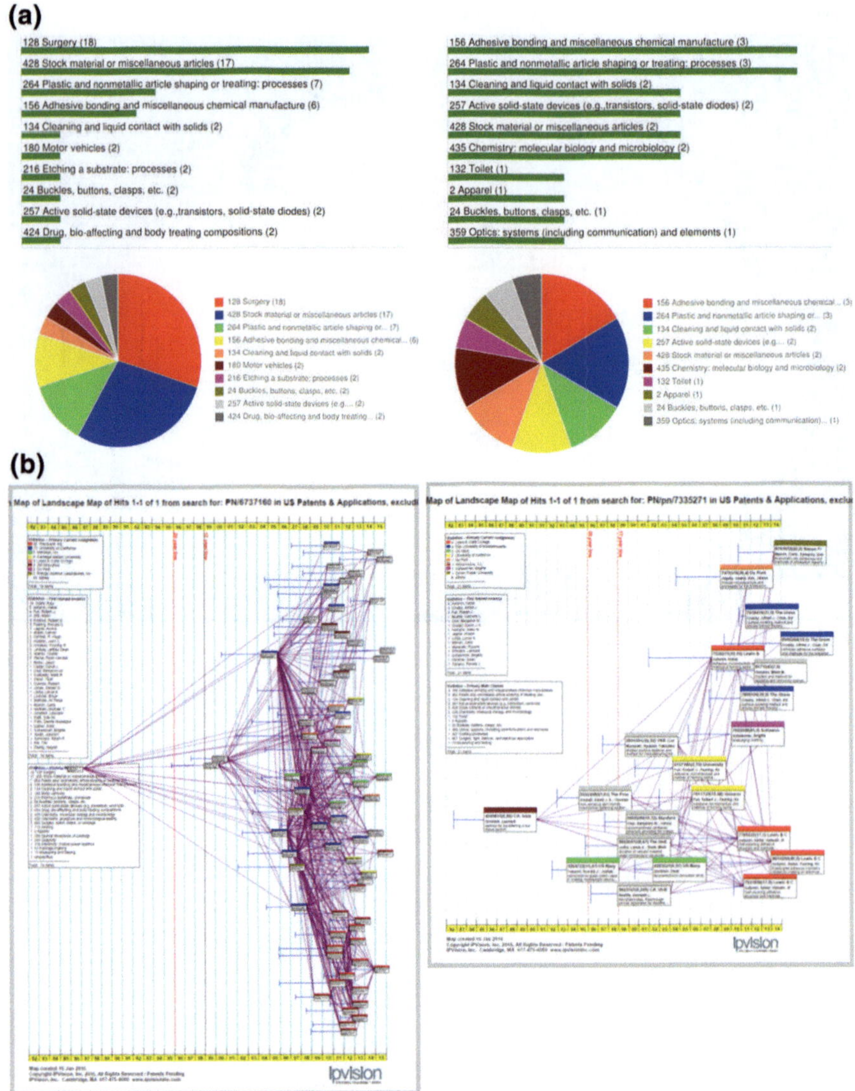

Fig. 1.9 **a** Patent citations and classifications of US6737160B1 (*left*) and US7335271B2 (*right*). **b** Landscape maps showing networks of citations

1.4 Discussion

While interest in gecko adhesion, from both a biological and application point of view, is ancient, detailed research in this emerging field is only about 15 years old. Nevertheless, research on the structure, function and translation of fibrillar adhesive structures found in biological systems in general and in geckos in particular, has

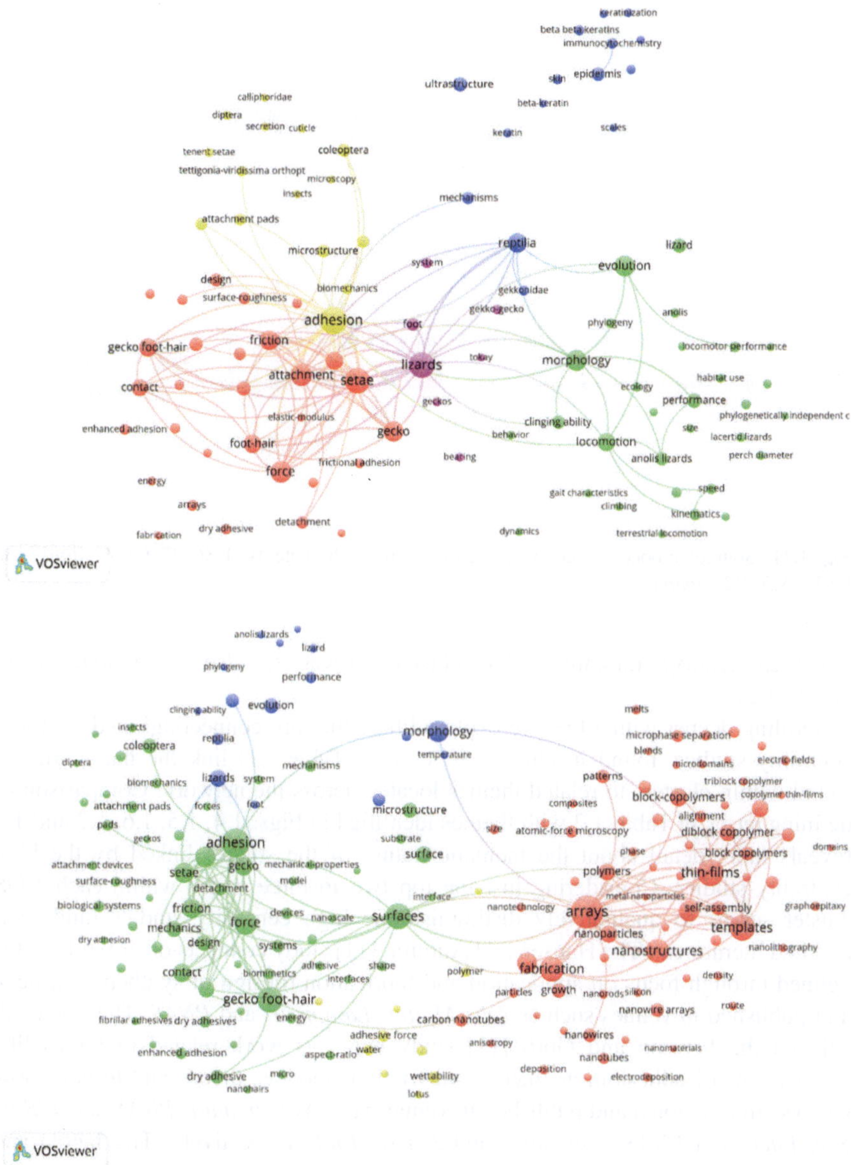

Fig. 1.10 Keyword co-occurrence maps associated with patents US6737160B1 (*top*) and US7335271B2 (*bottom*)

grown rapidly and extensively. Bibliometric analysis with CiteSpace III shows that while there are many dimensions to gecko adhesion research, much of it can be organized into approximately 6 main clusters accounting for more than 85% of the total co-citations (Table 1.2). Many of the terms that define the clusters are shared,

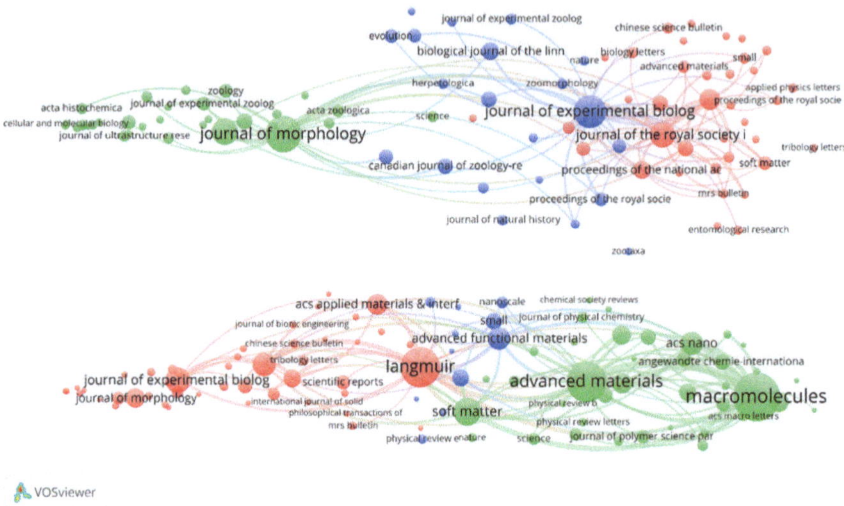

Fig. 1.11 Source co-occurrence maps associated with patents US6737160B1 (*top*) and US7335271B2 (*bottom*)

but their relative ranking within clusters serves to also differentiate them (Table 1.2).

Drilling deeper with VOS, we explored how the top 'connecting' studies (top 5 centrality studies, found in clusters 0, 1, 3, 5, Table 1.1) link the major themes found within clusters to related themes located across the network. Comparison of defining terms in Table 1.2 with themes identified in Figs. 1.4, 1.5, 1.6, 1.7 and 1.8 reveals more detail about the thematic nature of the studies linked by the high centrality studies. Considering just the top two to three nodes within each VOS cluster serves to qualitatively distinguish the high centrality studies (and their co-cited literatures). The Hansen and Autumn [11] study connects other work that is defined through focus on application and fabrication (design of synthetic mimics), and published in venues such as *Adv. Matter*, *Langmuir*, and *PNAS*. The Lee et al. [9] and the Persson and Gorb [12] study connects work related to topics like mechanisms of adhesion, geometry and friction (models, theory and tests of how the system functions) and published in venues such as *Langmuir*, *PNAS* and *J. Roy. Soc. Inter.* and *PNAS*, *Langmuir*, and *J. Exp. Biol.*, respectively. The Jagota and Bennison [13] study links studies focused on the micromechanics of setae and detachment mechanics and is published in venues such as *PNAS*, *J. Exp. Biol.* and *Adv. Matter*. The Glassmaker et al. [10] study links work on mechanics of adhesion like friction and application to synthetics, published in venues such as *Langmuir*, *PNAS*, *J. Roy. Soc. Interface and Adv. Matter.*

One of the primary objectives of bibliometric studies is to reveal sources and trends of knowledge developing in a chosen field [3]. The tools we employed here are designed for literatures that are indexed by fundamental research databases (e.g.,

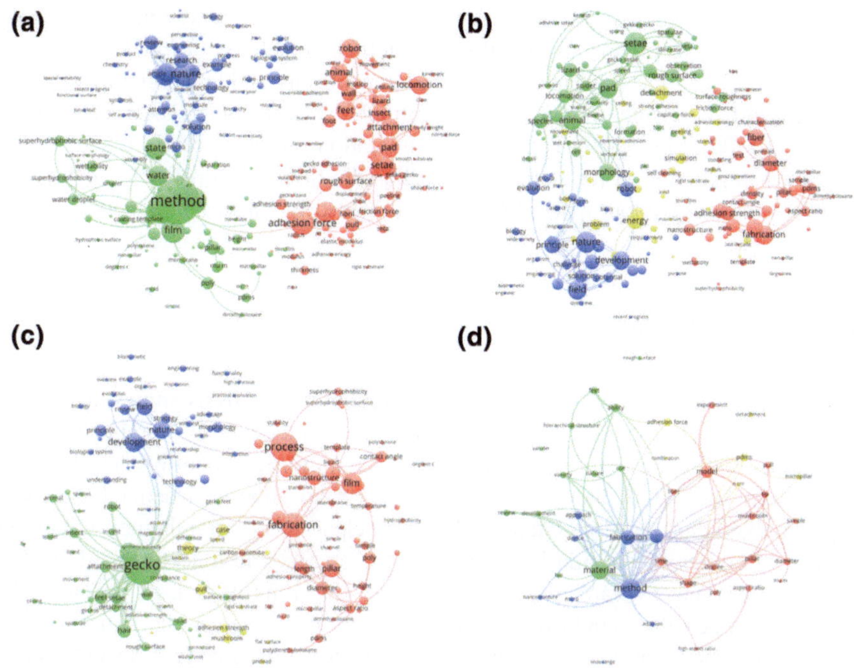

Fig. 1.12 Keyword co-occurrence maps of the top four burstness studies in the gecko fibrillar adhesion network. **a** Autumn et al. [8]. **b** Autumn et al. [14]. **c** Geim et al. [15]. **d** Arzt et al. [16]

WOS, Scopus, PubMed, etc.), rather than data indexed by databases such as the USPTO. In order to explore how the application and fundamental research side of gecko-inspired adhesion efforts are related, we examined patterns in the research literature cited by two early patents (Full et al. [23] and Autumn et al. [8]; Figs. 1.10 and 1.11). Both US6737160B1 and US7335271B2 networks show two relatively tightly linked clusters of nodes focusing on adhesion mechanics and performance on the one hand, and either the morphology of synthetic mimics or the natural system on the other. Another conspicuous feature of both patent cited literature networks is the weak linkage to other nodes representing work in the natural materials, ecological and evolutionary aspects of the natural systems, or diversity of analogous fibrillar systems (e.g., insects). As might be expected, the references in the original gecko fibrillar adhesion network with very high 'burstness' scores all show strong keyword co-occurrence patterns in 'fabrication' (red nodes; Fig. 1.12) and 'bio inspired design' or 'biomimetics' terms (blue nodes; Fig. 1.12). Each of these high 'burstness' studies show additional clusters of nodes, but overall, clusters of nodes are only weakly connected, suggesting that the major themes of knowledge development are not well integrated. Whether this reflects a constraint on future advancements or a limitation of previous work, or something else, is not clear. It also seems striking to us that none of the high 'burstness' studies, nor the cited literatures of the two patents show prominent nodes or clusters associated with the natural materials of the gecko adhesive system, apart from morphology [except the essentially isolated blue cluster of US6737160B1 (Fig. 1.10)]. In all the clusters of both the high centrality and high burstness, such a cluster is either missing or only weakly identifiable (but see blue cluster, patent map US6737160B1, Fig. 1.10). This gap in co-citation structure reinforces our own anecdotal impression of the state of the gecko adhesion literature [2, 18, 19]: focus on and knowledge about materials is overshadowed by structure/geometry.

1.5 Conclusions

Returning to our motivating questions, our bibliometric analysis shows that gecko adhesion research is part of a multifaceted research literature in biomimetic adhesion. Considering these in order, first (gecko vs. biomimetic adhesion), even within the broader context of biomimetic adhesion, 'gecko adhesion' defined the most highly ranked clusters of co-citations. It is also clear that several themes within the field of biomimetic adhesion have independent origins and trajectories. How that will change going forward is not clear, but we can think of at least two ways in which disparate topics could become more closely connected in the future: (1) integration or combination of independent biomimetic strategies into a single functional apparatus, and (2) findings of fundamental principles that unite previously distinct systems. As an example of the first, some have argued (e.g., robot locomotion cf. [20]) that one of the main advantages of biomimetic approaches comes from the ability for engineers to coopt multiple strategies into a single application even if those strategies don't exist in a single organism or system in nature, leading to functionality that goes beyond any single biological model (for new functionality [21]). Such a strategy could connect current research themes that are only loosely associated within gecko adhesion research, and also connect those themes to research areas that are currently external. An example of the second way can be seen from how a focus on van der Waals as a mechanism in the gecko system helped refine insights, and support clarification of mechanisms of adhesion in what were initially regarded as distinct systems, like the cribellate silks produced by some spiders [22].

Second, the major themes of gecko inspired adhesion are observed in the keywords identified by the high burstness and centrality citations: structure/function, fabrication, ecology/evolution, and biomimetics/bio inspiration. There is particularly tight integration among the themes structure/function with fabrication and weaker linkage among the other themes. We note that some of the performance capabilities of geckos that synthetic mimics have not yet captured, such as adhesion to wet or dirty surfaces and high repeatability, do not show up as major nodes in any of the networks. One conclusion to draw from this pattern is that these themes are either not the focus of intensive activity or are particularly hard to study, or both. With respect to designing and fabricating a more robust mimic of the gecko system, it would seem that activity in these areas would need to increase, and eventually connected to the other themes.

Third, weak linkage outside of the themes structure/function with fabrication, seem especially conspicuous at the interface between fundamental research and translational application. Whether this is a historical artefact of a relatively young and still emerging field or a constraint on future progress in both contexts is unclear, but perhaps worthy of further analysis and consideration. Again, the conspicuous areas including wet/dirty adhesion, self-cleaning, and resilience would be expected to include the features of a new adhesive platform that outperforms current pressure sensitive adhesive platforms, but there is very little evidence of research or progress as it might be reflected in the patent networks.

References

1. K. Autumn, P.H. Niewiarowski, J.B. Puthoff, Annu. Rev. Ecol. Evol. Syst. **45**, 445 (2014)
2. P.H. Niewiarowski, A.Y. Stark, A. Dhinojwala, J. Exp. Biol. **219**, 912 (2016)
3. C.M. Chen, J. Am. Soc. Inform. Sci. Technol. **57**, 359 (2006)
4. N.J. van Eck, L. Waltman, in *Advances in Data Analysis: Proceedings of the 30th Annual Conference of the Gesellschaft für Klassifikation e.V., Freie Universität Berlin, March 8–10, 2006*, eds. by R. Decker, H.J. Lenz (Springer, Berlin, 2007), p. 299
5. C. Chen, Inf. Process. Manag. **35**, 401 (1999)
6. D.J.D. Price, Science **149**, 510 (1965)
7. O. Persson, J. Am. Soc. Inf. Sci. **45**, 31 (1994)
8. K. Autumn, Y.A. Liang, S.T. Hsieh, W. Zesch, W.P. Chan, T.W. Kenny, R. Fearing, R. J. Full, Nature **405**, 681 (2000)
9. J. Lee, C. Majidi, B. Schubert, R.S. Fearing, J. R. Soc. Interface **5**, 835 (2008)
10. N.J. Glassmaker, A. Jagota, C.Y. Hui, J. Kim, J. R. Soc. Interface **1**, 23 (2004)
11. W.R. Hansen, K. Autumn, Proc. Natl. Acad. Sci. U. S. A. **102**, 385 (2005)
12. B.N.J. Persson, S. Gorb, J. Chem. Phys. **119**, 11437 (2003)
13. A. Jagota, S.J. Bennison, Integr. Comp. Biol. **42**, 1140 (2002)
14. K. Autumn et al., Proc. Natl. Acad. Sci. U. S. A. **99**, 12252 (2002)
15. A.K. Geim, S.V. Dubonos, I.V. Grigorieva, K.S. Novoselov, A.A. Zhukov, S.Y. Shapoval, Nat. Mater. **2**, 461 (2003)
16. E. Arzt, S. Gorb, R. Spolenak, Proc. Natl. Acad. Sci. U. S. A. **100**, 10603 (2003)
17. M. Sitti, R.S. Fearing, J. Adhes. Sci. Technol. **17**, 1055 (2003)
18. D. Jain, A.Y. Stark, P.H. Niewiarowski, T. Miyoshi, A. Dhinojwala, Sci. Rep. **5** (2015)
19. A.Y. Stark, D.M. Dryden, J. Olderman, K.A. Peterson, P.H. Niewiarowski, R.H. French, A. Dhinojwala, J. R. Soc. Interface **12**, 20150464 (2015)
20. R. Altendorfer, N. Moore, H. Komsuoglu, M. Buehler, H.B. Brown, D. McMordie, U. Saranli, R. Full, D.E. Koditschek, Auton. Robot. **11**, 207 (2001)
21. H. Lee, B.P. Lee, P.B. Messersmith, Nature **448**, 338 (2007)
22. A.C. Hawthorn, B.D. Opell, Biol. J. Lin. Soc. **77**, 1 (2002)
23. R.J. Full et al., *Adhesive microstructure and method of forming same*. U.S. Patent No. 6, 737, 160, 18 May (2004)

Chapter 2
Impact of Ambient Humidity on Traction Forces in Ladybird Beetles (*Coccinella septempunctata*)

Lars Heepe, Jonas O. Wolff and Stanislav N. Gorb

Abstract Many insects possess hairy adhesive foot pads, supplemented with a tarsal secretion, which allow them to securely and reversibly attach to various kinds of natural substrates, also under complex environmental conditions. Conditions such as ambient temperature and humidity have been shown to influence the attachment ability of the dry adhesive foot pads in spiders and geckos. For insects, however, the influence of the environment on attachment ability is lacking. In the present study, we therefore studied the attachment ability of the seven-spotted ladybird beetle (*Coccinella septempunctata*) at different humidities by traction force experiments. Our results indicate an optimal range of relative humidity with maximal traction forces. At both low (15%) and high (99%) relative humidity a decrease of attachment ability was found. At 60% humidity, the highest attachment forces were revealed. This relationship was found both in female and male beetles, despite of a deviating structure of adhesive setae and a significant difference in forces between sexes. These findings demonstrate that humidity similarly affects the

L. Heepe (✉) · J.O. Wolff · S.N. Gorb
Department of Functional Morphology and Biomechanics, Zoological Institute, Kiel University, Am Botanischen Garten 1–9, 24118 Kiel, Germany
e-mail: lheepe@zoologie.uni-kiel.de

J.O. Wolff
e-mail: jonas.wolff@mq.edu.au

S.N. Gorb
e-mail: sgorb@zoologie.uni-kiel.de

Present Address:
L. Heepe
Mads Clausen Institute, University of Southern Denmark,
NanoSYD Alsion 2, 6400 Sønderborg, Denmark

Present Address:
J.O. Wolff
Department of Biological Sciences, Macquarie University, Sydney, NSW 2109, Australia

L. Heepe et al. (eds.), *Bio-inspired Structured Adhesives*,
Biologically-Inspired Systems 9, DOI 10.1007/978-3-319-59114-8_2

function of both dry and wet adhesive pads, although both types of adhesive systems (wet and dry) are supposed to be based on different physical interactions (capillarity versus van der Waals forces).

2.1 Introduction

Substrate attachment plays an important role in the niche occupation of plant-dwelling insects, since it is substantial for resting and locomotion in a complex environment. Consequently, a high diversity of friction and adhesion enhancing structures has evolved among insects [1, 2]. Several studies showed that not only the intrinsic structure of an attachment organ determines its function, but also environmental parameters, such as the surface roughness or/and chemistry of the substrate [1, 3–9]. Also the ambient temperature and humidity may affect the attachment ability of adhesive organs, as it was shown in the dry adhesive pads of geckoes [10–13] and spiders [14]. For small arthropods these conditions may highly vary microspatially, especially in the boundary layer of plant leaves [15]. Furthermore, on most surfaces there is an adsorbed film of water molecules, with a thickness and mechanical properties highly influenced by ambient humidity [16, 17]. Due to its polarity it can work as a bonding agent between two surfaces, and therefore have a substantial effect on adhesion [18–21]. In insects the terminal contact elements of tenent setae are not dry, but rather wetted by a fluid secretion that is usually a microemulsion of water and oil, with varying fractions of substances [22–31]. The mixture of both polar and non-polar substances presumably helps to wet both hydrophobic and hydrophilic surfaces building a fluid meniscus between the seta and the substrate to yield high capillary forces (wet adhesion). The fluid can increase the contact area by filling minute cavities of micro and nano-rough surfaces, where setae otherwise cannot adapt to, and prevents slipping of the foot due to its specific rheological properties [31–35]. It was demonstrated that the presence of these secretions is crucial for the function of insect adhesive organs [6, 36]. Adhesion is affected, if the water content of the secretion is manipulated by a water adsorbing substrate [28]. However, it is not clear, how the thickness and composition of the secretion fluid film are affected by ambient humidity.

In the present study, force measurements with tethered seven-potted ladybirds (*Coccinella septempunctata*) were carried out in an environmental chamber at various controlled humidities. Experiments were carried out with female and male beetles, which differ in the structure of their adhesive pads [6]. Males possess tenent setae with discoidal tips that are assumed to be an adaptation to securely attach to the smooth elytrae of the female during both copulation and mate guarding [6, 37]. Two main questions were asked. (1) Does ambient humidity influence the attachment ability of the wet adhesive system of *C. septempunctata*? (2) Is the attachment ability in both sexes likewise affected by ambient humidity?

2.2 Experimental

2.2.1 Animals

For this study, the seven-spotted ladybird beetle *Coccinella septempunctata* LINNAEUS 1758 (Coleoptera, Coccinellidae) was chosen (Fig. 2.1a). This beetle species is a generalist, living on diverse plant surfaces [38, 39]. In their natural environment, beetles are exposed to various humidities, hence, their adhesive system must be well adapted to changing environmental conditions. Beetles were collected from bracken (*Pteridium aquilinum*) in the New Botanical Garden at Kiel University, Schleswig-Holstein, Germany. They were kept individually in plastic tubes at a temperature of 22–24 °C and relative humidity of 40–50%. These conditions were consistent with those at which the beetles were captured. The beetles were fed with honey. Figure 2.1 shows the hairy attachment devices of *C. septempunctata*. The tarsus is composed of three tarsomeres and two ventrally curved claws (Fig. 2.1b–d, h–j). Only the first two tarsomeres (T1 and T2 in Fig. 2.1) are ventrally covered by tenent setae. Different types of tarsal adhesive setae were distinguished: (1) setae with a pointed, filamentous tips (Fig. 2.1e), (2) setae with lanceolate tips (Fig. 2.1f), (3) setae with a flattened and rounded, spatula-shaped tips (Fig. 2.1g), and (4) setae terminated with discoidal terminal elements (Fig. 2.1k). All four types of tarsal adhesive setae were found in males. Females show the first three types only. Setae terminated with discoidal terminal elements were found in the centre of the first two tarsomeres (T1 and T2 in Fig. 2.1) for all legs in males (Fig. 2.1h–j), except for the first tarsomer (T1) of the hindleg (Fig. 2.1j).

2.2.2 Force Measurements in a Controlled Atmosphere

Traction force experiments were performed in a polymethylmethacrylate (PMMA) chamber (30 × 14 × 14 cm) in which relative humidity could be manipulated by the controlled mixture of dry and wetted air (Fig. 2.2, for details see [14]).

For this experiment, three levels of relative humidity (RH) were used: 15, 60 and 99%. The RH was monitored with a P330 digital hygrometer (Dostmann electronic GmbH, Wertheim-Reicholzheim, Germany). To generate humid conditions, air was passed through a water bottle, so that the vapour concentration in the air increased. To reach 99% of relative humidity, slight heating was used to increase the amount of water vapor. This did not affect the temperature within the experimentation chamber. To generate a dry environment, dry air from a pressurized air pipe was pumped into the experimentation box. To stabilize ambient humidity and the thickness of adsorbed water films, the procedure was started about 1 h before the experiments.

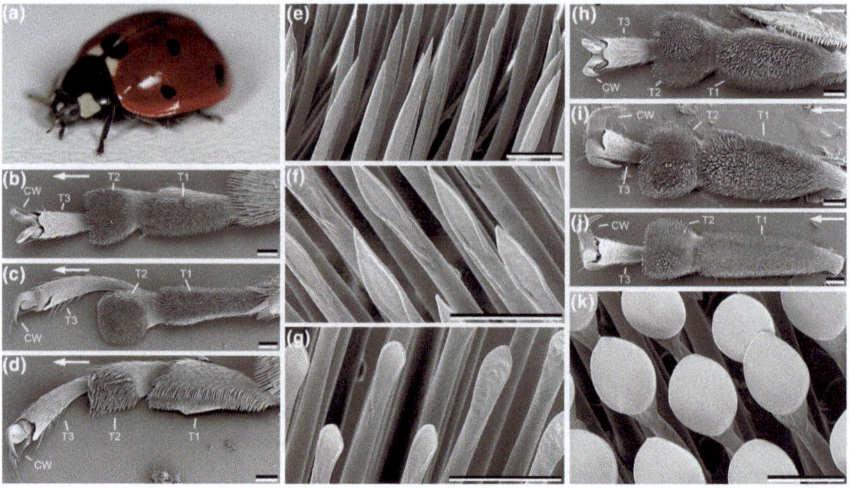

Fig. 2.1 Attachment devices of *Coccinella septempuctata* (**a**). Tarsi of forelegs (**b**), midlegs (**c**), and hindlegs (**d**) in females are ventrally covered by different types of tenent setae (**e–g**). Tarsi of forelegs (**h**), midlegs (**i**), and hindlegs (**j**) in males. Tarsi of males were also ventrally covered by tenet setae types shown in (**e**, **f**), but have an additional type, which is terminated with discoidal elements (**k**). *CW* claws; *T1* first proximal tarsomer; *T2* second proximal tarsomer; *T3* third proximal tarsomer. The *arrows* in **b–d** and **h–j** indicate distal direction. *Scale bars* in **b–d**, **h–j**, 100 μm. *Scale bars* in **e–g**, **k**, 10 μm

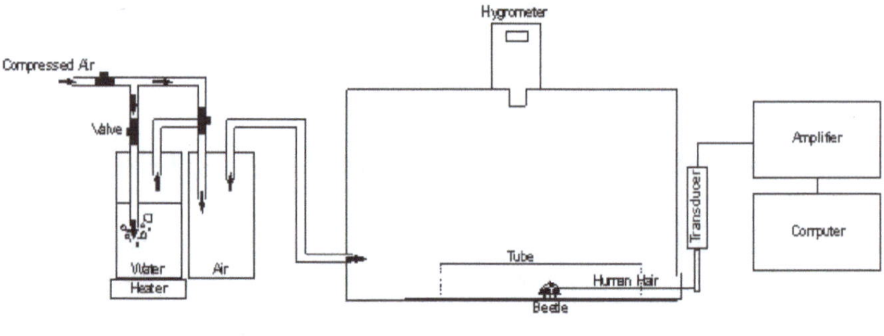

Fig. 2.2 Schematic of the experimental setup used for traction force experiments under controlled ambient humidities. Tethered beetles were connected to a force sensor by a 15–20 cm long human hair. Humidity was controlled by mixing dry compressed air (approx. 15% relative humidity) and wet air. Wet air was produced by heating water in a glass bottle, in order to increase the amount of water vapour. Temperature was continuously monitored in the chamber to ensure that the use of the heater did not lead to temperature increase within the chamber. For details, see [14]

A strain-gauge force transducer (10 g capacity; World Precision Instruments, Inc., FL, USA) connected to a Biopac MP100 amplifier (Biopac System Inc., Goleta, CA, USA) was used to measure forces generated by the beetles. Force data were recorded with the AcqKnowledge 3.7.0 software (Biopac Systems Inc.).

A half Plexiglas tube was installed inside the chamber, in order to guide beetle locomotion perpendicular to the force sensor and prevent it from climbing onto the lateral walls.

Prior to experiments the beetles were anesthetized with carbon dioxide and the elytra were glued together with a droplet of molten wax, to prevent them from flying (for details see [40]). Individual beetles were connected to the force transducer by a human hair (10–15 cm long) fixed to the elytra and to the force sensor with a droplet of molten wax. Beetles were placed on the Plexiglas bottom of the experimentation chamber and released such that they walk perpendicularly away from the force sensor. When the beetles were hold back by the strengthened hair they tried to pull forward, which led to transmission of traction forces to the force sensor. The traction force is dependent on the attachment ability of the pulling animal. This method has been used in previous studies [7, 14, 40]. We let the beetle pull for 1 min and then repeated the experiment in a second run, to make sure that the performance of individual animals was stable. In total, 10 runs for each individual beetle for 1 min at each of three humidity conditions were recorded.

The experiment was carried out in two different runs. The first run was done with 15 beetles (4 females and 11 males). All beetles were tested with one single level of humidity per day, further called Experiment 1. To exclude an influence of physiological conditions of individual animals on different experimental days, a second run was executed. In this run, 9 beetles (3 females and 6 males) were tested. Individual beetles were tested on one day at all three levels of relative humidity in a randomized fashion, further called Experiment 2. To evaluate the effect of the relative humidity on attachment force in both sexes of *C. septempunctata*, a Kruskal–Wallis One Way ANOVA on Ranks followed by an all pairwise multiple comparison procedure (Tukey Test, SigmaPlot 11.0, Systat Software Inc, San José, CA, USA) has been performed. In order to evaluate the effect of sex on the attachment force, a Mann–Whitney Rank Sum Test has been applied to the pooled data. Therefore, the data of the different relative humidity has been pooled individually for both experiments and sexes.

2.2.3 Observations of the Beetle Behavior at Different Relative Humidities

A Plexiglas sheet was fixed with double side adhesive tape at three different positions: horizontal (0°), vertical (90°) and upside-down (180°) inside the experimentation chamber. We observed the locomotion behavior of individual beetles on these surfaces. Additionally, to the three previously used humidity adjustments (RH of 15, 60 and 99%), we made some observations at RH higher than 99%, when water began to condense on the substrate.

2.3 Results

2.3.1 Observational Experiments

We observed that the beetle's attachment is affected, as the water begins to condense on the substrate, when the relative humidity exceeds 99%. In this case, beetles were not able to hold on the vertical Plexiglas slide and slid along its surface.

2.3.2 Experiment 1: One Level of Relative Humidity per Day

The attachment ability of beetles was highest at an RH of 60% and lower for RH of 15 and 99% (Fig. 2.3). Here and later data will be shown as median [minimum, maximum]. At an RH of 15%, females reached the median force of 6.9 mN [2.7, 12.4 mN]. At an RH of 60%, the median force was 11.2 mN [1.9, 18.6 mN]. At an RH of 99%, the median force was 5.5 mN [2.0, 17.5 mN]. The forces measured at different RH differed significantly (Kruskal–Wallis One Way ANOVA on Ranks: N = 40; H = 30.035; df = 2; $P \leq 0.001$). An all pairwise comparison (Tukey Test, see Table 2.1) revealed significant differences in measured traction forces between 15% RH and 60% RH as well as between 60% RH and 99% RH, but not between 15% RH and 99% RH.

At an RH of 15%, males reached the median force of 16.7 mN [8.1, 23.4 mN]. At an RH of 60%, the median force was 22.4 mN [11.7, 31.4 mN]. At an RH of 99%, the median force was 17.7 mN [8.2, 27.0 mN]. The forces measured at different RH differed significantly (Kruskal–Wallis One Way ANOVA on Ranks: N = 110; H = 133.1; df = 2; $P \leq 0.001$). An all pairwise comparison (Tukey

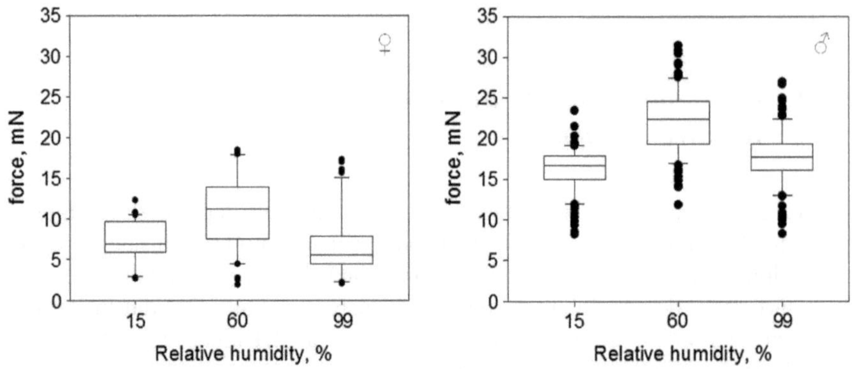

Fig. 2.3 *Box-and-whiskers plots*, based on the results of the first experiment, with one RH level tested per day. The *ends of the boxes* define the 25th and 75th percentiles, with a *line* at the median and *error bars* defining the 10th and 90th percentiles, *black dots* represent the outliers

Table 2.1 Statistics of the traction force measurements

Sex	Humidity	Experiment	q-value	Significance
Female[a]	15% versus 60%	1	5.200	Yes
Female[a]	15% versus 99%	1	2.377	No
Female[a]	60% versus 99%	1	7.577	Yes
Male[b]	15% versus 60%	1	15.634	Yes
Male[b]	15% versus 99%	1	3.776	Yes
Male[b]	60% versus 99%	1	11.858	Yes
Female[c]	15% versus 60%	2	6.003	Yes
Female[c]	15% versus 99%	2	3.473	Yes
Female[c]	60% versus 99%	2	2.530	No
Male[d]	15% versus 60%	2	4.311	Yes
Male[d]	15% versus 99%	2	3.806	Yes
Male[d]	60% versus 99%	2	8.117	Yes
Female versus male[e]	Pooled	1	$P < 0.001$	Yes
Female versus male[f]	Pooled	2	$P < 0.001$	Yes

[a–d]Kruskal–Wallis One Way ANOVA on Ranks with an all pairwise multiple comparison (Tukey Test)
[e,f]Mann–Whitney Rank Sum Test

Test, see Table 2.1) revealed significant differences in measured traction forces for all levels of RH.

2.3.3 *Experiment 2: Three Levels of Relative Humidity per Day*

Similar to the first experiment, highest median attachment forces were observed at 60% RH (Fig. 2.4). At an RH of 15%, females reached the median force of 7.1 mN [5.5, 9.9 mN]. At an RH of 60%, the median force was 8.8 mN [5.3, 11.0 mN]. At an RH of 99%, the median force was 8.5 mN [1.6, 11.3 mN]. The forces measured at different RH differed significantly (Kruskal–Wallis One Way ANOVA on Ranks: N = 30, H = 18.167, df = 2, $P \leq 0.001$). An all pairwise comparison (Tukey Test, see Table 2.1) revealed significant differences in measured traction forces between 15% RH and 60% RH as well as between 15% RH and 99% RH, but not between 60% RH and 99% RH.

At an RH of 15%, males reached the median force of 16.2 mN [13.2, 20.4 mN]. At an RH of 60%, the median force was 17.2 mN [13.1, 20.4 mN]. At an RH of 99%, the median force was 15.2 mN [11.6, 19.1 mN]. The forces measured at different RH differed significantly (Kruskal–Wallis One Way ANOVA on Ranks: N = 60, H = 32.984, df = 2, $P \leq 0.001$). An all pairwise comparison (Tukey Test, see Table 2.1) revealed significant differences in measured traction forces for all levels of RH.

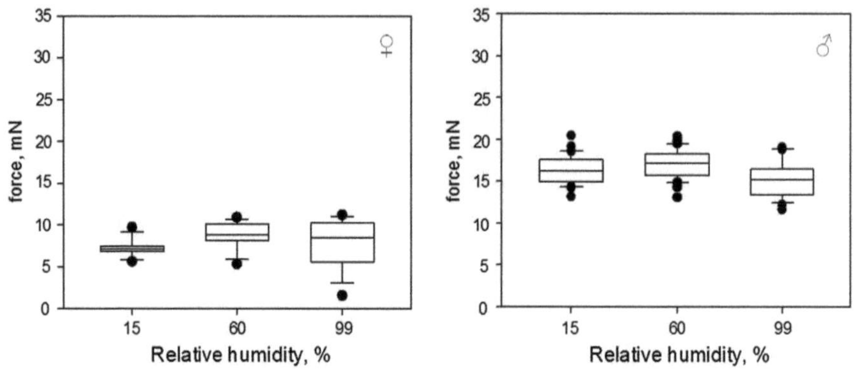

Fig. 2.4 *Box-and-whiskers diagrams* based on the results of the second experiment with all three levels of RH tested at the same day. The ends of the *boxes* define the 25th and 75th percentiles, with a *line* at the median and *error bars* defining the 10th and 90th percentiles, *black dots* represent the outliers

2.3.4 Effect of Sex

A significant difference was detected between sexes ($P < 0.001$, Mann–Whitney Rank Sum Test, see Table 2.1), where males achieved two times higher forces than females at all three levels of RH. Thereby, the attachment ability of both males and females was likewise affected by ambient humidity.

2.4 Discussion

The present study demonstrates, for the first time, that the performance of a wet (fluid supplemented) adhesive pad is influenced by ambient humidity. Our results indicate an optimal range of relative humidity with maximal traction forces in ladybird beetles. Similar observations have been made in dry adhesive pads of spiders, using a similar setup and method as in the present study [14], and geckos, revealed in measurements with living animals [11] and with isolated setae [10]. This is particularly interesting since both types of adhesive systems (wet and dry) are supposed to be based on different physical interactions (capillarity vs. van der Waals forces).

For the dry adhesive pads of spiders and geckos the humidity-related effects on adhesion can be explained by three different mechanisms: (1) capillary forces due to the formation of liquid bridges; (2) changes in the effective short-range interactions due to adsorbed monolayers of water on the substrate [11, 14]; (3) humidity-dependent material properties of insect cuticle and β-keratin (main constituent of gecko setae) [41–44]. In geckos, the effect of a RH on viscoelastic properties of the setal shaft was shown [13]. It was argued that with an increasing humidity the viscoelastic bulk energy dissipation increases within setae while being pulled off the substrate, leading to higher resistance of the adhesive contact [12, 45].

For *C. septempunctata* it was recently shown that the setal tips contain high amounts of the soft rubber-like protein resilin [46], which is a hygroscopic protein capable of binding high amounts of water [47]. Peisker et al. [46] showed a 6000-fold increase in the Young's modulus of the setal tips after drying. Consequently, the degree of hydration should also significantly affect the mechanical properties of the setae of the beetles in the present study and may also explain our results on ladybird beetles, at least the increase in traction forces from 15% RH to 60% RH.

Two other observations may further support the hypothesis that a change in material properties with increasing humidity is responsible for at least the increase in traction forces from 15% RH to 60% RH. Voigt et al. [48] observed significantly higher forces generated by beetles (*Leptinotarsa decemlineata*) on a dry surface, which were kept prior the experiment under humid conditions if compared to beetles kept prior the experiment under dry conditions. In the present study we observed a distinct difference between both types of experiments (Experiment 1 and 2). The differences between the measured traction forces, although significant, are smaller in case of all humidity levels measured on one day (Experiment 2), than in the case of one humidity level measured per day (Experiment 1). Since in Experiment 2 the humidity levels were randomized for individual runs, there were some beetles that were measured first at higher humidity and then at lower humidity. In combination with relatively short durations between individual runs, those individuals, according to the idea of changing material properties, should then show a better performance also at the lower humidity level if compared to Experiment 1, thus, reducing the observable differences between different humidity levels.

Although the hypothesis of changing material properties seems to be quite plausible, there are also several aspects speaking against it. The tarsal secretions in hairy attachment pads of beetles were found to be mainly lipid-based [22, 26, 27, 49, 50], with only a small volatile fraction (likely water) [30]. Peisker et al. [46] speculated that these secretions may have an additional function, apart from forming capillary bridges, which is to cover setal tips and thereby preventing the desiccation of setal tips, thus, keeping them in a soft rubber-like state. In this light, it seems unlikely that the setal tips absorb water from ambient humidity. However, this hypothesis has not been confirmed so far and little is known about where tarsal liquids are actually secreted and delivered in hairy attachment pads of beetles. Moreover, assuming capillarity to be responsible for the generated traction forces, it is not clear how the proposed increase in the viscoelastic bulk energy dissipation with increasing humidity should affect the attachment ability in this case, since the contact is mediated by the tarsal secretion and not by a solid–solid contact.

Coming back to the abovementioned mechanisms of capillary forces and/or changes in the effective short-range van der Waals interaction, which might influence the humidity dependent adhesion in geckos [10], it needs to be mentioned that both mechanisms require water to be present at the contact interface. It can be either absorbed water on the substrate, whose actual amount depends on the relative humidity and on the surface chemistry [19, 51] or capillary condensation [52]. Due to the high content of lipids in the tarsal secretions of the beetles, a direct effect of water capillary bridges can be neglected. Though, it cannot totally be excluded that the composition and properties of the tarsal secretion are altered by ambient humidity. Thus, for example, different contributions from viscous forces could be expected [31, 53]. For *C. septempunctata*, viscosity of the tarsal secretion was found to be of order 20 mPa s [31] and droplets of volumes around 1 fL of the tarsal secretion made contact angles smaller than 10° with freshly cleaved mica surfaces (calculation based on droplet geometries reported in [30]). Moreover, it is known that micro-emulsions may have complex structures depending on the particular water-to-oil ratio [54], which might also have significant effects on beetle adhesion. Indeed, it has been shown that the adhesive pad fluids of insects contain minute droplets of water that are crucial for the function of the adhesive system [28].

The reduction in traction forces observed in the present study at 99% RH (without visible water condensation) may be explained by a sufficient amount of adsorbed water on the substrate. It has been shown that the friction behaviour between two solids crucially depends on the water layer thickness between them [55]. At critical thickness of adsorbed monolayers water becomes liquid-like, which may render the onset of an "aquaplaning effect". Accordingly, in adhesion experiments of setal pads, adhesion rapidly dropped as soon as water condensation became visible on the substrate or moistening of substrate and/or pad surface ([14, 56] and this study).

In nature, animals are constantly exposed to certain humidity. Most of insects live on plants containing a humid boundary layer in the vicinity of the plant surface. The humidity close to leaf surfaces ranges between 30 and 75% [57]. It is probable that the insects' adhesive pad is adapted to work most efficient under these conditions.

Finally, comparing absolute values in the attachment forces of *C. septempunctata* beetles, we found evidence for a sexual dimorphism. Male beetles generated higher adhesion than females, which was previously reported and explained by the presence of specialized setae with discoidal tips that can generate higher adhesion on smooth substrates due to their advantageous, geometry-induced homogenous stress distribution [58–60]. Similar results were previously obtained for chysomelid beetles [5, 61]. The dimorphism is explained by the need of males to stay attached to the smooth elytrae of females during copulation and mate guarding. However, the attachment ability of males and females is similarly affected by humidity and, thus, underlying mechanisms of humidity dependent adhesion seem to be similar in both sexes.

2.5 Conclusions

The present study demonstrates, for the first time, the effect of ambient humidity on a wet adhesive pad that is supplemented by a lipid-based secretion. This shows that humidity similarly affects the function of both dry and wet adhesive pads. However, the underlying mechanisms leading to this effect remain unclear. To be able to distinguish between different mechanisms (changing material properties, water layer thickness, changing tarsal secretion properties), experiments with substrates of different surface free energy should be performed in the future.

Acknowledgements This book chapter is adapted from the publication Heepe, Wolff, and Gorb (2016) Influence of ambient humidity on the attachment ability of ladybird beetles (*Coccinella septempunctata*). Beilstein J. Nanotechnol. 7, 1332–1329. The authors like to thank J. Oesert, L. Reichel and H. Peisker for help with the experimental setup. V. Kastner (Max Planck Institute for Metals Research, Stuttgart, Germany) and U. Köhler are acknowledged for improvement of language of an early version of the manuscript. This work was partially supported by CARTRIB Project of The Leverhulme Trust to SG.

References

1. S.N. Gorb, *Attachment Devices of Insect Cuticle* (Kluwer Academic Publishers, Dordrecht, 2001)
2. S.N. Gorb, R.G. Beutel, Naturwissenschaften **88**, 530 (2001). doi:10.1007/s00114-001-0274-y
3. S.N. Gorb et al., Integr. Comp. Biol. **42**, 1127 (2002)
4. W.J. Federle, J. Exp. Biol. **209**, 2611 (2006)
5. D. Voigt, J.M. Schuppert, S. Dattinger, S.N. Gorb, J. Insect Physiol. **54**, 765 (2008)
6. E.V. Gorb, N. Hosoda, C. Miksch, S.N. Gorb, J. R. Soc. Interface **7**, 1571 (2010)
7. J.O. Wolff, S.N. Gorb, J. Exp. Biol. **215**, 179 (2012)
8. B. Prüm et al., Acta Biomater. **9**, 6360 (2013)
9. C. Grohmann, A. Blankenstein, S. Koops, S.N. Gorb, J. Exp. Biol. **217**, 4213 (2014)
10. G. Huber et al., Proc. Natl. Acad. Sci. U. S. A. **102**, 16293 (2005)
11. P.H. Niewiarowski et al., PLoS ONE **3**, e2192 (2008)
12. J.B. Puthoff, M.S. Prowse, M. Wilkinson, K. Autumn, J. Exp. Biol. **213**, 3699 (2010)
13. M.S. Prowse et al., Acta Biomater. **7**, 733 (2011)
14. J.O. Wolff, S.N. Gorb, Proc. R. Soc. Lond. Ser. B **279**, 139 (2011)
15. D.N. Ferro, E.E. Southwick, Environ. Entomol. **13**, 926 (1984)
16. P.B. Miranda, L. Xu, Y.R. Shen, M. Salmeron, Phys. Rev. Lett. **81**, 5876 (1998)
17. D.B. Asay, S.H. Kim, J. Phys. Chem. B **109**, 16760 (2005)
18. M. Binggeli, C.M. Mate, Appl. Phys. Lett. **65**, 415 (1994)
19. X. Xiao, L. Qian, Langmuir **16**, 8153 (2000)
20. J. Jang, G.C. Schatz, M.A. Ratner, J. Chem. Phys. **120**, 1157 (2004)
21. D.B. Asay, S.H. Kim, J. Chem. Phys. **124**, 174712 (2006)
22. S. Ishii, Appl. Entomol. Zool. **22**, 222 (1987)
23. T. Eisner, D.J. Anesnansley, Proc. Natl. Acad. Sci. U. S. A. **97**, 6568 (2000)
24. W. Vötsch et al., Insect Biochem. Mol. Biol. **32**, 1605 (2002)
25. M.G. Langer, J.P. Ruppersberg, S.N. Gorb, Proc. R. Soc. Lond. Ser. B **271**, 2209 (2004)
26. S.F. Geiselhardt, S. Geiselhardt, K. Peschke, Chemoecology **19**, 185 (2009)
27. S.F. Geiselhardt et al., J. Insect Physiol. **56**, 398 (2010)

28. J.-H. Dirks, C.F. Clemente, W.J. Federle, J. R. Soc. Interface **2010**(7), 587 (2010)
29. O. Betz, in *Biological Adhesive Systems. from Nature to Technical and Medical Application*, ed. by J. Byern, I. Grunwald (Springer, Vienna, 2010), p. 307
30. H. Peisker, S.N. Gorb, J. Exp. Biol. **215**, 1266 (2012)
31. H. Peisker, L. Heepe, A.E. Kovalev, S.N. Gorb, J. R. Soc. Interface **11**, 20140752 (2014)
32. B. Abou et al., J. R. Soc. Interface **7**, 1745 (2010)
33. J.-H. Dirks, W. Federle, Soft Matter **7**, 11047 (2011)
34. A.E. Kovalev, M. Varenberg, S.N. Gorb, Soft Matter **8**, 7560 (2012)
35. A.E. Kovalev, A.E. Filippov, S.N. Gorb, J. R. Soc. Interface **11**, 20130988 (2013)
36. P. Drechsler, W.J. Federle, Comp. Physiol. A **192**, 1213 (2006)
37. N.E. Stork, J. Nat. Hist. **17**, 583 (1983)
38. M.E.N. Majerus, *Ladybirds* (Harper Collins, London, 1994)
39. I. Hodek, J.P. Michaud, Eur. J. Entomol. **105**, 1 (2008)
40. E.V. Gorb et al., J. Exp. Biol. **208**, 4651 (2005)
41. A. Kitchner, J.F.V. Vincent, J. Mater. Sci. **1987**(22), 1385 (1987)
42. S.N. Collins et al., J. Mater. Sci. **33**, 5185 (1998)
43. A.M. Taylor, R.H. Bonser, J.W. Farrent, J. Mater. Sci. **39**, 939 (2004)
44. N. Barbakadze, S. Enders, S. Gorb, E. Arzt, J. Exp. Biol. **209**, 722 (2006)
45. B. Chen, H. Gao, Int. J. Appl. Mech. **2**, 1 (2010)
46. H. Peisker, J. Michels, S.N. Gorb, Nat. Commun. **4**, 1661 (2013)
47. T.J. Weis-Fogh, J. Exp. Biol. **37**, 889 (1960)
48. D. Voigt, J.M. Schuppert, S. Dattinger, S.N. Gorb, J. Zool. **281**, 227 (2010)
49. A. Kosaki, R. Yamaoka, Jpn. J. Appl. Entomol. Zool. **40**, 47 (1996)
50. O. Betz, J. Morphol. **255**, 24 (2003)
51. B. Bhushan, *Handbook of Micro/Nano Tribology* (CRC Press, New York, 1998)
52. M. Scherge, S.N. Gorb, *Biological Micro- and Nanotribology—Nature's Solutions* (Springer, Berlin, 2001)
53. J. Stefan, Ann. Phys. (Berlin, Ger.) **230**, 316 (1875)
54. P.G. De Gennes, C. Taupin, J. Phys. Chem. **86**, 2294 (1982)
55. M. Scherge, X. Li, J.A. Schaefer, Tribol. Lett. **6**, 215 (1999)
56. A.Y. Stark, T.W. Sullivan, P.H. Niewiarowski, J. Exp. Biol. **215**, 3080 (2012)
57. T. Boulard et al., Agric. For. Meteorol. **110**, 159 (2002)
58. G. Carbone, E. Pierro, S.N. Gorb, Soft Matter **7**, 5545 (2011)
59. L. Heepe, A.E. Kovalev, A.E. Filippov, S.N. Gorb, Phys. Rev. Lett. **111**, 104301 (2013)
60. L. Heepe, G. Carbone, E. Pierro, A.E. Kovalev, S.N. Gorb, Appl. Phys. Lett. **104**, 011906 (2014)
61. Y. Pelletier, Z. Smilowitz, Can. Entomol. **119**, 1139 (1987)

Chapter 3
Roughness Versus Chemistry: Effect of Different Surface Properties on Insect Adhesion

Elena V. Gorb, Matt W. England, Tomoya Sato, Makoto Yagihashi, Atsushi Hozumi and Stanislav N. Gorb

Abstract In order to study the effect of different surface properties on the attachment ability of seven-spotted ladybird beetles *Coccinella septempunctata*, traction forces of both male and female beetles were systematically measured on eight types of surface, each with different chemical and topographical properties. Three of these surfaces were smooth with different surface wettabilities, including two types of hydrophobic monolayers, with alkyl- and perfluoroalkyl-terminated functional groups. The third smooth surface was a hydrophobic/oleophilic alkylsilane-derived hybrid film, showing low contact angle hysteresis for water and *n*-hexadecane. In addition, we also tested three rough surfaces with different roughness values and different wettabilities. Smooth, hydrophilic silicon wafers (Si) and glass surfaces were used as two reference surfaces. The results of traction force tests clearly demonstrated that chemical surface properties, such as static/dynamic dewettability of water and oil caused by specific chemical compositions, had a certain effect on the beetle attachment: hydrophobic and oleophobic substrates tend to reduce the attachment forces. Surface roughness was found to be the dominant factor, strongly affecting the attachment ability of the beetles up to complete diminishing of attachment.

E.V. Gorb (✉)
Department of Functional Morphology and Biomechanics, Kiel University, Am Botanischen Garten 9, 24118 Kiel, Germany
e-mail: egorb@zoologie.uni-kiel.de

M.W. England · T. Sato · A. Hozumi
National Institute of Advanced Industrial Science and Technology (AIST), 2266-98, Anagahora, Shimoshidami, Moriyama, Nagoya 463-8560, Japan

M. Yagihashi
Nagoya Municipal Industrial Research Institute, 4-41, Rokuban, Atsuta, 456-0058 Nagoya, Japan

S.N. Gorb
Department of Functional Morphology and Biomechanics, Zoological Institute, Kiel University, Am Botanischen Garten 9, 24118 Kiel, Germany

L. Heepe et al. (eds.), *Bio-inspired Structured Adhesives*, Biologically-Inspired Systems 9, DOI 10.1007/978-3-319-59114-8_3

3.1 Introduction

The development of functional coatings that artificially mimic the properties of surfaces found in nature [1–4] to produce exceptional wetting/dewetting properties, such as superhydrophobicity, superhydrophilicity, and superoleophobicity (more commonly known as superomniphobicity), has been a major topic for research over the past decade [5–15].

There are countless examples of functional surfaces inspired by plants, such as lotus leaves [5, 13] and the pitchers of carnivorous plants [9, 14] that can be used to tune the wetting/dewetting properties of surfaces on various substrates. Certain of these natural surfaces can effectively prevent wetting by water, while simultaneously protecting against attachment by insects by taking advantage of the same or very similar surface features [16–23]. Unfortunately, these natural anti-attachment properties have received relatively little attention from researchers working on surface science and engineering [24, 25]. Another possible reason might be that the properties of unwettable biological surfaces, other than surface wetting/dewetting, have not been tested. The question of whether surface chemistry or surface roughness is primarily responsible for natural anti-attachment properties has also not been fully resolved. Therefore, a comparative study of the attachment behavior of insects on artificially designed (low/high surface energy) surfaces of varying surface roughness, has been postulated as an effective strategy to identify the most important parameters influencing insect attachment.

Many insects, including beetles, can attach to inverted surfaces using specific hairy adhesive pads, covered with tenent setae, which secrete an adhesive fluid which typically consists of a mixture of alcohols, fatty acids, and hydrocarbons [26–32]. Several hypotheses exist on how plant surfaces prevent insect attachment. These are typically based on (1) the reduction of the contact area between the substrate and the insect adhesive pad through surface micro-roughness, (2) a decrease in substrate surface energy that enhances dewetting of the insect attachment fluid, (3) fluid absorption by textured substrates, (4) contamination of the insect pads by easily erodible particles of the substrate, and (5) the reduction of wetting by pad fluid due to coverage of the substrate by another fluid (or solid which can be dissolved by the pad fluid) [22]. These mechanisms are to some extent conventional strategies utilized in functional surface design that gives us a unique chance to develop artificial surfaces with such properties, and test their anti-adhesive effects on insects [5–8, 10–13, 15]. However, recent studies on insect attachment have yielded contradicting results. For example, a previous experimental study on attachment of the beetle *Gastrophysa viridula* to the leaf surface of its host plant *Rumex obtusifolius*, and artificial micro-roughened and smooth (hydrophobic and hydrophilic) surfaces, has shown a stronger insect performance on smooth surfaces when compared to those with microroughness [33]. It was also found that surface hydrophobicity alone resulted in some decrease in the attachment force of the beetles, but when combined with surface micro-roughness it caused an even more pronounced reduction. Prüm et al. [17] measured the traction force of the beetle *Leptinotarsa decemlineata* on different plant surfaces and their artificial replicas and reported that surface roughness exerted a strong influence on attachment, whereas surface chemistry was found to have no significant influence, despite both of these affecting the magnitude of contact angles (CAs) of water. Additionally, the attachment of the leaf beetle *Gastrophysa viridula* did not strongly depend on the free energy of the surface of the substrate [34]. More recently, the attachment strength of the beetle *Galerucella nymphaeae* on surfaces with different surface energies, showing CAs in the range from 6° to 109°, was examined [35]. These beetles, both at their larval and adult stages, showed the highest forces on surfaces with water CAs close to 83° (similar to those of their host plant), while hydrophilic (CAs of 6° and 26°) and hydrophobic (CA of 109°) surfaces caused a reduction of their adhesive ability.

A strong dependence of adhesive ability on the chemistry of the substrates during locomotion underwater was recently found for the beetle *Gastrophysa viridula* [34]. Using air bubbles trapped between their adhesive setae, these beetles are able to walk on flooded substrates, including those under water. Their attachment to hydrophilic surfaces was reduced under water when compared to their attachment in air, whereas the attachment to hydrophobic surfaces under water was considerably stronger, and comparable to that observed in air. The oil-covered hairy pads on the feet of the beetle show a pinning effect, which retains air bubbles, and capillary attachment is produced by bubbles in contact with the hydrophobic substrate. Additionally, the liquid bridges between the foot and the substrate also produce capillary forces. Inspired by this idea, artificial silicone polymer structures with underwater adhesive properties were fabricated [34].

Thus, the relationship between surface structures and the attachment of insects, in combination with their particular chemical/physical properties, has not yet been fully resolved. Therefore, in order to obtain a deeper understanding of this bio-attachment phenomenon, it is crucial to systematically investigate the influence of both the surface chemistry and surface morphology on insect attachment properties using a greater range of surfaces with different surface wettabilities, in combination with both smooth and rough surface textures. Specifically, we focused our attention on both the static CAs of water and oil (*n*-hexadecane), and their dynamic [advancing (θ_A) and receding (θ_R)] CAs, especially CA hysteresis (Δ_θ, the difference between the values of θ_A and θ_R), which is a mass-independent measure of the resistance to macroscopic liquid drop movement on inclined surfaces.

In this study, we investigated eight different surfaces. Three of these were smooth with different surface wettabilities, including two types of hydrophobic monolayers, with alkyl- and perfluoroalkyl-terminated functional groups. The third smooth surface was a hydrophobic/oleophilic alkylsilane-derived hybrid film, showing low CA hysteresis for water and *n*-hexadecane. In addition, we also studied three rough surfaces with different wettabilities. For these, we used a commercially available superhydrophobic coating system, which was also used to prepare a rough superhydrophilic surface, by subjecting it to vacuum UV (VUV) light treatment. A superomniphobic surface (defined here as a surface exhibiting both superhydrophobicity and superoleophobicity) was created using candle soot as templates for SiO_2 nanoparticles, which were modified with a perfluoroalkylsilane monolayer. All surfaces were prepared on Si substrates. We measured traction forces of adult seven-spotted ladybird beetles *Coccinella septempunctata*, both males and females, on these six sample surfaces and two reference surfaces, i.e., smooth, hydrophilic silicon wafers (Si) and glass surfaces. Our sample surfaces displayed a wide range of surface chemical and topographical properties, and while both of these had a significant effect on the magnitude of CAs for probe liquids, the attachment abilities of the ladybird beetles were found to be predominantly influenced by the surface topography.

3.2 Experimental

3.2.1 Materials

Ethanol, 0.01 M HCl, and *n*-hexadecane were purchased from Wako Pure Chemical Industries Ltd. (Osaka, Japan). Tetramethoxysilane (TMOS, $Si(OCH_3)_4$) and *n*-octadecyltrimethoxysilane (ODS, $CH_3(CH_2)_{17}Si(OCH_3)_3$) were purchased from Tokyo Chemical Industry Co., Ltd. (Tokyo, Japan). (Heptadecafluoro-1,1,2,2-tetrahydrodecyl) trimethoxysilane (FAS17, $CF_3(CF_2)_7CH_2CH_2Si(OCH_3)_3$), (heptadecafluoro-1,1,2,2-tetrahydrodecyl)trichlorosilane (FAS17-Cl, $CF_3(CF_2)_7CH_2CH_2SiCl_3$), and decyltriethoxysilane (C_{10}, $CH_3(CH_2)_9Si(OC_2H_5)_3$) were purchased from Gelest Inc. (Morrisville, PA, USA). Never Wet™ (superhydrophobic coatings) was purchased from Rust-Oleum Corporation (Vernon Hills, IL, USA). All chemicals were used as received without further purification.

3.2.1.1 Preparation of Flat and Rough Sample Surfaces

Two smooth, hydrophobic monolayer-covered surfaces, terminated with octadecylsilyl ($CH_3(CH_2)_{17}$-) or perfluoroalkyl ($CF_3(CF_2)_7CH_2CH_2$-) groups, were prepared using chemical vapor deposition (CVD) of ODS or FAS17 [36], respectively. UV-ozone treated Si substrates (2×2 cm^2 and 5×5 cm^2) were placed on a heat-resistant glass plate ($18 \times 18 \times 0.4$ cm^3) with a small aluminum-made vessel, or alternately placed in a Teflon container with a glass vessel, containing 0.2 mL of organosilane (ODS or FAS17), in a dry N_2 atmosphere at less than 5% relative humidity. Another heat-resistant glass plate was then placed on top of it using an O-ring (approximately 150 mm diameter and 8 mm thickness) as a spacer, and the four corners of the glass plates were secured using four clamps. Alternately, the Teflon container was sealed with an airtight screw-on Teflon cap. The reaction container was then heated for three days in an oven maintained at 180 °C for ODS and 150 °C for FAS17. Finally, the treated samples were rinsed with *n*-hexane, then water, and finally blown dry with a stream of N_2. Besides these two flat monolayer-covered surfaces, we also prepared a smooth alkylsilane (C_{10})-derived hybrid film using conventional co-hydrolysis and co-condensation [10]. Briefly, precursor solutions were prepared by mixing C_{10} and TMOS in an ethanol/hydrochloric acid solution for 24 h at room temperature (25 ± 2 °C). The typical molar ratio of the precursor solution was 0.73 C_{10}:2.92 TMOS:32 EtOH:14 H_2O:7.7 × 10^{-3} HCl. The precursor solution was then spin-coated (500 rpm for 5 s and 1000 rpm for 10 s) onto UV–ozone-cleaned Si substrates (5 × 5 cm^2) at room temperature, under a relative humidity of 40 ± 5%. All samples were dried in air at room temperature for more than 24 h. Details of our preparation methods for the monolayers and hybrid film have been described elsewhere [10, 12]. Three rough surfaces, each showing different wetting properties (superhydrophobicity, superhydrophilicity, and superomnipobicity) and morphologies were prepared as follows. The commercially available Never Wet™ coating system was used to prepare two rough superhydrophobic surfaces. Base coats were first deposited onto UV-cleaned Si substrates (2 × 2 cm^2 and 5 × 5 cm^2), then dried in air at room temperature (25 ± 2 °C) for more than 30 min. Next, topcoats were deposited onto the surfaces and cured at 100 °C for 24 h, hereafter referred to as Never Wet. Superhydrophilic surfaces were prepared by exposing superhydrophobic Never Wet surfaces to VUV light generated from an excimer lamp (Ushio Inc., UER20-172 V; λ = 172 nm and 10 mW/cm^2) at 10^3 Pa for 2 min, hereafter referred to as VUV-Never Wet. Superomniphobic surfaces were prepared according to a method modified from a report previously published by Deng et al. [37]. The candle-soot-covered Si substrates (2 × 2 cm^2 and 5 × 5 cm^2) were exposed to VUV light at 10^3 Pa for 30 min. The samples were then exposed to TMOS vapor for 4 h at 80 °C using the CVD method described previously. Next, the samples underwent thermal calcination in air for 3 h at 600 °C in order to remove any organic components, and thus obtain SiO_2 nanostructures. After VUV irradiation at 10^3 Pa for 30 min, the samples were finally exposed to a FAS17-Cl vapor at room temperature (25 ± 2 °C) for more than 3 h under reduced pressure, hereafter referred to as Soot-TMOS-FAS17Cl.

3.2.1.2 Characterization of Sample Surfaces

The thicknesses of the ODS and FAS17 monolayers were measured using ellipsometry (Philips, PZ2000). The thicknesses of the superhydrophobic (Never Wet), superhydrophilic (VUV-Never Wet), and superomniphobic films (Soot-TMOS-FAS17Cl) were estimated from cross-sectional images acquired by a scanning electron microscope (SEM, Phenom Pro Scanning Electron Microscope, Phenom World). The surface morphologies of the samples were either observed using the same SEM system or by atomic force microscope in a tapping mode (AFM, XE-100, Park Systems), with a Si probe (910 M-NCHR; spring constant of 42 N/m and response frequency of 330 kHz, Park Systems). The surface roughness (root-mean square roughness, R_{rms}) was estimated using two separate techniques due to the huge disparity in the size of surface textures on smooth and rough samples. Our five smooth samples (glass, Si, ODS, FAS17, and C_{10}-hybrid) were estimated by AFM, while those of the three rough samples (Never Wet, VUV-Never Wet, and Soot-TMOS-FAS17Cl) were measured using a stylus profilometer (Surftest SJ-301, Mitutoyo Corp.).

Static and dynamic CA data (θ_S and θ_A/θ_R values) for water and oil (*n*-hexadecane) were recorded using CA goniometers (model CA-V150, Kyowa Interface Science). The CA data reported here were determined by averaging values measured at 5–10 different points on each surface of the sample. All values for each sample were in a range of ±2°.

Surface chemical properties of the samples were studied by applying X-ray photoelectron spectroscopy (XPS). Spectra were obtained using a Physical Electronics Quantum 2000 spectrometer with 200 μm spot size and monochromatic Al Kα radiation (1486.68 eV). The X-ray source was operated at 50 W and 15 kV with the pass energy of the analyzer at 29.35 eV. The pressure in the analysis chamber was around 6×10^{-9} Pa during all measurements. Core-level signals were obtained at a photoelectron takeoff angle of 15° (surface-sensitive mode) with respect to the sample surface. The binding energy (BE) scales were referenced to 284.6 eV, as determined by the locations of the peak maxima of the C 1s spectra of a hydrocarbon (CH_x). Surface compositions were determined by the corresponding core-level spectral area ratios, calculated using the relative sensitivity factor method. The relative error for all XPS data used to determine surface composition was estimated to be ±2%.

3.2.1.3 Insect Force Tests

Insect attachment ability was studied in traction experiments with tethered adult seven-spotted ladybird beetles *Coccinella septempunctata* (Coleoptera, Coccinellidae) by using a load cell force transducer (10 g capacity, Biopac Systems Ltd., Santa Barbara, CA, USA) as described by Gorb et al. [23]. Insects were collected near Stohl (surroundings of Kiel, Germany). Forces generated by both males and females walking horizontally on eight different surfaces [five smooth and three rough, each with different surface chemical and physical properties, (see above)] were measured. Obtained force–time curves were used to estimate the maximal traction force. Experiments were performed at 23 °C temperature and 26–29% relative humidity. We tested 10 male and 10 females and carried out 160 traction tests in total.

The experimental design included eight successive force tests with each insect individual: first on smooth, hydrophilic glass to check insect fitness, then on the other seven samples (Si, ODS, FAS 17, C_{10}-hybrid, Never Wet, VUV-Never Wet, and Soot-TMOS-FAS17Cl) in a random order. In order to regain their superhydrophilic properties, Si and VUV-Never Wet samples, prior to the force tests, were plasma treated for 8 min with compressed air (pressure = 2.0 mbar) by applying a low pressure plasma system (Zepto, Diener electronic, Ebhausen, Germany), working at 100 W and 40 kHz regime. The samples were used in traction force experiments during maximum of 60 min after the plasma treatment.

Statistical analyses of the force values were carried out with SigmaStat 3.5 (Systat Software Inc., Point Richmond, CA, USA). For one-way ANOVA (Kruskal-Wallis one-way ANOVA on ranks), data obtained with males and females were pooled together. The effects of the sex of insect individuals and the surface type on the traction force values were examined using two-way ANOVA. Pairwise comparisons of sexes and surfaces were performed with the Tukey test. Force values are given as the mean ± standard deviation (SD).

3.3 Results and Discussion

As shown in Fig. 3.1, all sample surfaces observed by AFM appeared to be smooth over the entire scanning areas (3 × 3 μm^2). The average R_{rms} values of ODS- and FAS17-monolayer-covered Si surfaces were very low (Table 3.1), indicating that the Si substrates could be modified uniformly (their thicknesses were between 1 and 2 nm) without any marked change in morphology (compared with the R_{rms} of the Si surface). Although the film thickness and R_{rms} of the C_{10}-hybrid film itself was up to 700 times thicker (700 nm) and approximately 10 times higher (1.24 nm), respectively, than those of ODS and FAS17 monolayers, its surface was also fairly smooth. In contrast, SEM images shown in Fig. 3.2 confirmed that films of the three rough surfaces, Never Wet, VUV-Never Wet, and Soot-TMOS-FAS17Cl, were substantially thicker (estimated by cross-sectional SEM images) than the monolayers and hybrid film, and also highly textured (Table 3.1). Due to these particulate film formation on these surfaces, their R_{rms} values were extremely high, around three to four orders of magnitude higher than those of the four smooth surfaces, as estimated by a stylus profilometer.

Surface dewetting properties were investigated by measuring static/dynamic CAs (θ_S, θ_A, and θ_R) of water and *n*-hexadecane droplets. As shown in Table 3.1, they were found to be strongly dependent on both surface chemical and topographical properties. Water CA measurements confirmed that the glass, Si and VUV-Never Wet surfaces were superhydrophilic with θ_S close to zero, while the other five surfaces exhibited either hydrophobic (ODS, FAS17 and C_{10}-hybrid surfaces; water CAs greater than 100°) or superhydrophobic (Never Wet and Soot-TMOS-FAS17Cl surfaces; water CAs larger than 150°) properties. Water droplets on the latter surfaces moved very easily and would roll off at very low tilt angles of the substrate because of their extremely large CAs and low CA hysteresis ($\Delta\theta$ = 2°–5°). On the other hand, while the static/dynamic CAs of smooth C_{10}-hybrid surfaces were considerably lower than those of the superhydrophobic surfaces, the CA hysteresis of water on the hybrid film was also low ($\Delta\theta$ = 8°, the lowest of the smooth surfaces). Thus, contrasting with the smooth ODS and FAS17 monolayer surfaces, water droplets similarly slid off the C_{10}-hybrid surfaces at low tilt angles, regardless of the magnitude of their static/dynamic CAs [10, 36].

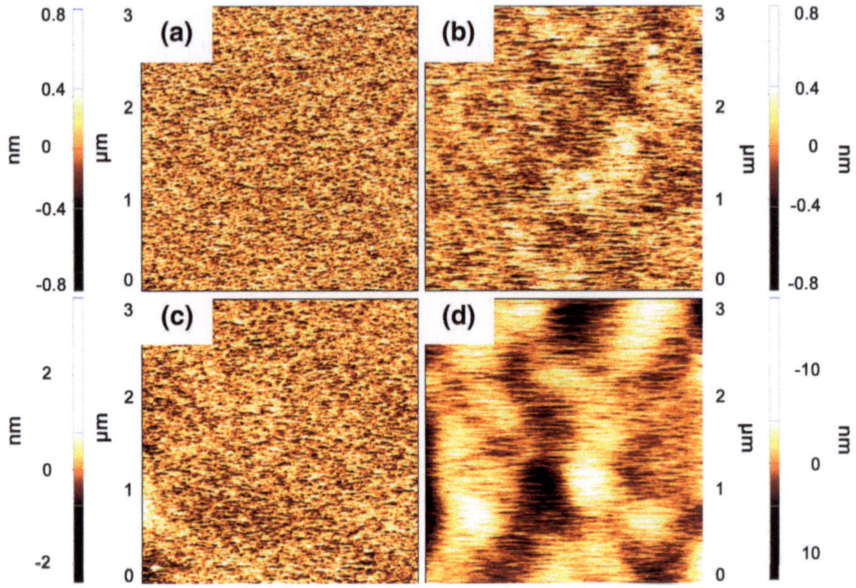

Fig. 3.1 Typical AFM images of Si substrates: **a** before and after **b** ODS-monolayer formation, **c** FAS17-monolayer formation, and **d** C_{10}-hybrid film formation (scan areas are 3 × 3 μm^2)

Table 3.1 Surface properties of the samples

Sample	θ_S water	θ_A/θ_R water	θ_S *n*-hexadecane	θ_A/θ_R *n*-hexadecane	Thickness	R_{rms}^a
Si	–	–	–	–	–	0.16 nm
ODS	103°	106°/96°	10°	12°/5°	1.9 nm	0.23 nm
FAS17	113°	121°/108°	70°	73°/60°	1.1 nm	0.27 nm
C_{10}-hybrid	111°	114°/106°	35°	36°/35°	700 nm	1.24 nm
Never Wet	155°	160°/158°	Dissolved	–	~35 μm	5.2 μm
VUV-Never Wet	–	–	Dissolved	–	~35 μm	6.3 μm
Soot-TMOS-FAS17Cl	163°	165°/160°	155°	161°/153°	<1.6 μm	0.2 μm

[a] R_{rms} values of Si with and without an ODS-monolayer, FAS17-monolayer, and C_{10}-hybrid film were estimated using AFM images (3 × 3 μm^2) shown in Fig. 3.1; those of Never Wet, VUV-Never Wet, and Soot-TMOS-FAS17Cl were estimated by a stylus profilometer

The CAs of *n*-hexadecane showed significantly greater variation. Monolayers of ODS and FAS17, and C_{10}-hybrid film surfaces were all oleophilic with θ_S of 10°, 70°, and 35°, respectively. Among them, the C_{10}-hybrid film surface in particular exhibited negligible CA hysteresis ($\Delta\theta$ = 1°), and excellent dynamic dewettability [10]. Unfortunately, the Never Wet surfaces, both before and after VUV irradiation, were dissolved by the *n*-hexadecane. However, the Soot-TMOS-FAS17Cl surface exhibited very high θ_S (155°) and low CA hysteresis ($\Delta\theta$ = 8°) for *n*-hexadecane,

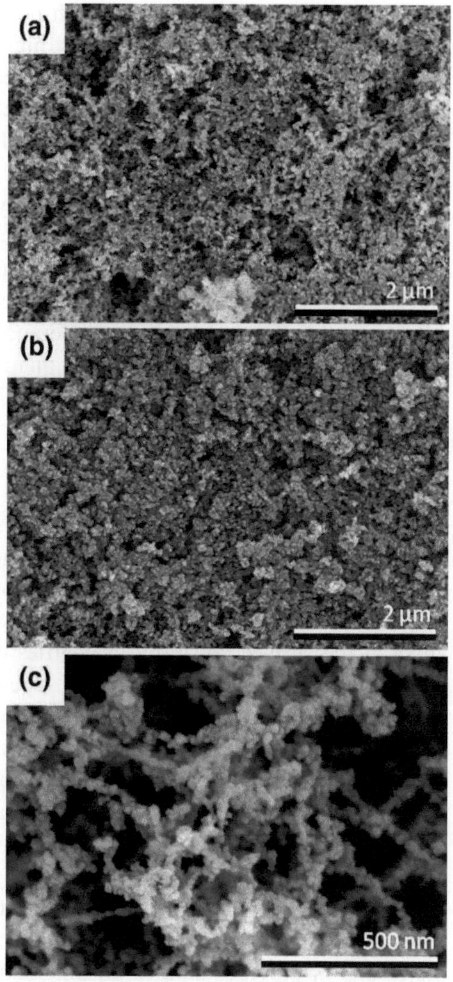

Fig. 3.2 Typical top-down-view SEM images of sample surfaces: **a** Never Wet, **b** VUV-Never Wet, and **c** Soot-TMOS-FAS17Cl

and thus *n*-hexadecane droplets on this surface could move, without pinning, more smoothly than they could on a smooth FAS17 surface ($\Delta\theta$ = 13°).

The relationships between surface wettability and chemical composition of our sample surfaces were established using XPS. The surface chemical compositions of each sample are summarized in Table 3.2. All non-perfluorinated sample surfaces were primarily composed of three elements [Si, oxygen (O), and carbon (C)]. As expected, the Never Wet surface showed a decrease in C concentration (ca. 4 atom%) and an increase in the surface O concentration (ca. 6 atom%) following only 2 min VUV irradiation, in agreement with its marked increase in hydrophilicity (superhydrophilicity). Interestingly, the XPS analysis also revealed that the Never Wet surface had the lowest concentration of C (ca. 24 atom%) of any non-perfluorinated sample surface and showed excellent static/dynamic dewetting behavior. In addition, in spite of both perfluorinated sample surfaces (FAS17 and Soot-TMOS-FAS17Cl) showing very high surface fluorine concentrations (above 40 atom%), the former surface displayed inferior static/dynamic dewettability compared to that of the latter one. This clearly indicated that surface roughness, rather than C and F concentrations, had the strongest influence upon surface wettability.

Based on the surface chemical and physical properties of our samples shown above, we next examined the attachment ability of *C. septempunctata* beetles by measuring their traction forces on these surfaces. The average traction force produced by the insects on test surfaces ranged from 0.56 mN to 9.82 mN (Fig. 3.3). All insects performed well on reference smooth, hydrophilic glass surfaces, showing force values either higher than (compared to C_{10}-hybrid, Never Wet, VUV-Never Wet, and Soot-TMOS-FAS17Cl; Tukey test, $P < 0.05$) or similar to (compared to Si, ODS, and FAS17; Tukey test, $P > 0.05$) those obtained on other samples (Fig. 3.3a). Among our samples (Kruskal-Wallis one-way ANOVA on ranks: $H_{6,139} = 123.062$, $P < 0.001$), two distinct groups can be clearly distinguished: smooth surfaces showing successful insect attachment (Si, ODS, and FAS17) and rough ones reducing the attachment (Never Wet, VUV-Never Wet, and Soot-TMOS-FAS17Cl). Interestingly, no statistical differences were detected between surfaces within each group (Tukey test, $P > 0.05$) indicating no significant effect of the surface dewetting properties on insect attachment. Only smooth C_{10}-hybrid samples showed intermediate results, differing ($P < 0.05$) from smooth Si and similar ($P > 0.05$) to rough Never Wet (both Tukey test). Both the sex of the insect individuals (Fig. 3.3b) and the surface type affected force values (two-way ANOVA: $F_{1,139} = 4.992$, $P = 0.027$ and $F_{6,139} = 40.878$, $P < 0.001$, respectively). However, there was no statistically significant interaction between these two factors (two-way ANOVA: $F_{6,139} = 0.763$, $P = 0.601$).

Our results show that the chemistry of smooth surfaces plays some role in the beetle attachment. The general trend is that hydrophobic and oleophobic substrates tend to reduce the attachment forces of beetles toward such surfaces. In addition, the results of previous studies on this matter have been fairly heterogeneous. In a recent publication, data from the literature dealing with force measurements of different insects on surfaces with different surface energies were carefully compared [35]. No significant dependence of insect attachment forces on water CAs was

Table 3.2 Sample surface compositions as estimated by XPS

Sample	Si (atom%)	O (atom%)	C (atom%)	F (atom%)
Si	60.1	39.9	–	–
ODS	27.7	41.5	30.8	–
FAS17	12.7	13.3	33.1	40.9
C_{10}-hybrid	18.2	29.1	52.7	–
Never Wet	27.7	46.1	24.2	–
VUV-Never Wet	27.6	52.0	20.4	–
Soot-TMOS-FAS17Cl	9.6	15.6	26.7	46.2

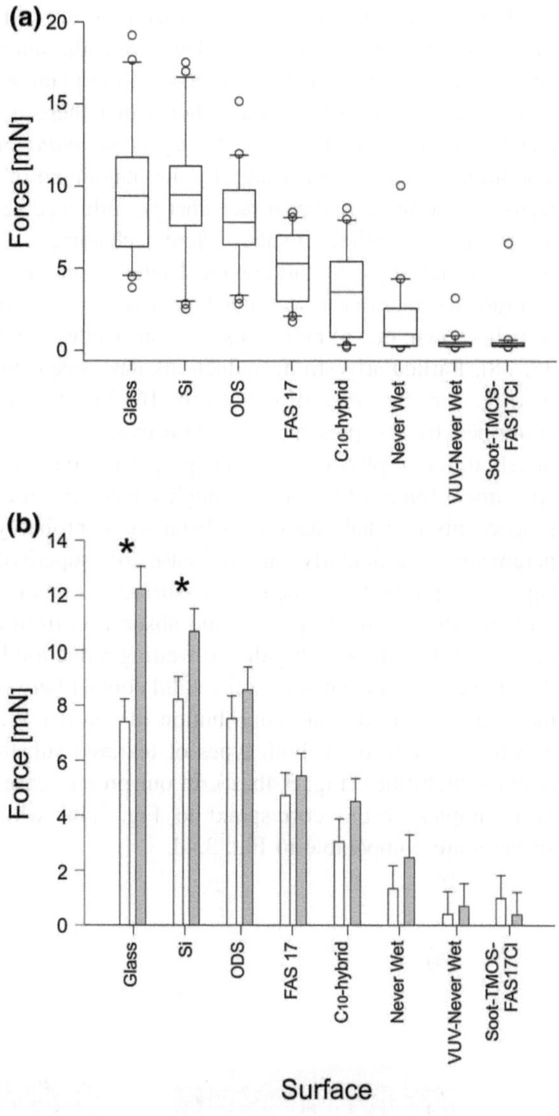

Fig. 3.3 Traction forces of *Coccinella septempunctata* beetles measured on different sample surfaces: **a** data on males and females pooled together; **b** males (*grey bars*) and females (*white bars*). *Asterisks* in **b** indicate statistically significant differences between forces generated by males versus females (Tukey test, $P < 0.05$)

shown in the five experiments recorded in the four studies compared. However, seven experimental set-ups from six different studies revealed dependence of attachment forces on water CAs. Generally in the latter studies, at surfaces with water CAs above 100°, attachment forces were lower than those on surfaces with water CAs below 40°. It is important to note that in all of these studies different species, developmental stages, sexes and experimental designs were used. In some of these studies, insect species that are strongly specialized to host plants whose leaf surfaces have very specific surface energies (water CA about 80°), such as the beetle *Galerucella nymphеae* which lives on the leaf surface of the water lily, the maximum attachment force was detected at the intermediate range of water CAs, approximately corresponding to those of the plant leaves [35].

Careful analysis of our data shows that there is no direct dependence of the attachment force of the beetle *C. septempunctata* on the surface energies of the smooth substrates, as measured by the magnitude of CAs. It also seems that other factors, in addition to the surface energy, influence beetle attachment. These factors are presumably related to the different characters of chemical substances, which may mediate physical and chemical interactions in contact to different extents. In contrast, the fact that insect attachment is very sensitive to the substrate roughness is well-known, due to numerous previous studies on the subject [16, 19–21, 23, 29, 33, 38]. Particularly strong reductions have been observed on textured substrates with R_{rms} in the range of 0.1–3 μm [16, 23, 29, 33, 38]. These results were also confirmed by the present study. However, in spite of our surfaces having completely different physical/chemical properties, there seem to be no differences in the attachment forces of beetles on rough substrates with different surface energies. We suspect this unusual attachment behavior is probably related to different physical parameters, particularly in the case of superhydrophilic/superoleophilic and superhydrophobic/superoleophobic surfaces. For the former types of surface, it is probably due to overly strong fluid absorption from the pads (Fig. 3.4a, b). In the latter case, it is presumably due to wetting reduction by the pad fluid (Fig. 3.4c, d). Additionally, at the sites of solid–solid contact between insect pads and substrates, the true contact area and contribution of van der Waals forces are believed to be effectively reduced for both types of textured substrate, despite them having different wettabilities (Fig. 3.4b, d). In our present case, Never Wet and VUV-Never Wet samples would correspond to Fig. 3.4b, while the Soot-TMOS-FAS17Cl surfaces are comparable to Fig. 3.4d.

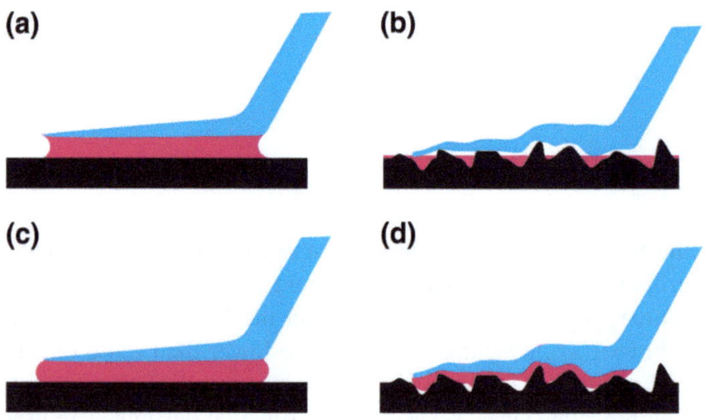

Fig. 3.4 Hypothetical wetting behavior of the pad fluid (*purple*) in the contact region between the tenent setal tip (*blue*) and different sample surfaces (*black*): **a**, **c** flat surfaces, **b**, **d** textured surfaces; **a**, **b** fluids that readily wet the surfaces, **c**, **d** fluids that poorly wet the surfaces

These effects were previously modeled to explain how the fluid flows on textured substrates with different roughness parameters and surface energies [39]. These numerical studies demonstrated that a higher density of geometrical surface structures of the rigid substrate results in a greater loss of fluid from the pad. The draining rate of the pad fluid is more rapid on fine roughness. A decreased affinity of the substrate to the pad fluid leads to significant reduction of the fluid loss. However, the substrate will not be wetted and will likely become slippery for the pad [39]. Our data demonstrated here provide clear experimental evidence for previous numerical predictions.

Attachment of male *C. septempunctata* was significantly stronger on the smooth hydrophilic surfaces. This effect was less pronounced or even vanished on our textured surfaces, which has been previously explained by the differences in the contact shape of the tenent setae [23, 38].

In summary, we have clearly shown that the insect anti-adhesive effect is due to both surface chemistry and texture, but it is primarily driven by the substrate roughness, and less by surface chemistry. It seems to be a universal effect for both dry [40, 41] and wet (but not glue-mediated) [23, 29, 38, 42] adhesive systems.

Acknowledgements This book chapter is adapted from the publication England, Sato, Yagihashi, Hozumi, Gorb, and Gorb (2016) Surface roughness rather than surface chemistry essentially affects insect adhesion, Beilstein J. Nanotechnol. 7, 1471 (2016). The authors thank Charlotte Bellmann and Ingo Grawe (both Kiel University) for their assistance with the force measurements and Emre Kizilkan (Kiel University) for the introduction into the plasma treatment device. This work was partially supported by CARTRIB Project of The Leverhulme Trust (E.V. Gorb and S.N. Gorb) and JSPS KAKENHI Grant Number JP24120005 (A. Hozumi) in Scientific Research on Innovative Areas "Innovative Materials Engineering Based on Biological Diversity".

References

1. B. Bhushan, Philos. Trans. R. Soc. Lond. A **367**, 1445 (2009)
2. K. Koch, W. Barthlott, Philos. Trans. R. Soc. Lond. A **367**, 1487 (2009)
3. A. Tuteja et al., MRS Bull. **33**, 752 (2008)
4. T.-S. Wong, T. Sun, L. Feng, J. Aizenberg, MRS Bull. **38**, 366 (2013)
5. N.A. Patankar, Langmuir **20**, 8209 (2004)
6. A. Tuteja et al., Science **318**, 1618 (2007)
7. A. Tuteja et al., Proc. Natl. Acad. Sci. U. S. A. **105**, 18200 (2008)
8. A. Hozumi, T.J. McCarthy, Langmuir **26**, 2567 (2010)
9. T.-S. Wong et al., Nature **477**, 443 (2011)
10. C. Urata, D.F. Cheng, B. Masheder, A. Hozumi, RSC Adv **2**, 9805 (2012)
11. C. Urata, B. Masheder, D.F. Cheng, A. Hozumi, Langmuir **29**, 12472 (2013)
12. C. Urata et al., Langmuir **30**, 4049 (2014)
13. S.S. Latthe, C. Terashima, K. Nakata, A. Fujishima, Molecules **19**, 4256 (2014)
14. D.F. Miranda et al., APL Mater. **2**, 056108 (2014)
15. J.T. Simpson, S.R. Hunter, T. Aytug, Rep. Prog. Phys **78**, 086501 (2015)
16. B. Prüm, R. Seidel, H.F. Bohn, T. Speck, J.R. Soc, Interface **9**, 127 (2012)
17. B. Prüm et al., Acta Biomater **9**, 6360 (2013)
18. E.V. Gorb et al., J. Exp. Biol **208**, 4651 (2005)

19. N.E. Stork, Entomol. Exp. Appl **28**, 100 (1980)
20. S.D. Eigenbrode, C. White, M. Rohde, C.J. Simon, Environ. Entomol **27**, 902 (1998)
21. S.D. Eigenbrode, N.N. Kabalo, Entomol. Exp. Appl **91**, 125 (1999)
22. E.V. Gorb, S.N. Gorb, Entomol. Exp. Appl **105**, 13 (2002)
23. E.V. Gorb, N. Hosoda, C. Miksch, S.N. Gorb, Soc. Interface **7**, 1571 (2010)
24. H. Peisker, S.N. Gorb, J. Exp. Biol **213**, 3457 (2010)
25. J. Purtov, E.V. Gorb, M. Steinhart, S.N. Gorb, Appl. Phys. A **111**, 183 (2013)
26. S.N. Gorb et al., Integr. Comp. Biol **42**, 1127 (2002)
27. S. Ishii, Appl. Entomol. Zool **22**, 222 (1987)
28. T. Eisner, D.J. Anesnansley, Proc. Natl. Acad. Sci. U. S. A **97**, 6568 (2000)
29. S.N. Gorb, *Attachment Devices of Insect Cuticle* (Kluwer Academic Publishers, Dordrecht, 2001)
30. W. Federle, M. Riehle, A.S.G. Curtis, R. Full, J. Integr. Comp. Biol **42**, 1100 (2002)
31. W. Vötsch et al., Insect Biochem. Mol. Biol **32**, 1605 (2002)
32. S.F. Geiselhardt et al., J. Insect Physiol **56**, 398 (2010)
33. E. Gorb, S. Gorb, Entomol. Exp. Appl **130**, 222 (2009)
34. N. Hosoda, S.N. Gorb, Proc. R. Soc. Lond. B **279**, 4236 (2012)
35. C. Grohmann, A. Blankenstein, S. Koops, S.N. Gorb, J. Exp. Biol **217**, 4213 (2014)
36. A. Hozumi, K. Ushiyama, H. Sugimura, O. Takai, Langmuir **15**, 7600 (1999)
37. X. Deng, L. Mammen, H.-J. Butt, D. Vollmer, Science **335**, 67 (2012)
38. D. Voigt, J.M. Schuppert, S. Dattinger, S.N. Gorb, J. Insect Physiol **54**, 765 (2008)
39. A.E. Kovalev, A.E. Filippov, S.N. Gorb, J.R. Soc, Interface **11**, 20130988 (2013)
40. G. Huber et al., Acta Biomater **3**, 607 (2007)
41. J.O. Wolff, S.N. Gorb, J. Exp. Biol **215**, 179 (2012)
42. A. Peressadko, S.N. Gorb, in *First International Industrial Conference Bionik 2004*, ed. by I. Boblan, R. Bannasch (VDI, Düsseldorf, 2004)

Chapter 4
Effect of Substrate Stiffness on the Attachment Ability in Ladybird Beetles *Coccinella septempunctata*

Lars Heepe, Dennis S. Petersen, Lisa Tölle, Jonas O. Wolff and Stanislav N. Gorb

Abstract In order to study the effect of different substrate stiffness on the attachment ability of seven-spotted ladybird beetles *Coccinella septempunctata*, we have performed friction experiments with female and male beetles on smooth silicone elastomer substrates of different stiffness, using a centrifugal force tester. Experiments were performed with intact beetles and those without claws. In addition, males were also tested on washed substrates to exclude effects of unpolymerized residues on the substrates. The results of the centrifuge experiment revealed that the attachment ability of females was not affected by the substrate stiffness within the range of tested stiffness. Males, however, showed a decreasing attachment ability with decreasing substrate stiffness. We argue that this sexual dimorphism may be explained by the presence of a specialized, discoidal seta type in males, which is not present in females. This type of setae, when softer than the substrate, presumably exhibits a homogeneous, peak-free interfacial stress distribution, accompanied with high pull-off forces necessary to detach these setae from the substrate. When the substrate, however, becomes softer than the discoidal setae, the interfacial stress distribution will change and exhibit stress peaks at the edges of

L. Heepe (✉) · D.S. Petersen · L. Tölle · J.O. Wolff · S.N. Gorb
Department of Functional Morphology and Biomechanics, Zoological Institute, Kiel University, Am Botanischen Garten 1-9, 24118 Kiel, Germany
e-mail: lheepe@zoologie.uni-kiel.de

Present Address:
L. Heepe
Mads Clausen Institute, University of Southern Denmark, NanoSYD Alsion 2, 6400 Sønderborg, Denmark

Present Address:
J.O. Wolff
Department of Biological Sciences, Macquarie University, Sydney, NSW 2109, Australia

L. Heepe et al. (eds.), *Bio-inspired Structured Adhesives*, Biologically-Inspired Systems 9, DOI 10.1007/978-3-319-59114-8_4

the contact. This, in fact, will lower the pull-off forces, in agreement with the experimental results. Surprisingly, the claws had almost no influence on the attachment ability of the beetles. The present study depicts, for the first time, the effect of substrate stiffness on the attachment ability in climbing animals, whose natural habitats provide a variety of substrate stiffness ranging from rigid rock surfaces to soft, biofilm covered substrates.

4.1 Introduction

Reliable attachment to diverse and unpredictable substrates for resting and locomotion plays an important role in the niche occupation of insects. Consequently, a high diversity of attachment devices for adhesion and friction control have evolved among insects [1, 2]. Independently, similar attachment devices have also been developed in arachnids [3], amphibians (e.g. [4]), and in reptiles (e.g. [5–9]). It has been shown that the performance of these attachment devices is not only determined by their intrinsic structural and material properties, but also by extrinsic environmental factors, such as surface roughness (e.g. [10–13]), surface chemistry (e.g. [14, 15]), ambient temperature and humidity [16–20]. Furthermore, within the habitats of those climbing animals, the stiffness of prevailing substrates may vary considerably. Compare, for example, the stiffness of a clean rock surface, which is of order several GPa, with a surface covered by a biofilm having a stiffness of only up to hundreds of kPa [21]. However, most experimental studies on the attachment ability of climbing animals have been performed on rigid substrates and so far, it is not clear how the attachment ability is affected by soft substrates.

To address this problem, we performed force measurements with seven-spotted ladybird beetles (*Coccinella septempunctata*) (Fig. 4.1a) on smooth silicone elastomer substrates of varying stiffness. The attachment devices of ladybird feet comprise a pair of claws and dense hairy soles (Fig. 4.1b). The adhesive setae have flattened tips of tapered or spatulate shapes (Fig. 4.1c–e). In males, additionally, setae with discoidal tips are present (Fig. 4.1f). These play a role for the attachment to the smooth elytrae of the mating partner. For a detailed description of the attachment devices of *C. septempunctata* see [22]. Experiments were carried out with female and male beetles. In order to determine the effect of substrate stiffness on different parts of the attachment system, the experiments were carried out with and without claws. Three main questions were asked: (1) Is the attachment ability of the beetles affected by the substrate stiffness? (2) How do claws and setal pads contribute to a possible effect of substrate stiffness on the attachment ability? (3) Given the differences in setal structures, is the attachment ability in both sexes likewise affected by substrate stiffness?

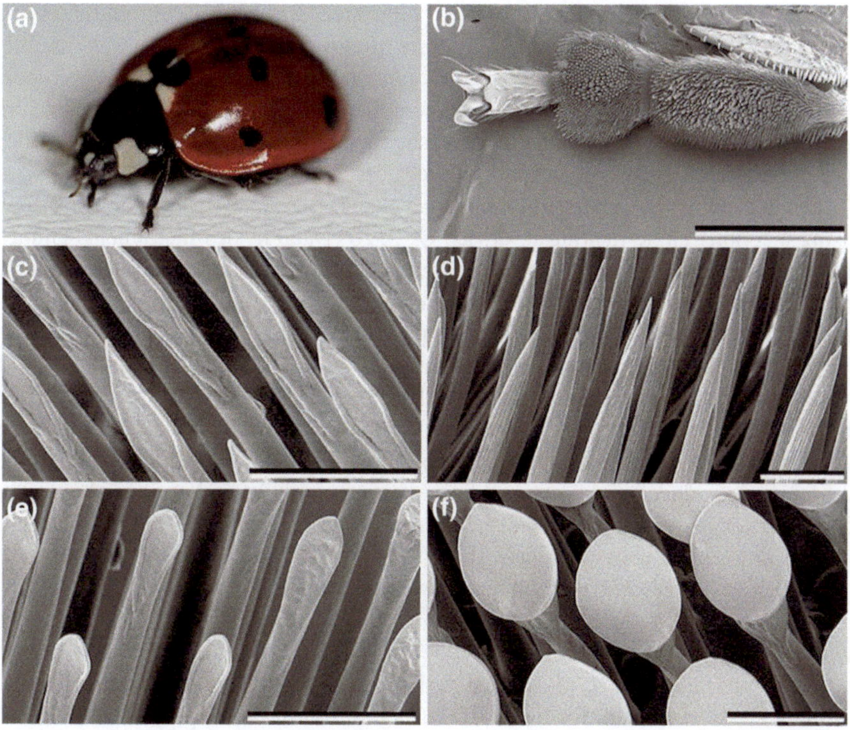

Fig. 4.1 Attachment devices of *C. septempunctata* (**a**). Tarsi are covered with different types of tenent setae, here the foreleg of a male beetle is shown (**b**). Setae types shown in scanning electron micrographs (**c–e**) are found in both males and females. Discoidal setae type in (**f**) only occurs in male attachment devices. *Scale bar* in (**b**) 500, and 10 µm (**c–f**). Adapted from [20]

4.2 Materials and Methods

4.2.1 Animals

For this study, the seven-spotted ladybird beetle *C. septempunctata* LINNAEUS 1758 (Coleoptera, Coccinellidae) was chosen (Fig. 4.1a). Adult beetles were either obtained from Katz Biotech AG (Baruth, Germany) or collected near Stohl (surroundings of Kiel, Germany). They were kept at 10 °C and fed with honey. The weight of the insects ranged from 21.6 to 37.4 mg.

4.2.2 Sample Preparation and Characterization

Three disc-like substrates with a diameter of 100 and 2 mm thickness were prepared from polydimethylsiloxane (PDMS, Sylgard® 184 silicone elastomer kit,

Dow Corning Inc., Midland, MI, USA) with three different mixing ratios (curing agent:base) of 1:10, 1:20, and 1:30. Mixture of curing agent and base was carefully mixed for 15 min. Then, the mixture was poured into Petri dishes made from poly (methyl methacrylate) (PMMA). PDMS was then cured in the Petri dishes for 12 h at 65 °C. Subsequently, samples were washed with liquid soap, rinsed in bi-distilled water, and dried under vacuum in a desiccator.

Contact angles of water, diiodomethane, and ethylene glycol were measured using a contact angle measuring device OCA 20 (DataPhysics Instruments GmbH, Filderstadt, Germany) using the sessile drop method. For each liquid five individual measurements were performed with a drop volume of 1 µl. Temperature and relative humidity were 21 °C and 30%, respectively.

Young's moduli of the substrates were determined in tensile tests using a zwickiLine Z0.5 materials testing machine (Zwick GmbH & Co. KG, Ulm, Germany) equipped with a Xforce HP 10 N load cell (Zwick GmbH). Test samples with a length of about 80 mm, a width of about 25 mm, and a thickness of about 2 mm were produced and processed from the same mixture as described above. The clamping length was 40 mm and the crosshead velocity was 1 mm/min. For each mixing ratio, three tests were performed.

Surface roughness of the substrates was analyzed with a scanning white-light interferometer Zygo NewView 6k (Zygo Corp., Middlefield, CT, USA) equipped with a 5× objective lens. Each substrate was analyzed at three randomly chosen positions with a scan size of 1.4 mm × 1.05 mm. Resulting 2-dimensional height profiles were corrected for waviness using a fourth-order polynomial and as roughness parameter the root mean square (rms) roughness and R_a were calculated.

In a second experiment, samples were washed in toluene for 24 h in order to remove uncured residues, especially from the 1:20 and 1:30 substrates. Samples were then dried for 6 h at 65 °C, subsequently washed in liquid soap, rinsed in bi-distilled water, and finally dried under vacuum conditions in a desiccator. Contact angles, Young's moduli, and surface roughness were measured again as described above.

4.2.3 Force Measurements

Prior to the experiment, each beetle was sexed and weighed using an analytical balance AG 204 Delta Range (Mettler Toledo GmbH, Greifensee, Switzerland). To assess attachment forces, the computer-controlled centrifugal force tester Tetra Zentri-01-P (Tetra GmbH, Ilmenau, Germany) was applied. For a detailed description of the experimental setup see [23]. Beetles were placed individually on the horizontal surface of the centrifuge drum, covered with either of the three sample surfaces. Beetles on the centrifuge drum were accelerated from 50 to

3000 rpm in 20 s. At a certain angular velocity, the growing centrifugal force acting on the horizontally placed beetles overcame the static friction force between attachment devices and the substrate, and the beetles started to slide towards the drums edge. Finally, the centrifugal force caused the beetles to detach from the drum surface. Using fiber optic sensors and the data acquisition software PC Fly (Tetra GmbH, Ilmenau, Germany), the position of the beetles on the rotating drum was recorded as a function of its angular velocity and used to calculate the maximum friction force in each individual trial. Each individual beetle was tested on each substrate three times in a row, allowing a recovery time of 30 s between individual trails. For data processing, only the maximum value of the three trails was used. Measurements of individual beetles on the different substrates were randomized. After these measurements, the claws of tested individuals were removed, (for details on the procedure see [24]). For that purpose, beetles were anesthetized with CO_2 and immobilized. Pretarsal claws were then carefully cut with micro-scissors so that virtually no stumps were left which could be potentially used for locomotion. Care was taken not to cut into the tarsus. The same force measurements were then repeated for the beetles without claws. In total, 23 females with claws and 20 without claws were tested. For males, 9 with claws and 7 without claws were tested. For this experiment beetles obtained from Katz Biotech AG were used (see Sect. 4.2.1). In order to account for potential effects resulting from uncured residues and/or changes in the surface chemistry by changing the mixing ratio of PDMS substrates, 13 male beetles with claws were tested in a second experiment on washed substrates for a comparison. For this experiment beetles collected in the field near Kiel, Germany were used. Temperature and relative humidity were 21 °C and 30%, respectively.

4.2.4 Statistical Analysis

The results obtained were statistically analyzed with SigmaPlot 12.5 (Systat Software Inc., San Jose, CA, USA). In order to assess differences between the attachment ability of beetles on different substrates, a One-way ANOVA, eventually followed by an all pairwise comparison of individual groups, was applied for females and males, with and without claws, respectively, if assumptions of normally distributed data and equal variance between individual groups were fulfilled. If these assumptions were not fulfilled, a Kruskal-Wallis One-way ANOVA on ranks was applied. In order to pairwise compare the attachment ability of beetles with and without claws on the same substrate, *t* tests have been applied. If required assumptions were not fulfilled a Mann-Whitney rank sum test was applied. The overall significance level was set to 0.05.

4.3 Results

4.3.1 Sample Properties

For the unwashed substrates, we obtained Young's moduli of approx. 2.2 MPa for the standard 1:10 mixture, approx. 0.7 MPa for the 1:20 mixture, and approx. 0.3 MPa for the 1:30 mixture (Fig. 4.2a, dark gray points) in agreement with previous studies [25]. For the washed substrates we obtained Young's moduli of approx. 1.9 MPa for the standard 1:10 mixture, approx. 0.7 MPa for the 1:20 mixture, and approx. 0.3 MPa for the 1:30 mixture (Fig. 4.2a, light gray points).

Figure 4.2b, c show contact angles of water (w), ethylene glycol (e), and diiodomethane (d) for the different substrates before and after washing the substrates, respectively. For unwashed substrates, median water contact angle increased from 112.5° on the 1:10 mixture to around 117° on the 1:30 mixtures. Median ethylene glycol contact angles increased from around 101° on the 1:10 standard mixture to approx. 105° on the 1:20 and 1:30 mixture. Median diiodomethane contact angles increased from around 90.5° on the 1:10 mixture to around 98.5° on the 1:30 mixture. High contact angles of all three liquids indicate the low surface free energy of all PDMS mixtures [26]. For washed substrates, median water contact angles were 112.5° on the 1:10 mixture, around 110° on the 1:20 mixtures, and around 117° on the 1:30 mixture. Median ethylene glycol contact angles decreased from around 100° on the 1:10 standard mixture to approx. 95.5° on the 1:30 mixture.

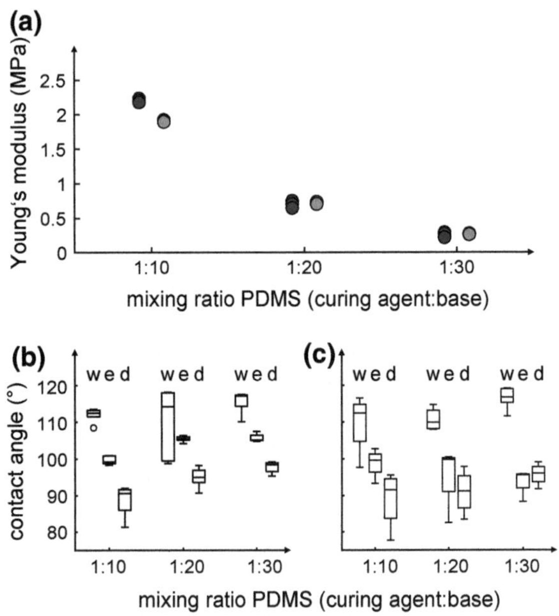

Fig. 4.2 **a** Young's moduli of unwashed (*dark gray*) and washed (*light gray*) PDMS substrates with mixing ratios (curing agent:base) 1:10, 1:20, 1:30. **b** *Boxplots* of measured contact angles of water (w), ethylene glycol (e), and diiodomethane (d) on unwashed substrates with different mixing ratios. **c** *Boxplots* of measured contact angles of water (W), ethylene glycol (E), and diiodomethane (D) on washed substrates with different mixing ratios

Median diiodomethane contact angles were 91.5° on the 1:10 mixture, around 91° on the 1:20 mixtures, and around 96° on the 1:30 mixture.

The surface roughness of all substrates (unwashed and washed) was similarly smooth with root mean square (rms) roughness ranging from 13 to 68 nm and R_a ranging from 6 to 9 nm, as expected for PDMS molded against a smooth Petri dish bottom.

4.3.2 Influence of the Substrate Stiffness on the Attachment Ability of Ladybird Beetles

In females, there was no significant difference in friction forces between the three substrates (Fig. 4.3a). Furthermore, there was no significant difference in friction forces of female beetles with and without claws, except on the 1:20 substrate (Mann-Whitney rank sum test, $U = 149$, $T = 521$, $n(\text{small}) = 20$, $n(\text{big}) = 23$, $P = 0.050$), where claw-less beetles showed a higher attachment ability. This difference is still present, when forces are normalized on body weight (safety factor, Fig. 4.3b, t test, $t = -2.383$, $dof = 41$, two-tailed $P = 0.022$), but not, when friction forces were normalized by the average friction force of all beetles obtained on the 1:10 substrate (Fig. 4.3c). However, there is a trend of higher attachment ability of beetles without claws for the friction forces and the safety factors on all substrates.

For males, there was a tendency of decreasing attachment ability with decreasing substrate stiffness as indicated by the dashed line in Fig. 4.3d. This trend is supported by a significant difference in the attachment ability of beetles with claws between the 1:10 and the 1:30 substrate (Mann-Whitney rank sum test on original and normalized friction forces, $U = 17$, $T = 109$, $n(\text{small}) = 9$, $n(\text{big}) = 9$, $P = 0.042$). However, these significances only appeared when both substrates were tested separately using a Mann-Whitney rank sum test. Within performed ANOVAs, including all substrates, these significances did not appear, potentially as a result from a very low power of performed tests due to the small sample sizes: $n = 9$ for male beetles with claws and $n = 7$ for male beetles without claws. All other comparisons did not show statistical significances. Males showed, in general, a higher attachment ability as females in accordance to previous studies on *C. septempunctata* on smooth substrates [20, 22].

After washing the substrates with toluene, intact male beetles showed significantly higher friction forces on the 1:10 if compared to the 1:30 substrate (Fig. 4.4a, Mann-Whitney rank sum test, $U = 37$, $T = 115$, $n(\text{small}) = 12$, $n(\text{big}) = 13$, $P = 0.028$). The same behavior was observed for the safety factors safety factor (Fig. 4.4b, One-way ANOVA, $n = 13$, $dof = 2$, $SS = 2144.646$, $MS = 1072.323$, $F = 7.067$, $P = 0.003$; Holm-Sidak posthoc test, $t = 3.628$, $P = 0.003$) and the normalized friction forces (Fig. 4.4c, Kruskal-Wallis One-way ANOVA on ranks, $n = 13$, $H = 7.191$, $dof = 2$, $P = 0.027$, Tukey posthoc test, $q = 3.764$).

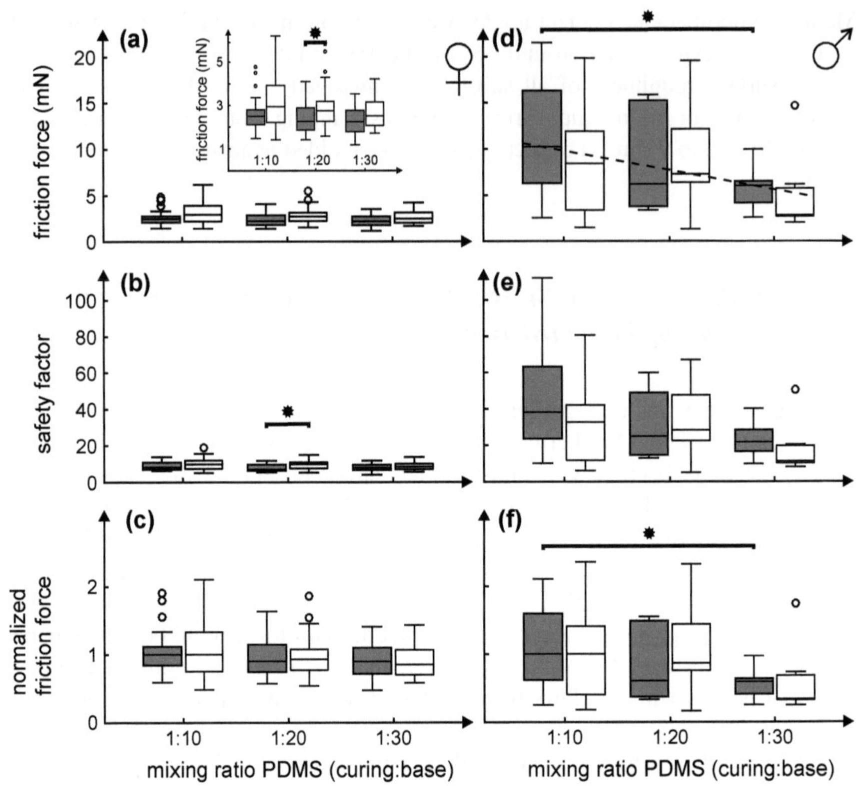

Fig. 4.3 Attachment ability of female (**a–c**) and male (**d–f**) ladybird beetles with (*gray boxes*) and without (*unfilled boxes*) claws on PDMS substrates with different mixing ratios (curing agent: base), i.e. decreasing substrate stiffness from 1:10, 1:20, to 1:30. Substrate stiffness are given in Fig. 4.2. **a, d** *Boxplots* show the friction forces the animals sustained before detachment from the substrate. *Inset* in **a** shows the same data, but at different scale. **b, e** *Boxplots* show the corresponding safety factors, i.e. friction force per individual beetle divided by weight force of the individual. **c, f** *Boxplots* show normalized friction forces; all data is normalized to the average friction force obtained on 1:10 substrate. Bars with asterisks in (**a, b, d, f**) indicate statistical differences, see Sect. 4.2.4. Dashed line in (**d**) indicate decreasing attachment ability of male beetles with decreasing substrate stiffness. Females with claws, $N = 23$; females without claws, $N = 20$; males with claws, $N = 9$; males without claws, $N = 7$

Moreover, intact male beetles had a significantly higher safety factor on the 1:10 substrate if compared to the 1:20 substrate (One-way ANOVA, $n = 13$, $dof = 2$, $SS = 2144.646$, $MS = 1072.323$, $F = 7.067$, $P = 0.003$; Holm-Sidak posthoc test, $t = 2.638$, $P = 0.025$). There are no significant differences in beetles' attachment ability (i.e. friction forces, safety factors and normalized friction forces) on each substrate before and after washing.

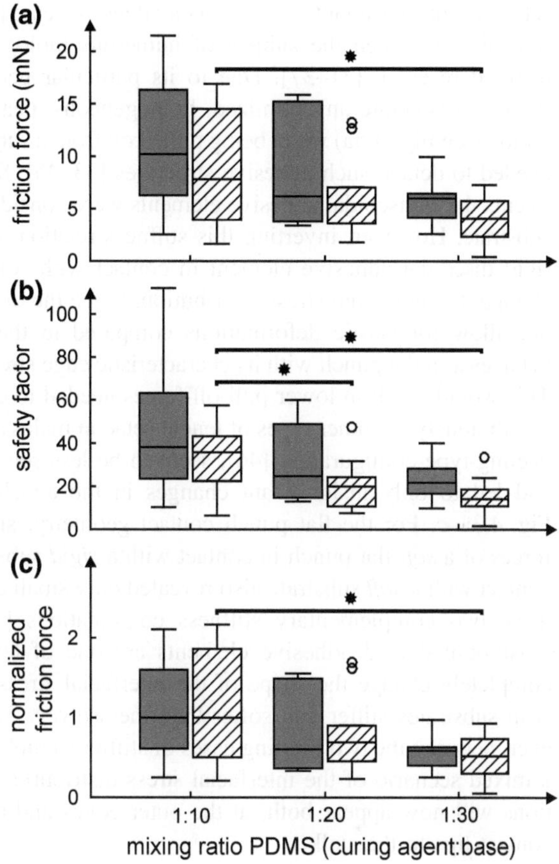

Fig. 4.4 Attachment ability of male beetles with claws (*gray boxes*, same data shown in Fig. 4.3) before washing the substrates in toluene and after washing (*hatched boxes*). *Boxplots* show the maximum friction forces (**a**), the corresponding safety factors (**b**), and the normalized friction forces (**c**), which were normalized to the 1:10 substrate. *Bars* with *asterisks* indicate statistical differences. Males with and without claws, $N = 13$

4.4 Discussion

4.4.1 *Sexual Dimorphism in the Attachment Ability of* C. septempunctata *on Substrates with Different Stiffness*

The present study demonstrates, for the first time, the effect of substrate stiffness on the attachment ability of climbing animals. We have observed a sexual dimorphism in the attachment ability of *C. septempunctata* on soft substrates with different stiffness. Females, with and without claws, were not affected by reducing the substrate stiffness from around 2 MPa to around 0.3 MPa. Whereas males, with and without claws, showed decreasing attachment ability with decreasing stiffness of the substrates. This may be explained by different types of tenent setae present in the attachment devices in males compared to females. Males possess setae with discoidal tips (see Fig. 4.1f) in addition to the usual setal equipment. These specialized setae account for about 50% of tarsal adhesive setae [22]. This contact geometry of setal tips has been the subject of numerous biological [27–30] and biologically inspired research [31–37]. Due to its particular geometry, the discoidal tip was shown to generate an optimized, homogeneous, peak-free interfacial stress distribution (see Fig. 4.5a) when being pulled off from a substrate. High pull-off forces are needed to detach such adhesive structures [33–45]. So far, in theory and in experiments these discoidal adhesive elements were considered soft in comparison to the substrate. However, inverting this stiffness relation as depicted in Fig. 4.5b, i.e. a rigid discoidal adhesive element in contact with soft substrate, would drastically change the interfacial stress distribution. Since the discoidal adhesive element does not allow for elastic deformations compared to the soft substrate, it essentially behaves as a flat punch with its characteristic edge stress concentrations [38–40, 43]. This would result in lower pull-off forces needed to detach the structure.

Obviously, all other types of tenent setae in males and females, which resemble a peeling-type configuration [46], seem to be less sensitive to the substrate stiffness and led to only insignificant changes in the attachment ability in females, see Fig. 4.3a–c. For the flat punch contact geometry, simulations on the detachment forces of a *soft* flat punch in contact with a *rigid* substrate and a *rigid* flat punch in contact with a *soft* substrate also revealed only small difference in pull-off forces for these two complementary stiffness combinations [47]. Potentially, discoidal or mushroom-shaped adhesive elements are one of few contact geometries, which completely change the shape of the interfacial stress distribution when in contact with substrates stiffer and softer than themselves. At an intermediate stage, however, between the two limiting cases of stiffness combinations, as shown in Fig. 4.5, a mixed scenario of the interfacial stress distribution will occur. Stress concentrations will now appear both, at the outer edges and at the transition from the thin contact lips to the stalk.

Unfortunately, for *C. septempunctata* the stiffness of discoidal setal tips in males is not known yet, but will, of course, depend on the actual material properties as well as on its ultrastructure. Recently, setae with pointed tips in females of *C. septempunctata* were investigated and shown to contain high amounts of the soft rubber-like protein resilin with a Young's modulus of about 1 MPa, determined by

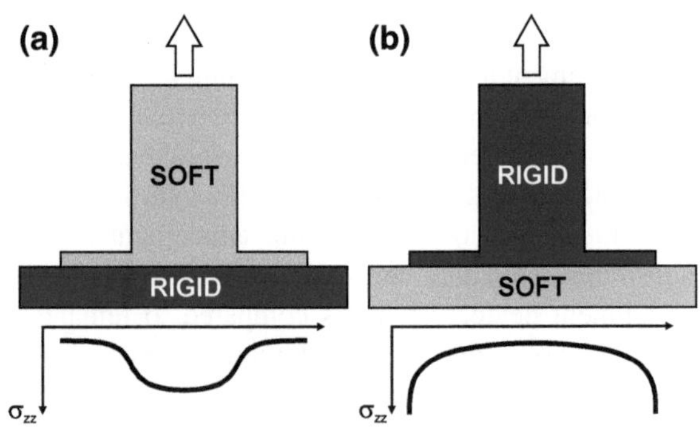

Fig. 4.5 Qualitative representation of the interfacial stress distribution σ_{zz} of a *soft* discoidal adhesive element pulled from a *rigid* substrate (**a**) and a *rigid* discoidal adhesive element pulled from a *soft* substrate (**b**)

nanoindentation using an atomic force microscopy [48]. However, for discoidal setae in males, it was hypothesized that this soft rubber-like material is incorporated only at the transition from the setal shaft to the terminal contact element [42]. This hypothesis seems to be confirmed by preliminary observations on the autofluorescence of the cuticle in male discoidal setae of *C. septempunctata* (personal observation). The actual discoidal terminal element which makes contact with the substrate seems to be mainly composed of chitin and tanned proteins, similar to the composition of the setal shaft (L. Heepe unpublished data). Young's moduli measured by Peisker et al. [48] for a similar region in female setae was of order 1 GPa. For *Gastrophysa viridula*, having similar types of tenent setae, Young's moduli in the range from 5 to 73 MPa were reported [49], whereas the estimated Young's modulus of discoidal setal tips was distinctly higher (46–73 MPa) than for the spatulate (8–32 MPa) and pointed tip shape (5–8 MPa).

Above mentioned observations on discoidal setae suggest that the stiffness of all substrates used in the present study are lower by at least an order of magnitude if compared to the stiffness of this discoidal seta type.

Using the safety factor, we may compare the attachment ability of *C. septempunctata* with previous experiments on rigid substrates. For females actively pulling in a traction force experiments on a smooth PMMA substrate (stiffness of order 1 GPa) an average safety factor (assuming a weight of 30 mg, see Sect. 2.1) of about 30 is obtained, for males in the same experiment about 60 [20]. In fact, the surface chemistry of PMMA and PDMS is different, but traction force tests of *C. septempunctata* revealed that surface chemistry only has a marginal effect on the attachment ability of beetles [15]. In the present experiment, on the 1:10 substrate, an average safety factor of about 10 and about 40 was obtained for females and males respectively. Those safety factors are consistently lower if compared to the results obtained on the rigid substrate. In fact, the actual differences might be much higher. Here, the attachment ability of beetles was tested on a centrifuge, which potentially measures the maximum attachment ability of the animals. Whereas in a friction force experiment in which animals actively pull, as in the experiment on the PMMA substrate reported above, the maximum attachment ability observed is limited by the actual muscular power of the animal.

Indeed, discoidal setae in beetles differ slightly in their morphology compared to the idealized, axis symmetric case shown in Fig. 4.5. Here, the setal shaft is eccentrically connected and tilted under an angle smaller than 90° relatively to the discoidal terminal contact element. However, high-speed video recordings of the detachment process of individual discoidal-shaped setae of male beetles walking upside down on a smooth glass slide, indicate two different types of detachment. Figure 4.6a–d shows still images of an individual seta detaching by crack nucleation inside the contact area (a). The crack grew radially (b) until partially the contact edge has been detached (c). Then, detachment changed to a peeling-type of behavior with a straight delamination edge (d). This kind of detachment is in agreement with the interfacial stress distribution shown in Fig. 4.5a. Figure 4.6e–h shows still images of an individual seta detaching by crack nucleation at one point of the contact edge (f), followed by a peeling-type of crack propagation (g, h). This kind of detachment implies an edge stress concentration, similar to the case shown in Fig. 4.5b. Based in this observation, we speculate that male beetle are able to control the type of detachment and thus the pull-off resistance, by changing the angle of an applied detachment force.

4.4.2 The Role of Claws

We have compared the attachment ability of beetles with and without claws based on the hypothesis, that with decreasing substrate stiffness, claws would be able to locally deform the substrate and to mechanically interlock with it which would result in an increase of forces with decreasing substrate stiffness. However, such a mechanism was not observed. Unexpectedly, females showed consistently higher average attachment ability without claws than with them. In males, no such tendency was observed. At this point, we can only speculate whether these results are just an artifact or indicate a certain claw behavior during attachment and detachment. One potential explanation could be, that, during acceleration on the centrifuge, the beetles tried to find surface asperities and pressed their claws into the substrate, but with this actually impede attachment ability of the adhesive pad nearest to the claws. However, these are just speculations and require an optical assessment of claws during locomotion on soft substrates.

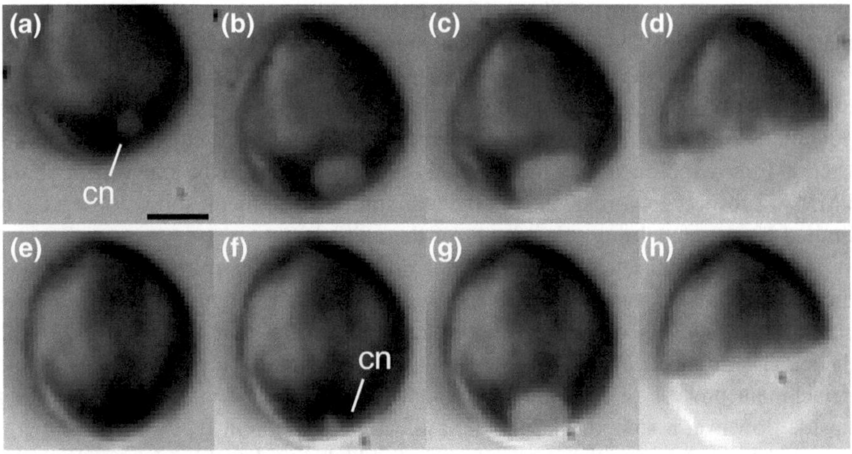

Fig. 4.6 Discoidal seta of a male ladybird *C. septempunctata* in contact with a glass substrate (*dark gray area*). Image sequence **a–d** shows the detachment process initiated by crack nucleation (cn) inside the contact area (**a**). Subsequently, crack propagation changed to a peeling-type of detachment (**b–d**). Image sequence **e–h** shows the detachment process initiated by crack nucleation (cn) at the edge of the contact (**f**) followed a peeling-type of crack propagation (**g**, **h**). **a** $\Delta t = 0$ ms, **b** $\Delta t = 209$ ms, **c** $\Delta t = 259$ ms, **d** $\Delta t = 439$ ms, **e** $\Delta t = 0$ ms, **f** $\Delta t = 135$ ms, **g** $\Delta t = 139$ ms, **h** $\Delta t = 182$ ms. *Scale bar*, 3 μm

4.4.3 Substrate Condition

We have produced substrates with different stiffness made of a two component silicon elastomer by changing the mixing ratio of the curing agent to the base. By changing the mixing ratio, i.e. using less curing agent than the recommended standard ratio 1:10, in order to soften the polymerized substrate, uncured residues are expected. These residues may have potentially changed the surface chemistry and could thus distort our obtained results. However, since males and females were both tested on the same unwashed substrates only the relation between the different substrates could be altered, but neither the comparison of beetles with and without claws on the same substrate nor the comparison of males and females on the same substrate. Due to the fact, that females did not show any significant differences in the attachment ability on the different substrates, we assume any potential difference in surface chemistry present in the different substrates to be of minor importance, in accordance with previous observations [15]. Moreover, different viscoelastic properties of PDMS with varying mixing ratio [50] could have increased the attachment ability of beetles on the softer substrates due to higher viscoelastic energy dissipation [51]. However, females' attachment ability did not change on the softer substrates and males' attachment ability even decreased. Therefore, we assume viscoelastic effects also to be of minor importance for the present study. For males (with claws), we have repeated the experiments on toluene washed substrates, a solvent for uncured PDMS residues [52]. No statistical significant differences were observed for the attachment ability of male beetles before and after washing the substrates. Since washing did not changed distinctly the stiffness, surface roughness, and contact angles of the samples, we are confident that we have essentially probed the effect of substrate stiffness rather than effects related to surface chemistry or roughness.

4.5 Conclusions

The present study demonstrates that the substrate stiffness may have an effect on the attachment ability of climbing animals. However, this effect seems to depend on the geometry of the contact elements. Although the presence of different substrates, which may additionally by covered with biofilms, in the habitats of climbing animals allow for a large variety of substrate stiffness, it was so far largely neglected in animal adhesion studies. With this experimental approach we shed, for the first time, light on the importance of the substrate stiffness for the attachment ability of climbing animals, which may also be of importance of technical and medical applications. However, there remain open questions. Females without claws showed the tendency of higher attachment ability if compared to females with claws. The underlying mechanisms leading to this effect remain unclear and require further investigations. In contrast, claw ablation did not have a significant effect in males. Thus, the role of claws is ambiguous.

Acknowledgements This book chapter is based on the original publication L. Heepe, D. S. Petersen, L Tölle, J.O. Wolff, and S.N. Gorb, Sexual dimorphism in the attachment ability of the ladybird beetle *C. septempunctata* on soft substrates, Appl. Phys. A 123, 34 (2017), with permission of Springer. We thank J. Schnoor for providing high-speed observations of setae detachment and Titus Pelzl for assistance during force measurements. This work was partially supported by CARTRIB Project of The Leverhulme Trust (S. N. Gorb) by projects CP 1550 and 1623 by a grant of the Cluster of Excellence 80 The Future Ocean (L. Heepe and S. N. Gorb). The Future Ocean is funded within the framework of the Excellence Initiative by the Deutsche Forschungsgemeinschaft (DFG) on behalf of the German federal and state governments. J.O. Wolff is funded by a Macquarie Research Fellowship of Macquarie University, Sydney.

References

1. S. Gorb, *Attachment Devices of Insect Cuticle* (Kluwer Academic Publishers, London, 2001)
2. S.N. Gorb, R.G. Beutel, Naturwissenschaften **88**, 530 (2001)
3. J.O. Wolff, S.N. Gorb, *Attachment Structures and Adhesive Secretions in Arachnids* (Springer, New York, 2016)
4. G. Hanna, W.J.P. Barnes, J. Exp. Biol. **155**, 103 (1991)
5. A. Haase, Arch. Naturgesch. **66**, 321 (1900)
6. R. Ruibal, V. Ernst, J. Morphol. **117**, 271 (1965)
7. U. Hiller, Z. Morphol, Tiere **62**, 307 (1968)
8. D.J. Irschick, C.C. Austin, K. Petren, R.N. Fisher, J.B. Losos, O. Ellers, Biol. J. Linn. Soc. **59**, 1169 (1996)
9. K. Autumn, P.H. Niewiarowski, J.B. Puthoff, Annu. Rev. Ecol. Evol. Syst. **45**, 445 (2014)
10. G. Huber, S.N. Gorb, N. Hosoda, R. Spolenak, E. Arzt, Acta Biomater. **3**, 607 (2007)
11. D. Voigt, J.M. Schuppert, S. Dattinger, S. Gorb, J. Insect Physiol. **54**, 765 (2008)
12. J.M.R. Bullock, W. Federle, Insect Sci. **18**, 298 (2011)
13. J.O. Wolff, S.N. Gorb, J. Exp. Biol. **215**, 179 (2012)
14. I. Badge, A.Y. Stark, E.L. Paoloni, P.H. Niewiarowski, A. Dhinojwala, Sci. Rep. **4**, 6643 (2016)
15. M.W. England, T. Sato, M. Yagihashi, A. Hozumi, S.N. Gorb, E.V. Gorb, Beilstein J. Nanotechnol. **7**, 1471 (2016)
16. W. Federle, W. Baumgartner, B. Hölldobler, J. Exp. Biol. **207**, 67 (2004)
17. P.H. Niewiarowski, S. Lopez, L. Ge, E. Hagan, A. Dhinojwala, PLoS ONE **3**, e2192 (2008)
18. J.B. Puthoff, M.S. Prowse, M. Wilkinson, K. Autumn, J. Exp. Biol. **213**, 3699 (2010)
19. J.O. Wolff, S.N. Gorb, Proc. R. Soc. B **279**, 139 (2012)
20. L. Heepe, J.O. Wolff, S.N. Gorb, Beilstein J. Nanotechnol. **7**, 1322 (2016)
21. M. Böl, A.E. Ehret, A.B. Albero, J. Hellriegel, R. Krull, Crit. Rev. Biotechnol. **33**, 145 (2013)
22. E.V. Gorb, N. Hosoda, C. Miksch, S.N. Gorb, J.R. Soc, Interface **7**, 1571 (2011)
23. S. Gorb, E. Gorb, V. Kastner, J. Exp. Biol. **204**, 1421 (2001)
24. P. Bußhardt, D. Kunze, S.N. Gorb, Sci. Rep. **4**, 6998 (2014)
25. B. Trappmann, J.E. Gautrot, J.T. Connelly, D.G.T. Strange, Y. Li, M.L. Oyen, M.A. Cohen Stuart, H. Boehm, B. Li, V. Vogel, J.P. Spatz, F.M. Watt, W.T.S. Huck, Nat. Mater. **11**, 642 (2012)
26. J.E. Mark, *Polymer Data Handbook* (Oxford University Press, Oxford, 1999)
27. P.H. Tsang, G. Li, Y.V. Brun, L.B. Freund, J.X. Tang, Proc. Natl. Acad. Sci. USA **103**, 5764 (2006)

28. S.N. Gorb, M. Varenberg, J. Adhes. Sci. Technol. **21**, 1175 (2007)
29. J.M.R. Bullock, W. Federle, Naturwissenschaften **98**, 381 (2011)
30. E. Hennebert, R. Santos, P. Flammang, Zoosymposia **7**, 25 (2012)
31. S. Gorb, M. Varenberg, A. Peressadko, J. Tuma, J.R. Soc, Interface **4**, 721 (2007)
32. S. Kim, B. Aksak, M. Sitti, Appl. Phys. Lett. **91**, 221913 (2007)
33. A. del Campo, C. Greiner, E. Arzt, Langmuir **23**, 10235 (2007)
34. J. Davies, S. Haq, T. Hawke, J.P. Sargent, Int. J. Adhes. Adhes. **29**, 380 (2009)
35. W.G. Bae, D. Kim, M.K. Kwak, L. Ha, S.M. Kang, K.Y. Suh, Adv. Healthc. Mater. **2**, 109 (2013)
36. K. Dening, L. Heepe, L. Afferrante, G. Carbone, S.N. Gorb, Appl. Phys. A **116**, 567 (2014)
37. L. Heepe, A.E. Kovalev, S.N. Gorb, Beilstein J. Nanotechnol. **5**, 903 (2014)
38. A.V. Spuskanyuk, R.M. McMeeking, V.S. Deshpande, E. Arzt, Acta Biomater. **4**, 1669 (2008)
39. G. Carbone, E. Pierro, S.N. Gorb, Soft Matter **7**, 5545 (2011)
40. G. Carbone, E. Pierro, Meccanica **48**, 1819 (2013)
41. L. Heepe, A.E. Kovalev, A.E. Filippov, S.N. Gorb, Phys. Rev. Lett. **111**, 104301 (2013)
42. L. Heepe, G. Carbone, E. Pierro, A.E. Kovalev, S.N. Gorb, Appl. Phys. Lett. **104**, 011906 (2014)
43. L. Heepe, S.N. Gorb, Annu. Rev. Mater. Res. **44**, 173 (2014)
44. B. Aksak, K. Sahin, M. Sitti, Beilstein J. Nanotechnol. **5**, 630 (2014)
45. J.W. Neubauer, L. Xue, J. Erath, D.M. Drotlef, A. del Campo, A. Fery, A.C.S. Appl, Mater. Interfaces **8**, 17870 (2016)
46. M. Varenberg, N.M. Pugno, S.N. Gorb, Soft Matter **6**, 3269 (2010)
47. S.N. Khaderi, N.A. Fleck, E. Arzt, R.M. McMeeking, J. Mech. Phys. Solids **75**, 159 (2015)
48. H. Peisker, J. Michels, S.N. Gorb, Nat. Commun. **4**, 1661 (2013)
49. S. Gernay, W. Federle, P. Lambert, T. Gilet, J.R. Soc, Interface **13**, 20160371 (2016)
50. J. Nase, A. Lindner, C. Creton, Phys. Rev. Lett. **101**, 074503 (2008)
51. B.N.J. Persson, O. Albohr, G. Heinrich, H. Ueba, J. Phys. Condens. Matter **17**, R1071 (2005)
52. J.N. Lee, C. Park, G.M. Whitesides, Anal. Chem. **75**, 6544 (2003)

Chapter 5
Structural Effects of Glue Application in Spiders—What Can We Learn from Silk Anchors?

Jonas O. Wolff

Abstract Silk is a key innovation in spiders that equally fascinates biologists and material scientists. The anchoring of silken threads is a critical, yet poorly understood aspect of web-building—one that could lead to the development of advanced adhesives. To anchor silk, most spiders produce a structurally unique and rather unexplored bio-adhesive: the two-compound piriform gland secretion. The secretion is spun into elaborate patterns, so called attachment discs that anchor silken threads to substrates. The piriform gland secretion is a glue-coated fibre that combines the high toughness of silk with strong adhesion, even to highly repellent surfaces like Teflon. The glue is used highly economically, dries within less than a second after extrusion and can remain stable for years. Its hierarchical organization, discontinuous contact area and the embedding of compliant fibres may explain the high adhesive performance and flaw tolerance of such a composite product coming from a single, rather simple, type of silk glands. These principles contradict, in many regards, the paradigms of adhesives design and application, like aimed homogeneity of the bonding. Understanding the function of silk anchors may therefore trigger the development of novel industrial adhesives with embedded fibres, and methods of gaining higher bonding strength with less material consumption by smart glue application.

Present Address:
J.O. Wolff (✉)
Department of Biological Sciences, Macquarie University,
Sydney, NSW 2109, Australia
e-mail: jonas.wolff@mq.edu.au

J.O. Wolff
Department of Functional Morphology and Biomechanics, Zoological Institute,
University of Kiel, Am Botanischen Garten 9, 24098 Kiel, Germany

L. Heepe et al. (eds.), *Bio-inspired Structured Adhesives*,
Biologically-Inspired Systems 9, DOI 10.1007/978-3-319-59114-8_5

5.1 Introduction

Spider silk is one of the toughest biological materials, surpassing artificial fibres by its combination of high strength and extensibility [1–3]. Spider silk and webs are therefore of utmost interest to a range of research disciplines (e.g., biologists, material scientists and engineers) who use very different approaches in order to understand the secrets of its superior material performance, which has evolved over millions of years. The vivid and interdisciplinary field of silk research aims not less than an integrative understanding of the structure-function relationships from the molecular level to whole webs. Here I focus on one of the least understood, yet substantial, parts of a spider's web: the anchorages of threads. I will review the state of knowledge on thread anchorages and the material they are made of, and illustrate their potential to elicit new impulses and innovations for adhesives technology.

5.2 How to Glue a Rope

Silk is used for nearly every aspect in a spider's life: locomotion, prey capture and reproduction [4]. However, a thread can only fulfil its specific function if it is combined with a strong anchorage that attaches it to substrates or other threads. Spiders have evolved a special type of secretion, the piriform (or pyriform) gland silk to fulfil this purpose [5, 6]. Of all (up to 7) types of silks that web-building spiders produce, piriform (PI) gland silk is (despite its importance) the least studied, with highly unknown material properties [7].

The PI silk gland is fed by two types of secretory cells: one produces a silk thread and the other a glue coating that consists of nanofibrils and an amorphous matrix [5, 8, 9]. The spigots (nozzle-like gland openings) of the piriform glands stand in a dense array on the anterior lateral spinnerets (ALS). They are adjacent to the single or few major ampullate (MA) gland spigots that produce the dragline. This specific configuration is rather conserved among spiders and enables a direct agglutination of the dragline to substrates or to other threads [6, 10]. This is achieved by rubbing the spinnerets against each other and on the substrate through an innate movement programme [5, 6]. The resulting adhesive plaque (also called attachment disc) is a network of the minute glue-coated piriform silk fibres that forms an envelope around the dragline and laterally spreads on the substrate (Fig. 5.1b). The attachment discs are discernible as a tiny white dots against a dark background. These anchors must resist high forces, such as the weight of the spider or wind drag acting on the web.

Thus, silk anchors play a critical mechanical role and presumably have evolved under strong selection for extraordinary material performance. For example, the aerial webs of spiders have evolved in response to the evolution of insect flight [11]. These webs are typically attached to vegetation, which is often characterised by anti-adhesive and anti-wetting properties [12]. Like any other adhesive the anchor

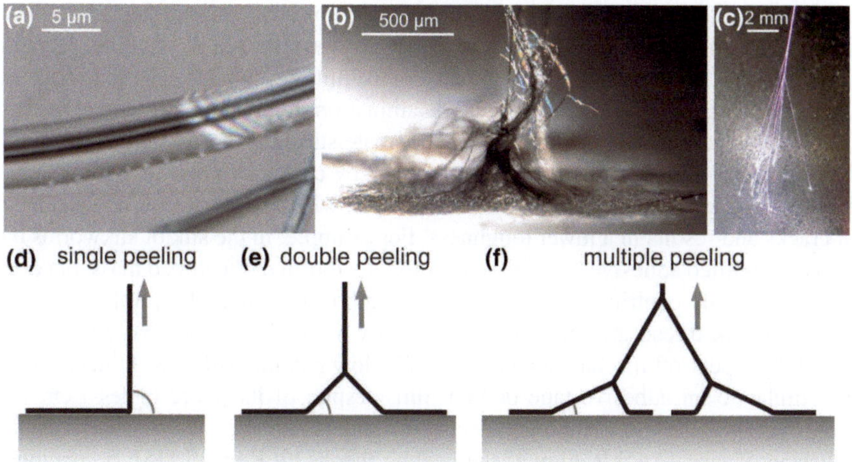

Fig. 5.1 Silk anchors with different levels of hierarchy. **a** 1 level: single glue-coated thread (here: PI thread of a gnaphosid spider). **b** 2 levels: attachment disc, with the main thread (dragline) indirectly glued to the substrate by means of numerous PI threads. **c** 3 levels: support thread of an orb web. These threads are bundles of several draglines, each one anchored to the substrate by an attachment disc. Photo by Jonathan Neumann, with kind permission. **d–f** Schematic illustrations of corresponding contact-mechanical theories of tape-like thin films in contact with a smooth surface. Note the difference in the detachment (peeling) angle, resulting from hierarchical branching. **d** Peeling-model after Kendall [15], **e** double-peeling model after Pugno [16], and **f** multiple (branched) peeling model after Bosia et al. [17]

glue is affected by a low surface free energy [9, 13]. Still, the strength of an attachment disc spun on a highly anti-adhesive surface, like Teflon, can support the spider's body weight [13]. The key for such an outstanding performance is a specific hierarchical structure [14] and composition that is highly conserved among spiders.

5.3 The Hierarchical Structure of Silk Thread Anchorages

5.3.1 Single Thread

Many silks, like that of caterpillars [18, 19] or other insects [20, 21], exhibit a layered structure, including a fibrous core and an adhesive coat that is initially liquid but solidifies quickly after emergence from the gland. A similar composition is found in the silks of primitive spiders, the Mesothelae and Mygalomorphae [22]. The silk of these spiders is not very tough if compared with the silk of derived orb web spiders [23, 24], but can be glued on substrates or to each other, in order to build a lining that stabilizes the spider's burrow [25]. A single thread here acts both as a structural element and as an adhesive. However, these two functions have divergent demands on the material properties. The structural fibre should be resistant against pulling forces. In many silks this is achieved by an anisotropic semicrystalline structure permitting both high tensile strength and extensibility [2, 26, 27]. The glue coat, on the other hand, must resist shearing and delamination. It must initially spread well on the surface and quickly solidify in order to create strong inherent material strength. In biological adhesives this is usually achieved by curing or cross-linking [28–30]. The resulting glue coat is probably more isotropic, which facilitates the advancement of cracks and results in a lower toughness. For example, in the silk of silkworms the coat of solidified adhesive (sericin) makes about a fourth of the thread mass, but does not significantly contribute to the tensile strength of the thread [31, 32].

A thin silk thread attached to the surface by a solidified glue coat will usually detach by a peel-off mechanism, because of its low bending stiffness. It, hence, may act similar to an adhesive tape or thin film, despite of its (more or less) circular cross-section. The most important theory to understand the factors governing the adhesion of tape-like thin films is the peeling theory by Kendall [15]. It concludes that the width, thickness and elasticity of the tape, and the angle under which it is pulled from the rigid substrate have a great effect on its resistance against delamination. It should be noted, that the strength of the attachment is determined by the peak force, which is reached before the delamination crack is initiated and the tape is starting to detach. Once delamination is started and the crack in the interface propagates, the force drops. This means that the length of the tape does not have an effect on the pull-off strength, but the work that is necessary to detach the tape. Greater width of the tape and a lower peel-off angle result in a higher resistance against detachment, because the stress is more widely distributed in the delamination zone. When the peeling angle is close to zero, which means, that the tape (or the thread) is pulled nearly parallel to the substrate surface, the elasticity of the tape plays an important role [15]. However, in a single glue-coated anchor thread the peeling angle can hardly be controlled as many natural substrates have surfaces with complex topography and the thread may be stressed in variable directions. Moreover, only substrates with good wetting properties will allow a high width (relative to the height) as the glue can spread laterally. Hence, the load acting on the adhesive bond is poorly controllable, and this, together with the unpredictable adhesion due to inhomogeneous substrate surfaces, makes a thread's own glue coat an inefficient anchorage.

5.3.2 Attachment Disc

When we look at the thread anchorages of modern spiders, the Araneomorphae, we find that the two functions, structure and adhesion, are performed by independent threads coming from different glands. The structural thread, the dragline, is a product of the major ampullate (MA) gland. It is one of the strongest and toughest of all silk fibres a spider can produce [7]. The dragline is a bundle of at least two fibres, each coming from one of the paired anterior lateral spinnerets (ALS). The adhesion function is performed by the threads of the piriform (PI) glands, which are usually short glue-coated fibres with a sub-micron diameter. Each of both ALSs bears 30–50 times more PI spigots than MA spigots [5, 33–35]. While the MA gland is one of the largest of all silk glands of a web-building spider, the numerous PI glands are minute and only store very little amount of material. The PI glands are activated only for a short moment, and it is still unknown, how the secretion emergence can start and stop so suddenly. It is probable that the rubbing of the spinnerets against each other plays a role in this mechanism [6]. The material is then applied onto the substrate with quick lateral sweep movements of the spinnerets (Fig. 5.2c). In between each sweep the spinnerets are rubbed against each other, which pulls the ends of the PI threads from the substrate surface in an elevated position and conglutinates them with the dragline [36]. The number and track shape of sweeps in such a sequence may greatly vary between species, individuals and also between spinning events of a single individual. It is unclear, if the resulting variation in anchor size is an adaptation to different substrates or loading situations.

Anchoring the structural threads with multiple sticky threads has several advantages. First, it enhances the flaw tolerance and fracture toughness of the anchorage. The substrate surfaces of the anchorages are highly unpredictable and often inhomogeneous. Furthermore, attachment discs are spun very quickly, and the filling degree of single PI glands may vary. Therefore flaws, failures and pre-existing cracks cannot be prevented. Under such conditions hierarchical branching is the way to create a robust anchorage, in which the failure of a single element does not affect the integrity of the whole structure. Second, crack induction in or near the substrate joint is retarded, because stress is distributed over great parts of the branches. Third, in such a structure the detachment mechanism may be optimized. As discussed above a single agglutinated thread suffers from a lack of control of the peeling angle, which is crucial, since the adhesive bonding itself is similarly unpredictable. Gorb and Varenberg [37] first pointed out that spider attachment discs exhibit a self-stabilising geometry, followed by Pugno and co-workers, who modelled the attachment disc with the help of the, so-called, multiple peeling theory [38–40]. Here the basic element of a thread anchorage is two contra-lateral tapes that are simultaneously peeled off [16] (Fig. 5.1e). In such a configuration the peel-off angle always stabilizes in an optimum, regardless of the initial pulling angle [38–40].

Sahni et al. [14] proposed the comparable ‘staple-pin model’ to describe the function of the attachment disc. Here the dragline is agglutinated to the substrate by a set of parallel tapes, which subsequently detach by contralateral peeling. The authors assume that the initial peeling angle is close to zero, because the dragline lays flat on the substrate when unloaded. The assumption differs from the model by Pugno [16], who comes to the conclusion that the peeling angle must be rather steep due to a high elasticity of the PI fibres. Indeed, Sahni et al. [14] did not consider that the dragline is usually not directly agglutinated to the substrate but slightly suspended above [9, 13]. This has profound effects on peeling angle and stress distribution. Derived from their observations the team of Sahni produced artificial attachment discs, using an electrospinning technique [41]. In these models the main

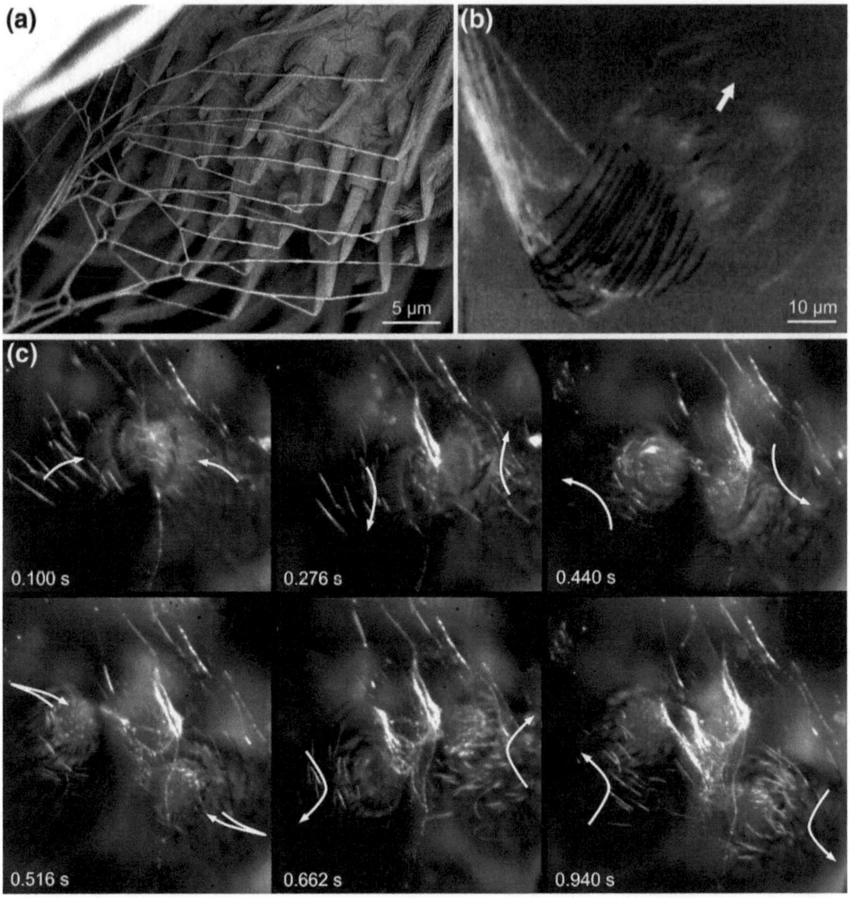

Fig. 5.2 Spinning of attachment discs. **a** Cryo-scanning electron microscope image of the anterior lateral spinneret (ALS) of an *Uloborus plumipes* spider snap frozen while attachment disc spinning (emerging PI silk slightly brighten up, for clearness). The numerous piriform gland spigots and the emerging PI silk are visible. From Wolff et al. [9]. **b** High speed video frame of the ALS of *U. plumipes* spinning PI silk onto a smooth glass slide, as observed with means of a reflection interference contrast microscope (view from below). The *black lines* represent PI threads in contact with the glass slide. *Arrow* indicates the spinneret movement. From Wolff et al. [9]. **c** Frames from a high speed video sequence (500 fps) of the paired ALS of an *Eratigena atrica* spinning an attachment disc onto a glass slide (viewed from below). *Arrows* indicate the spinneret movements. The spinnerets are initially rubbed against each other, followed by multiple alternating lateral sweeps

(nylon) thread, was directly attached to the substrate by an overlay of polyurethane nanofibers. The models were spun on aluminium surfaces, some of which were either strongly hydrophobic or hydrophilic by a previous plasma treatment. As expected the pull-off force of the nylon thread was reduced on the hydrophobic surface. However, the force reduction was even higher on the strongly hydrophilic surfaces, because the fibres could not peel off due to their strong adhesion, but fractured instead. Because of the direct agglutination of the nylon thread onto the substrate, the polyurethane fibres could hardly deform and the fracture occurred in a narrow band around the main thread. In contrast, in the spider attachment disc a substantial part breaks off, when it is pulled from a hydrophilic substrate [9, 13, 14] (Fig. 5.4a). Experimental observations of pulled attachment discs show that during high load before the start of detachment or fracture the PI threads around the suspended dragline deform and their final angle with the substrate is rather steep [9] (e.g. Fig. 5.4a).

Interestingly, Sahni et al. [14] describe another architecture, called dendritic, in which the PI threads are radially arranged and only agglutinated in the outermost part. This architecture is used by cobweb spiders in, so-called, gumfoot-traps, which consist of a tacky pre-stressed thread that is detached and lifted, when a prey walks against it [42]. For this purpose it must detach when a certain stress threshold has reached. Sahni et al. [14] show that the breaking force of the attachment discs of gumfoots is much lower than that of the usual (scaffolding) disc. They explain the difference with the dissimilarity of the resulting peeling angles in both architectures, however their experimental data shows a breakage of the PI threads before a peel-off detachment is initiated. This may indicate that peeling does not play a role in the gumfoot disc. I studied both types of attachment discs with a light microscope using different illumination techniques, and found that in the gumfoot disc the dragline suspension is much more elevated (Fig. 5.3c, compare with Fig. 5.3a), like suggested by Sahni et al. [14]. However, the arrangement of PI threads around the loading point does not obviously differ between the gumfoot and the scaffolding disc (Fig. 5.3b, d). The main difference is that in the gumfoot disc the bridge consist of only few PI threads and therefore represents a predetermined breaking point.

In contrast to the staple-pin models by Pugno et al. and Sahni et al., Gorb and Varenberg [37] describe attachment discs as mushroom-like microstructures, where the stress is radially distributed and the delamination zone is circular. Indeed, such a morphology with a (more) medially attaching dragline may be present in many spider species, and occasionally results in a mushroom-like circular peel-off [36] (Fig. 5.5).

While, these simple models are helpful to understand some basic structural features that contribute to the high strength of the attachment disc, the actual, experimentally observed mechanical behaviour may highly differ from the theoretical presumptions.

First, the detachment mechanism highly differs on different substrates. As for all adhesives, both the polarity and the micro-topography of the substrate surface determine how good the silk anchor glue can spread and create a bonding [9, 13]. Kendall's peeling-model presumes that the breaking energy of the thin film is much higher than its adhesion to the substrate. However, pull-off tests from usual, hydrophilic glass always result in an internal fracture of the attachment disc [9, 13, 14] (Fig. 5.4a, b). This means, that the breaking force of the PI threads is lower than the detachment force in these situations. For such conditions the use of the peeling model is problematic. In contrast to Sahni et al. [14], Pugno et al. [38–40] take the

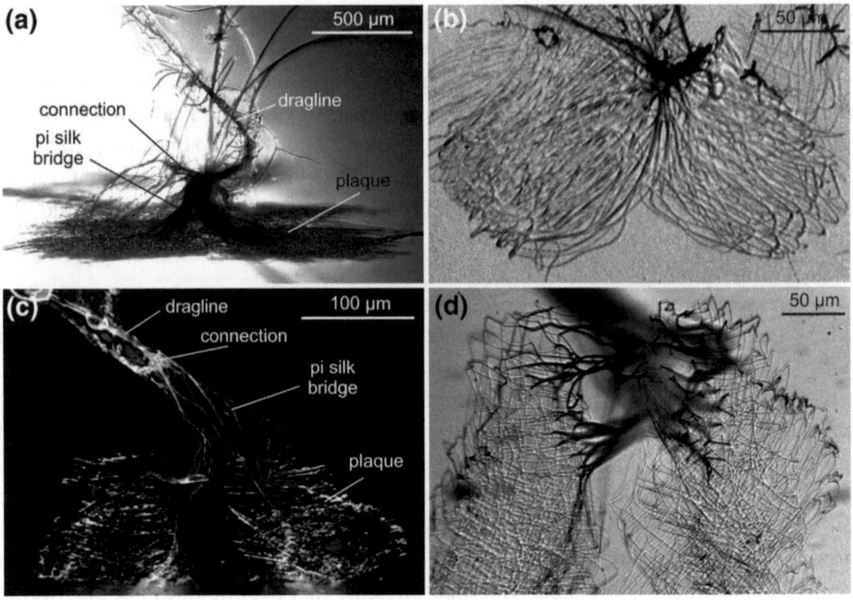

Fig. 5.3 Two different attachment disc architectures (**a** *Nephila senegalensis*, **b–d** *Parasteatoda tepidariorum*). **a**, **b** Scaffolding disc (usual thread anchor), **a** lateral view, **b** view from below, micrograph by Marina Wirth, with kind permission. **c**, **d** Gumfoot disc (anchor of a cobweb capture thread), **c** lateral view, **d** view from below

internal fracture mechanics into account and exactly define the boundary conditions of their numerical model: the structure is optimal, when adhesion and material strength are balanced (Fig. 5.4d). Such an optimum is presumed on substrate surfaces the spider is adapted to. Sahni et al. [14] stated that attachment discs are stronger than the dragline, and that during pull-off tests from glass always the dragline failed. From the biological point of view this would make no sense since this would mean an unused excess of adhesive capacity and material, which may have never evolved because of a missing selective advantage. Grawe et al. [13] performed pull-off tests of attachment discs spun by five different species, including the cobweb spider species used by Sahni and co-workers. They found that in attachment discs spun on glass the dragline failed only in 6–37% of all cases (depending on the species), when pulled. In the majority of trials there was an internal breakage of the attachment discs, in the PI thread network of the dragline suspension. Interestingly, they observed another failure mode that occurred with a frequency of 22–36% in the three web building spiders, but never in attachment discs of the two hunting spider species tested. In this case the bonding between the dragline and the enveloping PI threads fails, leading to the dragline being pulled through an otherwise intact attachment disc. This seems to be a specific trait of orb and cobweb spiders that use the PI secretion in so-called sliding junctions that flexibly combine different types of threads in a web [43, 44]. Total delamination of attachment discs can be observed on hydrophobic substrates, including polymer foils, silane-treated glass and tree leaves [9, 13] (Fig. 5.4b, c). It can therefore be assumed that such spiders are adapted to attach their threads on more hydrophilic surfaces, such as rocks and wood.

Second, the spinning pattern of a spider is more complex than a set of simple lateral sweeps. Derived spiders, such as orb- and cob-web spiders change the angle at the turning point of a lateral sweep, such that the single PI threads are not glued in parallel, but cross each other, resulting in a mesh-like pattern [9, 36] (e.g. Fig. 5.3b, c). Following the models by Pugno et al. [38–40] and Sahni et al. [14] one may assume that lateral connections are disadvantageous, because they permit a transmission of the delamination crack to neighbouring threads. However, experimental observations indicate that crossing fibres may temporarily stop the propagation of the delamination crack, and yield a further rise of the detachment force before delamination proceeds [9]. This may be explained with a sudden increase of the peeling line, which means a lateral stress distribution, resulting in a higher resistance. In artificial electrospun attachment disc models it was found that an increased density of the glue fibres resulted in higher peel-off forces [41], which further indicates a beneficial role of interconnections within the attachment disc.

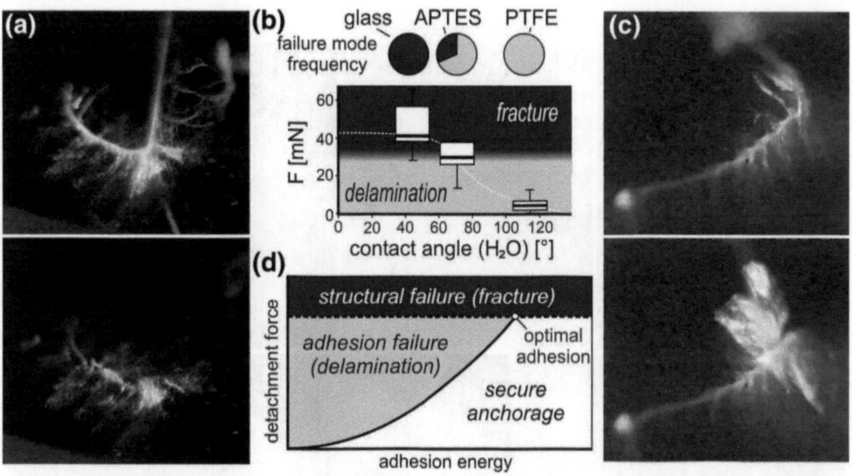

Fig. 5.4 Substrate dependent failure of attachment discs. **a** Video frames of a pull-off test of a *Nephila senegalensis* attachment disc from glass, above shortly before and below shortly after failure. In this case the adhesion is so strong that the anchor fails above the plaque, namely in the PI silk bridge. **b** Experimental pull-off strength and failure mode of *N. senegalensis* attachment discs spun against three substrates differing in their surface free energy (glass slides, untreated and silane-coated (APTES), and Teflon sheets (PTFE)). The *white dotted line* indicates the hypothesized relationship between surface polarity and pull-off force. Data from Grawe et al. [13] and Wolff et al. [9]. **c** Video frames of a pull-off test of a *N. senegalensis* attachment disc from Teflon, above before and below during failure. In this case the adhesion is weaker than the breaking strength of the attachment disc, and the plaque completely detaches from the substrate. **d** Failure mode phase diagram after the numerical model by Pugno et al. [39], relating the adhesion energy of the glue with the detachment force of the anchorage. Adapted from Cranford et al. [45]

Additionally, many spiders do not perform the sweep movements rectangularly towards the median line along which the dragline is usually orientated. On the front and rear end of the anchor, where the dragline emerges and the main load acts, the PI threads may be arranged axisymmetrically due to an alteration of movement directions between subsequent sweeps [36] (Fig. 5.5a–c). This means that load does not initially act on single threads at the front edge of the disc like in the staple-pin models, but is widely distributed over a network of fibres [9]. Presumably, a much higher initial force is required to overcome the adhesion of such a structure and initiate a crack in the interface or within the structure. Further, this configuration may lead to a circular or axisymmetric delamination edge where the peeling line gradually increases [9], like in mushroom-like microstructures [46] or tapes with an axisymmetric shape [47].

Eventually, it can be observed that many spiders deploy more material in the median part of the discs by applying a second or third layer of PI threads. This may enhance the stability of this part, which is important in high load situations on substrates with good glue adhesion. Further, this could result in a higher bending stiffness of the median part, which may have an effect on the detachment mechanism by peeling.

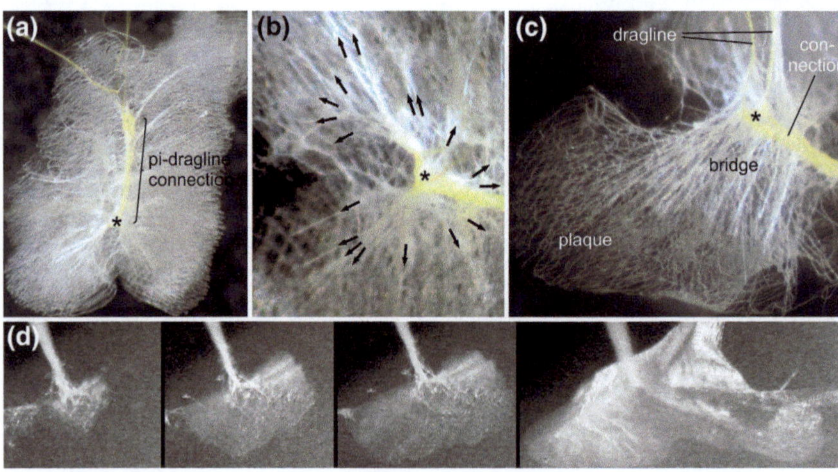

Fig. 5.5 Mushroom-like microstructure and detachment behaviour of attachment discs. **a–c** Reflection light micrographs of attachment discs of *Nephila plumipes*, detached. The dragline is frequently *yellow coloured* in individuals of this species, making it easy to distinguish between the MA and PI silks. The insertion point of the upstream dragline is marked by an *asterisk*. **a** Total attachment disc, viewed from above. Note that the PI-dragline connection piece is located in the median part and does not cover the entire length of the disc. Accordingly, the dragline insertion point is shifted from the front edge towards the median part. **b** Detail of the main loading region around the insertion point of the upstream dragline. *Arrows* indicate the orientation of PI threads in the bridge. **c** Detail of the loading region, viewed from more lateral. **d** Frames of a video recording of a pull-off test of an attachment disc of *Cupiennius salei* spun against Teflon, showing a *circular crack* propagation (detached part masked and slightly brighten up for clearness)

These examples depict that we are far from a comprehensive understanding of the mechanical behaviour of attachment discs. Different spiders may have evolved a set of modifications and adaptations that specifically enhance the function of thread anchors in different situations, and are yet to be studied.

5.3.3 *Multiple Attachment Discs*

When looking at suspended orb webs one may note that structural frame threads are often not anchored by a single, but multiple attachment discs that are radially arranged on the substrate surface (Fig. 5.1c). Such a structure reminds of the byssus threads of mussels that are attached by adhesive plaques and always applied in a radial network. The radial arrangement and specific length of the thread determines a load angle, under which the attachment of the plaques acts optimal [48, 49]. Bosia et al. [17] modelled—inspired from the branched adhesive setae of geckoes—the effect of hierarchical branching on adhesion strength, and found that it reduces the load on single attachments and lowers the peel-off angle. Thus in such a configuration the attachment discs may act synergistically. An interesting question for future studies is if the arrangement and angles of the branches are optimally adjusted in such a multiple anchor.

5.4 Spend Less: Yield More

A striking question is, why spiders apply numerous tiny glue trails instead of simply extruding a blob of glue to anchor their threads. One may assume that it is the nature of the spinning apparatus to produce fibre-like products. However, orb- and cobweb spiders also produce another adhesive which comes from a modified silk gland, but is not fibre-like: the viscid glue of the capture threads used to retain prey in the web. It comes from enlarged nozzles and forms a thick glue coat on the thread, which breaks up into regularly arranged droplets due to its inherent viscoelastic properties [50]. Comparison of the sticky droplets of a capture thread with a thread anchorage shows several differences, which provide an explanation for the discrepancy in glue use. Above all, both glues serve as a means of attachment over a different period. The viscid glue must retain the prey only temporarily, over a short period, until the spider has approached and overwhelmed it. It must be permanently tacky, because the catching event is highly unpredictable. This implicates an extreme softness and elasticity of the glue [51, 52], which has the drawback of reduced material strength, and hence substantial amounts of glue are necessary to resist the forces elicited by the struggling prey. The thread anchor, on the other hand, must be durable and stable. Its glue solidifies, which creates a higher material strength and demands less material. While the tacky capture threads can be used multiple times, thread anchors cannot be re-used for other thread attachments. It is an immanent rule of nature that the individual that uses the restricted resources most efficiently, will more efficiently reproduce than its wasteful competitors. In large orb web spiders only 2–10 μg of material is used in an attachment disc, which corresponds to ~10 ppm of the spider weight [9].

The nano-fibrillar glue application pattern is highly unique to spider silk anchors. A comparable structure, which can resist enormous pulling and shear stresses, is the byssus thread of a mussel used to attach to rocks or other mussels in the surf zone. Byssus threads attach by a mushroom-like adhesive plaque that is continuous [49, 53, 54]. A sessile mussel may afford to produce a massive plaque that attaches it durably. On the other hand, the thread anchor of a spider has a much higher demand of economy, because it is newly spun every time the spider moves forward and needs to secure itself, and in high frequency when building a web or a cocoon. Furthermore, the surfaces a spider encounters during locomotion and web-building may be extremely diverse in terms of their physical and chemical surface properties, whereas the substrates of mussels may be more defined. As discussed above, I assume that the hierarchical structure of the spider silk anchorage is of main importance to create an outstanding flaw tolerance and fracture toughness of the bonding. Further, it may save a substantial amount of material, if compared to a continuous adhesive plaque.

For technical applications, the idea of creating better bonds by using less adhesive material is tempting. It is a matter of both saving money and fossil resources. This is of high significance, since in industrial and medical applications often expensive elaborated, tailored adhesives are necessary. To date a homogeneous and continuous application of glue is regarded as a key for a reliable bonding. Therefore the highly divergent glue use by spiders is so intriguing. An attempt to mimic the structure and properties of attachment discs was undertaken by Jain et al. [41]. They used the electrospinning technique to apply sticky polyurethane nanofibres onto a nylon fibre. Unfortunately, their statement that this technique produces higher adhesion per unit of adhesive consumption cannot be evaluated, because comparative measurements using a continuous glue film of polyurethane have not been performed. The challenge of future biomimetic approaches is the implementation of the glue application pattern into joints of adherends or into sticky tapes.

5.5 Embedding Fibres

But silk anchors can teach us more than that. The embedding of ductile polymer fibres is an established method to enhance the fracture toughness of strong, but brittle materials, like concrete [55]. In contrast, a paradigm in adhesives design is the aim of a bond as homogeneous as possible, because little defects, such as enclosed air bubbles, can quickly induce cracks and fractures. Thus the specific nano-composite structure of the PI glue is highly divergent from conventional artificial adhesives.

Based on microscopy images published by Blasingame et al. [56] and testing different parameters with their model, Pugno et al. [38] estimated that the PI threads must be extremely extensible (8×) and soft (Young's modulus of 1 MPa). In contrast, the model by Sahni et al. [14] assumes a rather rigid material. However, they later found an extreme effect of the elasticity of the glue fibres in their biomimetic, electrospun attachment discs: when nylon nanofibers were used instead of polyurethane nanofibres to glue down the main thread the resulting structure was very brittle and easily broke apart [41]. Tensile tests of isolated strands of PI silk from a non-web building spider revealed that this silk is more extensible and softer than other silks [57], though the numbers are much lower than predicted from Pugno's model. On the ultrastructural level, the core–shell structure of silks may include layers of different stiffness [58, 59]. It is conceivable that the solidified glue coat is much stiffer than the embedded fibre. A fractographic analysis of an attachment disc showed an even fracture face of the glue and an irregular one of the embedded fibre [9, 57] (Fig. 5.6b). This speaks for a less ordered structure and high extensibility of the fibre proteins, and a more stiff and brittle, highly crosslinked material in the glue coat. Such a nano-composite of two materials with totally different properties may be one of the key functions within a spider's thread anchorage. If the embedded fibre is highly ductile it will distribute tensile stress over a high length of the fibre, which reduces stress concentration and the probability of crack induction in the glue coat.

In order to understand the micromechanics of the composite PI silk, and how it interacts with different substrates, we have to look at its chemical composition and nano-structure. Unfortunately, our knowledge on these matters is very sparse. This is mainly due to the small size of the PI glands and the tiny amounts of the glue in the silk anchor, which is challenging for the analysis of its composition and macro-molecular structure.

Most previous studies focussed on the proteinaceous fraction of the core fibre. Already Apstein [5] noted that the two components, core fibre and glue coat, already build two distinct phases in the gland lumen and stain differently. Kovoor

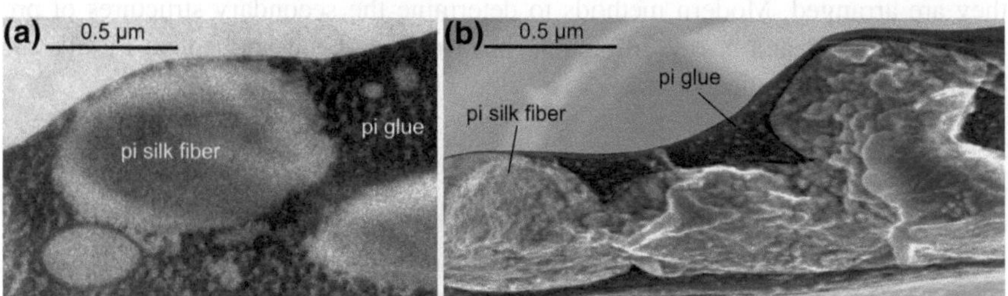

Fig. 5.6 Ultrastructure of PI silk. **a** Transmission electron microscopy image of a cross section of PI silk that had been attached to a smooth polymer surface (former contact side at the *bottom*). **b** Cryo-scanning electron microscopy image of transverse fracture of piriform fibre bundle. (After Wolff et al. [9]. In both images the background, and additionally in (**b**) the fracture faces of the fibres, were masked and slightly brighten up for more clearness.)

and Zylberberg [8] did different staining and digestion experiments and found hints that the glue consists of proteins rich in acidic groups and also includes carbohydrate compounds which may indicate glycoproteins. In contrast, the core fibre is made of protein rich in basic amino acids [8, 60]. The first comprehensive amino acid analysis of the PI gland content was performed by Andersen [61]. He noted a high proportion of amino acids with polar and charged side chains, which may support a good interactions with substrates. Further, he remarked that the low content of small side chain amino acids and high content of proline speak for an amorphous structure of the proteins forming a stretchy fibre. In contrast the observation that the PI secretion is birefringent [62] indicates that at least a fraction of the material should be highly ordered.

Six years ago, single genes and the corresponding amino acid sequences for the fibre forming proteins (so-called spidroins) have been analysed in a cobweb spider [56] and four species of orb web spiders [63, 64]. Spidroins are very large proteins that contain highly repetitive, short sequences of amino acids. These repetitive elements form secondary structures that affect the material properties of the silk [65]. This is best understood for the dragline silk, where poly-alanine motifs form crystal-like β-sheets that strengthen the material, and interconnecting amorphous glycine-rich domains contribute to a high extensibility [66–68]. The amino acid sequences in the PI spidroin differ from other spider silks by having unique repetitive elements [63, 64]. It is unclear, what kind of secondary structures are built by these and which material properties they cause. A proline-rich repetitive element in the PI silk of orb web spiders may produce a random-coil configuration with elastomeric properties [64, 69, 70] and contribute to compliancy, as the role for proline in other silks [71]. However, the PI spidroin found in cobweb spiders exhibits a very low content in proline [56], and yet there is no obvious difference in the macro-structure and performance of their attachment discs if compared to orb web spiders [13]. It cannot be excluded that PI fibres are composed of more than one spidroin and the different studies did not catch the homolog [64]. A study using Raman spectroscopy indicated a mixture of α-helices and β-sheets in the PI gland content [72], but it is unclear if these structures occur in the same protein and how they are arranged. Modern methods to determine the secondary structures of proteins, like NMR and X-ray diffraction, have not yet been applied on PI silk, and could help to get a clearer image.

5.6 Conclusions and Outlook

There are substantial differences in how spiders and humans apply glue. In spider silk anchors numerous glue-coated nanofibers, applied in a thin, discontinuous film provide a remarkable bonding performance, which is unachieved by artificial adhesives. This is an intriguing starting point to investigate silk thread anchors in order to exploit and evaluate their potential for the development of novel unconventional high-end adhesives and smart glue applications.

Acknowledgements I thank Mariella Herberstein, Joshua Madin and Stanislav Gorb for comments on this manuscript. This study was founded by a scholarship of the German Merit Foundation (Studienstiftung des Deutschen Volkes) and a Macquarie Research Fellowship of Macquarie University.

References

1. J.M. Gosline, M.W. Denny, M.E. DeMont, Spider silk as rubber. Nature **309**(5968), 551–552 (1984)
2. F. Vollrath, D.P. Knight, Liquid crystalline spinning of spider silk. Nature **410**(6828), 541–548 (2001)
3. I. Agnarsson, M. Kuntner, T.A. Blackledge, Bioprospecting finds the toughest biological material: extraordinary silk from a giant riverine orb spider. PLoS ONE **5**(9), e11234 (2010)
4. R.F. Foelix, *Biology of Spiders*, 3rd edn. (Oxford University Press, New York, 2011)
5. C. Apstein, Bau und Funktion der Spinndrüsen der Araneida. Arch. Naturg. **55**, 29–74 (1889)
6. W.G. Eberhard, Possible functional significance of spigot placement on the spinnerets of spiders. J. Arachnol. **38**(3), 407–414 (2010)
7. T.A. Blackledge, C.Y. Hayashi, Silken toolkits: biomechanics of silk fibers spun by the orb web spider *Argiope argentata* (Fabricius 1775). J. Exp. Biol. **209**(13), 2452–2461 (2006)
8. J. Kovoor, L. Zylberberg, Fine-structural aspects of silk secretion in a spider (*Araneus diadematus*). 1. Elaboration in the pyriform glands. Tissue Cell **12**(3), 547–556 (1980)
9. J.O. Wolff, I. Grawe, M. Wirth, A. Karstedt, S.N. Gorb, Spider's super-glue: thread anchors are composite adhesives with synergistic hierarchical organization. Soft Matter **11**(12), 2394–2403 (2015)
10. J.A. Murphy, M.J. Roberts, *Spider Families of the World and Their Spinnerets* (British Arachnological Society, York, 2015)
11. F. Vollrath, P. Selden, The role of behavior in the evolution of spiders, silks, and webs. Annu. Rev. Ecol. Evol. Syst. **38**, 819–846 (2007)
12. H.M. Whitney, W. Federle, Biomechanics of plant-insect interactions. Curr. Opin. Plant Biol. **16**(1), 105–111 (2013)
13. I. Grawe, J.O. Wolff, S.N. Gorb, Composition and substrate-dependent strength of the silken attachment discs in spiders. J. R. Soc. Interface **11**(98), 1742–5662 (2014)
14. V. Sahni, J. Harris, T.A. Blackledge, A. Dhinojwala, Cobweb-weaving spiders produce different attachment discs for locomotion and prey capture. Nat. Commun. **3**, 1106 (2012)
15. K. Kendall, Thin-film peeling—elastic term. J. Phys. D Appl. Phys. **8**(13), 1449–1452 (1975)
16. N.M. Pugno, The theory of multiple peeling. Int. J. Fract. **171**(2), 185–193 (2011)
17. F. Bosia, S. Colella, V. Mattoli, B. Mazzolai, N.M. Pugno, Hierarchical multiple peeling simulations. RSC Adv. **4**(48), 25447–25452 (2014)
18. P. Poza, J. Perez-Rigueiro, M. Elices, J. Llorca, Fractographic analysis of silkworm and spider silk. Eng. Fract. Mech. **69**(9), 1035–1048 (2002)
19. P. Jiang, H. Liu, C. Wang, L. Wu, J. Huang, C. Guo, Tensile behavior and morphology of differently degummed silkworm (*Bombyx mori*) cocoon silk fibres. Mater. Lett. **60**(7), 919–925 (2006)
20. M.S. Engster, Studies on silk secretion in the Trichoptera (F. Limnephilidae). Cell Tissue Res. **169**(1), 77–92 (1976)
21. T. Hatano, T. Nagashima, The secretion process of liquid silk with nanopillar structures from *Stenopsyche marmorata* (Trichoptera: Stenopsychidae). Sci. Rep. **5**, 9237 (2015)
22. J.M. Palmer, F.A. Coyle, F.W. Harrison, Structure and cytochemistry of the silk glands of the mygalomorph spider *Antrodiaetus unicolor* (Araneae, Antrodiaetidae). J. Morphol. **174**(3), 269–274 (1982)

23. B.O. Swanson, S.P. Anderson, C. DiGiovine, R.N. Ross, J.P. Dorsey, The evolution of complex biomaterial performance: the case of spider silk. Integr. Comp. Biol. **49**(1), 21–31 (2009)
24. T.A. Blackledge, J. Pérez-Rigueiro, G.R. Plaza, B. Perea, A. Navarro, G.V. Guinea, M. Elices, Sequential origin in the high performance properties of orb spider dragline silk. Sci. Rep. **2**, 782 (2012)
25. J.O. Wolff, S.N. Gorb, *Attachment Structures and Adhesive Secretions in Arachnids* (Springer, Cham, 2016)
26. Y. Termonia, Molecular modeling of spider silk elasticity. Macromolecules **27**(25), 7378–7381 (1994)
27. D. Porter, F. Vollrath, Z. Shao, Predicting the mechanical properties of spider silk as a model nanostructured polymer. Eur. Phys. J. E **16**(2), 199–206 (2005)
28. C. Fant, K. Sott, H. Elwing, F. Hook, Adsorption behavior and enzymatically or chemically induced cross-linking of a mussel adhesive protein. Biofouling **16**(2–4), 119–132 (2000)
29. M. Wiegemann, B. Watermann, Peculiarities of barnacle adhesive cured on non-stick surfaces. J. Adhes. Sci. Technol. **17**(14), 1957–1977 (2003)
30. M.J. Sever, J.T. Weisser, J. Monahan, S. Srinivasan, J.J. Wilker, Metal-mediated cross-linking in the generation of a marine-mussel adhesive. Angew. Chem. **116**(4), 454–456 (2004)
31. J. Pérez-Rigueiro, C. Viney, J. Llorca, M. Elices, Mechanical properties of single-brin silkworm silk. J. Appl. Polym. Sci. **75**(10), 1270–1277 (2000)
32. J. Pérez-Rigueiro, M. Elices, J. Llorca, C. Viney, Effect of degumming on the tensile properties of silkworm (*Bombyx mori*) silk fiber. J. Appl. Polym. Sci. **84**(7), 1431–1437 (2002)
33. J.A. Coddington, Spinneret silk spigot morphology: evidence for the monophyly of orbweaving spiders, Cyrtophorinae (Araneidae), and the group Theridiidae plus Nesticidae. J. Arachnol. **17**(1), 71–95 (1989)
34. L. Yu, J.A. Coddington, Ontogenetic changes in the spinning fields of *Nuctenea cornuta* and *Neoscona theisi* (Araneae, Araneidae). J. Arachnol. **18**(3), 331–345 (1990)
35. P. Dolejš, J. Buchar, L. Kubcová, J. Smrž, Developmental changes in the spinning apparatus over the life cycle of wolf spiders (Araneae: Lycosidae). Invertebr. Biol. **133**(3), 281–297 (2014)
36. J.O. Wolff, M.E. Herberstein, 3D-printing spiders: back-and-forth glue application yields silk anchorages with high pull-off resistance under varying loading situations. J. R. Soc. Interface **14**(127), 20160783 (2017).
37. S.N. Gorb, M. Varenberg, Mushroom-shaped geometry of contact elements in biological adhesive systems. J. Adhesion Sci. Technol. **21**(12–13), 1175–1183 (2007)
38. Pugno N, Vanzo J, Cranford S, Buehler M, Simultaneous material and structural optimization in the spider web attachment disk, in *Atti del XX Cong. Nazionale dell' Associazione Italiana di Meccanica Teorica ed Applicata, Minisymposium Micro- or Nano-mechanics*, Bologna, Italy, 12–15 Settembre 2011
39. N.M. Pugno, S.W. Cranford, M.J. Buehler, Synergetic material and structure optimization yields robust spider web anchorages. Small **9**(16), 2747–2756 (2013)
40. L. Brely, F. Bosia, N.M. Pugno, Numerical implementation of multiple peeling theory and its application to spider web anchorages. Interface Focus **5**(1), 20140051 (2015)
41. D. Jain, V. Sahni, A. Dhinojwala, Synthetic adhesive attachment discs inspired by spider's pyriform silk architecture. J. Polym. Sci. B Polym. Phys. **52**, 553–560 (2014)
42. S. Argintean, J. Chen, M. Kim, A.M.F. Moore, Resilient silk captures prey in black widow cobwebs. Appl. Phys. A **82**, 235–241 (2006)
43. W.G. Eberhard, Physical properties of sticky spirals and their connections: sliding connections in orb webs. J. Nat. Hist. **10**(5), 481–488 (1976)
44. S.N. Gorb, M.A. Landolfa, F.G. Barth, Dragline-associated behaviour of the orb web spider *Nephila clavipes* (Araneioidea, Tetragnathidae). J. Zool. Lond. **244**, 323–330 (1998)

45. S.W. Cranford, N.M. Pugno, M.J. Buehler, Silk and web synergy: the merging of material and structural performance, in *Biotechnology of Silk. Biologically-Inspired Systems*, vol. 5, ed. by T. Asakura, T. Miller (Springer, Dordrecht, 2014), pp. 219–268
46. L. Heepe, A.E. Kovalev, A.E. Filippov, S.N. Gorb, Adhesion failure at 180 000 frames per second: direct observation of the detachment process of a mushroom-shaped adhesive. Phys. Rev. Lett. **111**(10), 104301 (2013)
47. L. Afferrante, G. Carbone, G. Demelio, N. Pugno, Adhesion of elastic thin films: double peeling of tapes versus axisymmetric peeling of membranes. Tribol. Lett. **52**(3), 439–447 (2013)
48. Z. Qin, M.J. Buehler, Impact tolerance in mussel thread networks by heterogeneous material distribution. Nat. Commun. **4**, 2187 (2013)
49. K.W. Desmond, N.A. Zacchia, J.H. Waite, M.T. Valentine, Dynamics of mussel plaque detachment. Soft Matter **11**(34), 6832–6839 (2015)
50. V. Sahni, T.A. Blackledge, A. Dhinojwala, Viscoelastic solids explain spider web stickiness. Nat. Commun. **1**, 19 (2010)
51. C. Dahlquist, An investigation into the nature of tack. Adhes. Age **2**(10), 25–29 (1959)
52. C. Creton, Pressure-sensitive adhesives: an introductory course. MRS Bull. **28**(06), 434–439 (2003)
53. A. Tamarin, P. Lewis, J. Askey, The structure and formation of the byssus attachment plaque in *Mytilus*. J. Morphol. **149**(2), 199–221 (1976)
54. H.G. Silverman, F.F. Roberto, Understanding marine mussel adhesion. Mar. Biotechnol. **9**(6), 661–681 (2007)
55. V.C. Li, On engineered cementitious composites (ECC). J. Adv. Concr. Technol. **1**(3), 215–230 (2003)
56. E. Blasingame, T. Tuton-Blasingame, L. Larkin, A.M. Falick, L. Zhao, J. Fong, V. Vaidyanathan, A. Visperas, P. Geurts, X.Y. Hu, C. La Mattina, C. Vierra, Pyriform Spidroin 1, a novel member of the silk gene family that anchors dragline silk fibers in attachment discs of the black widow spider, *Latrodectus hesperus*. J. Biol. Chem. **284**(42), 29097–29108 (2009)
57. J.O. Wolff, M. Řezáč, T. Krejčí, S.N. Gorb, Hunting with sticky tape: functional shift in silk glands of araneophagous ground spiders (Gnaphosidae). J. Exp. Biol. *In press* (2017)
58. A. Sponner, W. Vater, S. Monajembashi, E. Unger, F. Grosse, K. Weisshart, Composition and hierarchical organisation of a spider silk. PLoS ONE **2**(10), e998 (2007)
59. C.P. Brown, J. MacLeod, H. Amenitsch, F. Cacho-Nerin, H.S. Gill, A.J. Price, E. Traversa, S. Licoccia, F. Rosei, The critical role of water in spider silk and its consequence for protein mechanics. Nanoscale **3**(9), 3805–3811 (2011)
60. J. Kovoor, Etude histochimique et cytologique des glandes sericigenes de quelques Argiopidae. Annales des Sciences naturelle Zoologie and biologie animale 12ème série **14**, 1–40 (1972)
61. S.O. Andersen, Amino acid composition of spider silks. Comp. Biochem. Physiol. **35**, 705–711 (1970)
62. R.W. Work, Web components associated with the major ampullate silk fibers of orb-web-building spiders. Trans. Am. Microsc. Soc. **100**(1), 1–20 (1981)
63. P. Geurts, L. Zhao, Y. Hsia, E. Gnesa, S. Tang, F. Jeffery, C. La Mattina, A. Franz, L. Larkin, C. Vierra, Synthetic spider silk fibers spun from Pyriform Spidroin 2, a glue silk protein discovered in orb-weaving spider attachment discs. Biomacromolecules **11**(12), 3495–3503 (2010)
64. D.J. Perry, D. Bittencourt, J. Siltberg-Liberles, E.L. Rech, R.V. Lewis, Piriform spider silk sequences reveal unique repetitive elements. Biomacromolecules **11**(11), 3000–3006 (2010)
65. C.Y. Hayashi, N.H. Shipley, R.V. Lewis, Hypotheses that correlate the sequence, structure, and mechanical properties of spider silk proteins. Int. J. Biol. Macromol. **24**(2), 271–275 (1999)
66. A.H. Simmons, C.A. Michal, L.W. Jelinski, Molecular orientation and two-component nature of the crystalline fraction of spider dragline silk. Science **271**(5245), 84–87 (1996)

67. J. Gosline, P. Guerette, C. Ortlepp, K. Savage, The mechanical design of spider silks: from fibroin sequence to mechanical function. J. Exp. Biol. **202**(23), 3295–3303 (1999)
68. S. Keten, M.J. Buehler, Nanostructure and molecular mechanics of spider dragline silk protein assemblies. J. R. Soc. Interface **7**(53), 1709–1721 (2010)
69. S. Rauscher, S. Baud, M. Miao, F.W. Keeley, R. Pomès, Proline and glycine control protein self-organization into elastomeric or amyloid fibrils. Structure **14**(11), 1667–1676 (2006)
70. R.C. Chaw, C.A. Saski, C.Y. Hayashi, Complete gene sequence of spider attachment silk protein (PySp1) reveals novel linker regions and extreme repeat homogenization. Insect Biochem. Mol. Biol. **81**, 80–90 (2017).
71. Y. Liu, Z. Shao, F. Vollrath, Elasticity of spider silks. Biomacromolecules **9**(7), 1782–1786 (2008)
72. T. Lefèvre, S. Boudreault, C. Cloutier, M. Pézolet, Diversity of molecular transformations involved in the formation of spider silks. J. Mol. Biol. **405**(1), 238–253 (2011)

Chapter 6
Optimal Adhesion Control via Cooperative Hierarchy, Grading, Geometries and Non-linearity of Anchorages and Adhesive Pads

Lucas Brely, Daniele Liprandi, Federico Bosia and Nicola M. Pugno

Abstract Optimization of dry adhesion in biological organisms is achieved using various strategies at different scale levels. In the past, studies have shown how contact splitting is used effectively by animals such as geckos and insects to increase the total peeling line of contacts and therefore the adhesion force. Also, tapering of contacts or grading of their mechanical properties has been shown to be instrumental in the achievement of improved adhesion efficiency. On a more macroscopic scale, structures such as spider web anchorages exploit hierarchical structure or nonlinear constitutive material properties to improve resilience and to achieve tunability in adhesion/detachment characteristics. Here, we analyse some of these properties and propose some mechanisms for the optimization of adhesion that have thus far been neglected in modelling approaches, and could be potentially exploited for the design of bioinspired adhesives. We consider hierarchical structure, contact tapering, grading of mechanical properties, and their interaction. It emerges that these mechanisms contribute on various size scales to the achievement of optimal adhesive properties through structural complexity and hierarchical organization.

L. Brely · D. Liprandi · F. Bosia
Department of Physics and "Nanostructured Interfaces and Surfaces" Centre, Università di Torino, via P. Giuria 1, 10125 Turin, Italy
e-mail: federico.bosia@unito.it

N.M. Pugno (✉)
Laboratory of Bio-Inspired and Graphene Nanomechanics, Department of Civil, Environmental and Mechanical Engineering, Università di Trento, via Mesiano, 77, 38123 Trento, Italy
e-mail: nicola.pugno@unitn.it

N.M. Pugno
School of Engineering and Materials Science, Queen Mary University of London, Mile End Road, E14NS London, UK

N.M. Pugno
Ket Lab, Edoardo Amaldi Foundation, Italian Space Agency, Via del Politecnico snc, Rome 00133, Italy

L. Heepe et al. (eds.), *Bio-inspired Structured Adhesives*, Biologically-Inspired Systems 9, DOI 10.1007/978-3-319-59114-8_6

6.1 Introduction

Natural structural materials have been widely used as a source of inspiration for advanced materials, due to their outstanding mechanical properties. This applies to the field of adhesives [1], where notable examples are fibrillar structures such as those found in gecko toe pads or in spiders or insects [2, 3]. Another interesting example is that of spider web attachment discs, in which different types of spider silk threads and hierarchical branching are employed to tune adhesion for specific applications [4–6]. In all of these examples, material properties and structural topology concur in determining adhesive functionality. As for many natural materials, in many cases hierarchical structures are observed ("hierarchical branching"), and it is thought that these are essential in determining optimal adaptation to the surface, load redistribution without self-bunching and possibly tunability in adhesive properties [7]. This hierarchical branching allows so-called "contact splitting", whereby larger contacts are split into smaller ones (fibrils), with benefits deriving from fibril deformation, adaptability to rough surfaces, size effects due to surface-to-volume ratio, uniformity of stress distributions [8]. Contact splitting also contributes to increasing the so-called "peeling line", i.e. the sum of the contact tape widths, also increasing adhesion [7]. Additionally, it has been shown that in biological structures adhesion can be optimized by variable contact unit geometry [9] and spatial variation of mechanical properties, e.g. in the tarsal setae of the ladybird beetle, allowing it to achieve adaptation to rough surfaces while simultaneously ensuring sufficient stability [10]. These concepts are illustrated in Fig. 6.1, highlighting the multiscale nature of these mechanisms.

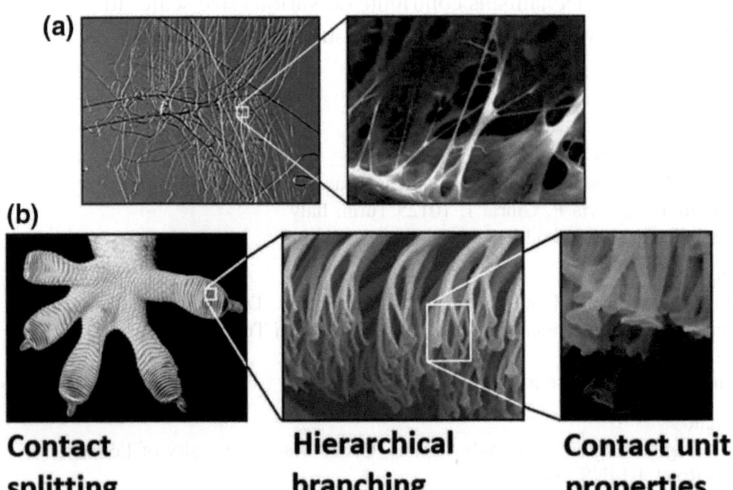

Fig. 6.1 Multiscale mechanisms contributing to fibrillar adhesion: contact splitting, hierarchical branching, contact unit properties (i.e. tapering or grading of mechanical properties): a) Spider web anchorages; b) gecko pads (from [25, 26])

Artificial dry adhesives mimicking natural systems have recently been introduced [11]. For example, creating pillar (or mushroom)-shaped patterns at micro (and nano) scale has allowed to successfully activate adhesion based on van der Waals interactions. The first artificial "mushroom-tape" or "gecko-tape" are based on a punch-like structure [12, 13], which is designed to provide an adhesive force mainly normal to the substrate. The optimization of these structures in terms of adhesive strength has been attempted using contact mechanics models [14]. As symptotically, the models predict an unlimited increase in adhesive strength. Using nanoscale contact units, the adhesive strength tends to the theoretical strength of the Van der Waals interaction. On the other hand, most natural designs are based on a tape-like geometry, which can be described using Kendall's "single peeling" theory [15] developed in the 1970s, recently extended to "multiple peeling" cases [16, 17] and applied to complex geometries [18]. Tape-like structures have also been introduced in artificial adhesives in order to optimize the shear mode adhesion of bioinspired tapes [19, 20]. The study of how nature organizes these basic constitutive units could lead to further optimization in the field of bioinspired adhesives. In this respect, introducing hierarchical structures instead of using regular patterns is the new challenge [21, 22]. To design optimal solutions, adequate modelling of all mechanical mechanisms is required, and thus reliable analytical/numerical approaches need to be developed [23, 24].

In this chapter, we analytically and numerically analyse various mechanisms that can contribute to the improvement of adhesion at different scale levels, at present observed experimentally in the literature but not fully explained theoretically. These include the creation of favourable delamination stress distributions thanks to spatial stiffness variation of the fibrillar elements and increasing the peeling line and adhesion force through contact tapering, i.e. width or thickness variations along the pad length, and hierarchical organization of multiple peeling geometries. The study is performed through energy-based analytic calculations and stress distributions analysis. Finite Element Method (FEM) numerical simulations and their multiscale interaction in complex hierarchical multiple peeling geometries are finally discussed.

6.2 Single and Multiple Peeling Theories Applied to Attachment Structures

Adhesion problems in biological structures can usually be investigated by resorting to thin film peeling theories [15, 16], due to the typical tape-like shape of the terminal contact units observed in systems such as gecko spatulae, insect setae, spider attachment discs, mussel adhesive plaques, etc. These typically involve multiple contact units, so that the detachment of adhesive biological structures must be treated as a multiple peeling problem. In this type of formulation, Griffith energy balance is used to determine whether the *k*th tape of a structure delaminates:

$$-\partial \Pi / \partial l_k = 2\gamma_k w_k \tag{6.1}$$

where Π *is* the potential energy, given as the difference between the elastic energy E and the external work W, i.e. $\Pi = E - W$, γ is the surface energy between tape and surface, w_k the tape width and l_k its detached length. For a thin film (i.e., neglecting the bending stiffness), the variation of external work ΔW can be expressed as:

$$\Delta W = F\Delta\eta \tag{6.2}$$

where F is the applied force and $\Delta\eta$ the increment in displacement. The elastic energy variation can be calculated as:

$$\Delta E = \frac{1}{2}\sum_{k=1}^{N} Y_k b_k w_k \left(l'_k \varepsilon'^2_k - l_k \varepsilon^2_k\right) \tag{6.3}$$

where Y_k is the kth tape elastic modulus, b_k its thickness, w_k its width and ε_k its strain. l'_k and ε'_k correspond to the length and the strain in the delaminated configuration (see Fig. 6.2a). For a single tape (Fig. 6.2b), the force necessary to peel the film from a substrate can be analytically obtained [15]:

$$T(1 - \cos\theta) + \frac{T^2}{2bwY} - 2\gamma w = 0 \tag{6.4}$$

where T is the tape tension, θ is the tape angle with respect to the substrate, and the subscript k has been dropped for simplicity. This leads to:

$$T = wbY\left(\cos\theta - 1 + \sqrt{(1 - \cos\theta)^2 + \frac{4\gamma}{bY}}\right) \tag{6.5}$$

which highlights the proportionality of the peeling force with the tape width w. This geometrical parameter is linked to the width of the peeling line, i.e. the line along which delamination occurs. As discussed in [7], a well-known strategy adopted by biological attachments systems consists in increasing the length of peeling line by simply splitting the contacts. This result is illustrated by the log relationship between the body mass and the sum of the peeling lines in animal fibrillar attachment structures [7].

In the case of multiple peeling problems [16], we first consider symmetrical double peeling, as shown in Fig. 6.2c, since asymmetrical cases can be treated as combined single peeling problems [18]. The applied force is the decomposition of the tape tensions, which leads to:

$$F = 2T\sin\theta \tag{6.6}$$

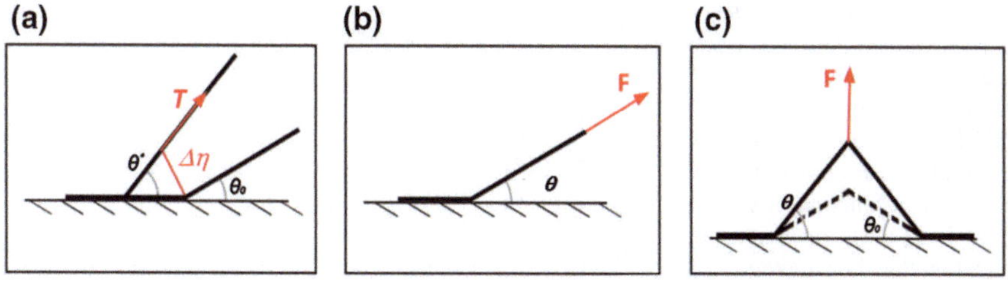

Fig. 6.2 **a** Tape angle before deformation ($\theta = \theta_0$) and at delamination ($\theta = \theta^*$) due to a tape tension T. **b** Single peeling configuration with an applied force F (in which case $\theta_0 = \theta^*$). **c** Symmetrical double peeling configuration with a vertically applied Force F: undeformed (θ_0) and deformed ($\theta < \theta^*$) tape angle

so that at delamination $(F = F^*,\ \theta = \theta^*)$:

$$F^* = 2wbY \sin\theta^* \left[\cos\theta^* - 1 + \sqrt{(1-\cos\theta^*)^2 + \frac{4\gamma}{bY}}\right] \tag{6.7}$$

In this case, the angle between the tapes and the substrate changes with the tape elastic deformation. Solving (6.7), one finds that an optimal peeling angle exists at which the peeling force is maximal. For angles below and above this value, the peeling force decreases. However, in (6.7) the peeling force is expressed as a function of the delamination (or "peeling") angle θ^*, so that peeling angle values below the optimal value correspond to negative "undeformed" initial angles θ_0, which is physically meaningless. The tape tension before delamination (Fig. 6.2c) can be expressed from the tape strain as $l(1+\varepsilon)\cos\theta = l\cos\theta_0$, so that:

$$T = bwY\varepsilon = bwY\left(\frac{\cos\theta_0}{\cos\theta} - 1\right) \tag{6.8}$$

while the external load applied to the system is given by (6.6). Considering as an example a tape with properties $Y_k = 1000$ MPa, $b_k = 1$ µm, $w_k = 2$ µm, $\gamma_k = 0.01$ MPa µm or $\gamma_k = 0.001$ MPa µm, calculations shown in Fig. 6.3a demonstrate that the peeling force F reaches a maximum when its intersects the peeling force F^* calculated as a function of angle variation when the initial tape angle is $\theta_0 = 0$ [18]. An initial undeformed angle of $\theta_0 = 0$ degrees is therefore the optimal peeling angle for maximizing the adhesion force.

The optimal asymptotic peeling angle is a function of the global deformability of the system, i.e. the deformation that the system can sustain before delamination

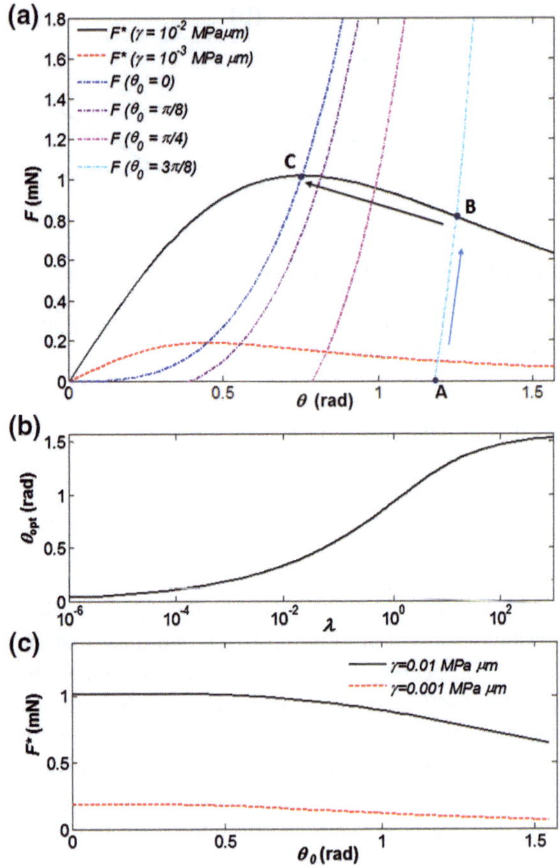

Fig. 6.3 **a** Peeling force F^* as a function of deformed tape angle θ, calculated with Multiple Peeling Theory (MPT), (6.7), compared to the external load F applied to the system as a function of θ, (6.6) and (6.8), for various initial angles. For a given initial angle, the tape deforms from point A=(θ_0;0) to point B=(θ^*;F^*), then deforms and delaminates until point C= (θ_{opt};F_{opt}) where it reaches the optimal asymptotic angle. The $\theta_0 = 0$ curve is the one that intersects the MPT curve for maximal peeling force values. **b** Optimal asymptotic peeling angle θ_{opt} for symmetric double peeling as a function of the ratio between adhesion energy and elasticity of the system. **c** Initial peeling force F^* as a function of the initial angle θ_0. In all cases, the peeling force then tends towards the maximum delamination force F^*=1 mN, for all initial angles.

occurs. This property can be quantified using the non-dimensional parameter $\lambda = \frac{4\gamma}{bY}$ representing the ratio between adhesion energy and elasticity [16]. From (6.7) and (6.8), it is possible to determine the optimal peeling angle $\theta^{opt} = \theta^*|_{\theta_0=0}$ at which the structure delaminates:

$$2\cos^3\theta^{opt} - (3+\lambda)\cos^2\theta^{opt} + 1 = 0 \tag{6.9}$$

The corresponding θ^{opt} versus λ curve is shown in Fig. 6.3b, showing a monotonically increasing nonlinear behaviour. More in general, we have:

$$\cos \theta_0 = \cos^2 \theta^* + \cos \theta^* \sqrt{(1 - \cos \theta^*)^2 + \lambda} \tag{6.10}$$

Using this relation, it is possible to plot the initial peeling force as a function of the undeformed angle θ_0 (Fig. 6.3c). These results highlight the fact that in multiple peeling cases, delamination is dependent from the deformability of the system, and therefore from the type of structure and the mechanical properties of the attachment.

The adhesion force of attachment structures can be deduced from planar multiple peeling models such as those discussed above. One example are so-called “dendritic” geometries in spider web anchorages, which can be described as radial branching structures where the pyriform silk fibres converge to a single point at a distance from the substrate. This results in a cone-like morphology (Fig. 6.1) which is involved in spider prey capture. It has been shown that in order to fulfil its function, the dendritic anchorage must exhibit a reduced pull off force [4, 27]. Due to its conical symmetry, the relationship between force and extension can be directly obtained from the symmetric double peeling configuration described above, modified by multiplying results by the appropriate number of tapes or membrane geometries [17]. Figure 6.4a shows simulation results for non membrane-like dendritic attachments when varying the initial contact angle θ_0. For $\theta_0 = \pi/4$, the system is firstly deformed without delamination, with a linear force-displacement relationship (Hooke’s law). Then, the tapes begin to delaminate and the peeling angle starts to vary, which results in an increase of the peeling force. This explains the elasto-plastic-like behaviour obtained in experimental [5] and in numerical results. The peeling force saturates when the peeling angle is asymptotically optimal. In the case of a limited available delamination length, however, full delamination could be achieved before the optimal peeling angle, and thus before the load plateau, is reached. This would allow to tune the adhesive strength of the structure. For a small peeling angle ($\theta_0 = \pi/16$), the initial elastic deformation displays hyperelastic behaviour due to the geometrical non linearity of the system, but then saturates to the maximum peeling force as soon as delamination begins. These two cases illustrate how by varying the peeling angle and attached tape lengths, it is possible to control the properties of the anchorage, inducing linear/nonlinear deformation and varying detachment loads. The possibility of tuning load response by selecting suitable multiple peeling structures has been observed experimentally, e.g. in spider web anchorages [5]. In more complex loading scenarios, such as when applied loads are not normal to the surface, asymmetric deformation states are induced (as shown in Fig. 6.4b), and FEM approaches are useful to evaluate the performance of these structures.

6.3 Hierarchical Branching in Adhesive Structures

Adhesive fibrillar structures found in Nature display much more complex geometries than the double peeling problem discussed above. One example is the branched structure found in spider attachment discs, where a high number of contacts

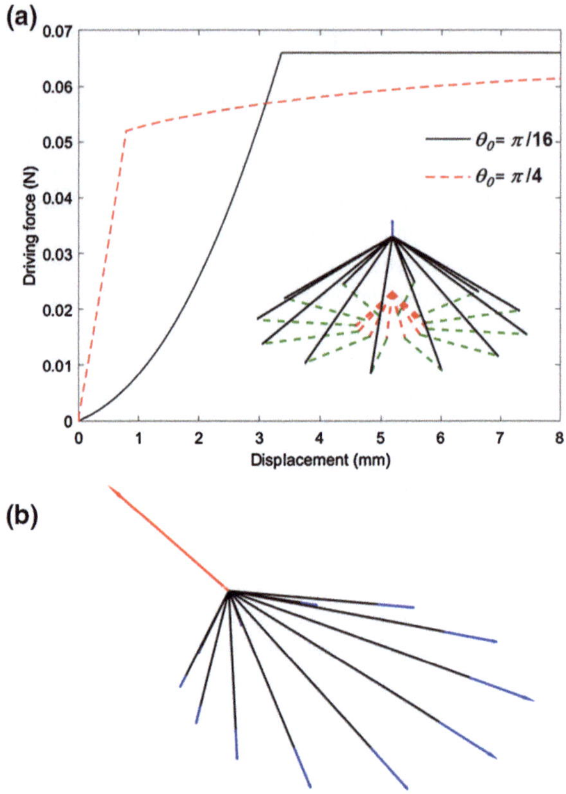

Fig. 6.4 **a** Force-extension curves for a non membrane-like dendritic anchorage, varying the initial peeling angle θ_0 of the anchorage. **b** Model of dendritic anchorage in an asymmetric loading scenario

with the substrate are obtained thanks to hierarchical structure and branching at several levels [5, 28], as illustrated in Fig. 6.1. To evaluate 2-D branched hierarchical multiple peeling configurations, FEM simulations have been carried out using the Structural Mechanics module of COMSOL Multiphysics 4.3. The interface between the tape and substrate is modelled adopting a Cohesive Zone Model (CZM) [29], based on a stress-softening constitutive law before delamination [24]. Simulation parameters are $b = 10^{-2}$ m, $t = 10^{-3}$ m, $Y = 3$ MPa, $\nu = 0.45$ (where ν is the Poisson's ratio). The hierarchical configurations are compared to the symmetrical double peeling case. A second-level hierarchical geometry is considered ("2"), where the two tapes branch out in a self-similar manner into two further tapes at equal distances from the centre. The third-level hierarchical geometry ("3") replicates this to a further level. These structures are shown in Fig. 6.5a, with the corresponding calculated load-displacement curves in Fig. 6.5b. The three configurations are compared for the same peeling line, length, thickness, and all three start from a fully adhered tape. It is apparent that increasing hierarchical branching increases contact splitting and the number of delamination points in the tape, thus distributing and reducing the stresses at the interface. This helps in avoiding stress concentrations and an early onset of tape delamination. On the other hand, tape

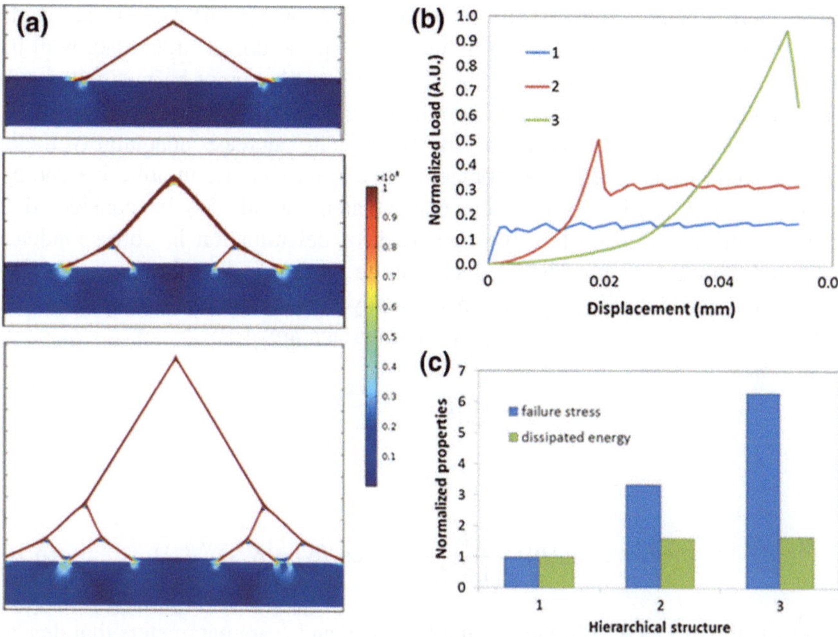

Fig. 6.5 Hierarchical peeling configurations: **a** FEM simulation results, pictured for different imposed displacement δ values. *1* first level (non-hierarchical) configuration; *2* second-level hierarchy; *3* third-level hierarchy. Colour scale represents von Mises stresses during delamination (*scale bar* shown on the *left*). **b** Corresponding Load versus displacement curves: the peaks in the curves correspond to full delamination of single tapes; **c** Comparison between the normalized delamination stress and dissipated energy at delamination for the 3 configurations

deformation and internal stresses are generally greater. The variation in normalized adhesive properties of the three structures is reported in Fig. 6.5c. There is an increase in adhesive failure stresses with hierarchy, i.e. the geometry with the highest hierarchical level achieves the best adhesive strength. Also, dissipated energy, which can be obtained as the area underlying the load-displacement curves, also increases for 2 and 3, showing how hierarchy favours an increase in both the strength and toughness of the adhesive interface.

6.4 Geometry and Mechanical Properties of Contact Units

At a lower scale level, the peeling force of a single tape depends on the mechanical properties and the geometry of the contact unit itself. It is well known that the gecko spatulae or insect setae are not simple uniform tapes, but display gradients in the mechanical properties along both the width and the thickness of the attached length [10]. It has been shown that due to the concentration variation of a softener, the resilin protein, the elastic modulus in the adhesive tarsal setae of the ladybird beetle varies approximately from 1 MPa to 6 GPa, with two gradients, one along the length of the setae, the other along its thickness, in the dorsal (in contact with the substrate) and ventral part of the setae [10]. This result cannot be justified simply using Kendall's theory, which predicts that the peeling force increases with the elastic modulus. The softening has the function to increase adaptability to the surface, which leads to a greater contact area, and an increase in adhesive energy. However, stress distributions on the contact area should also be considered to determine the possibility of early-stage onset of delamination in correspondence with stress concentrations. These effects can be captured using the stress distributions in the tape/interface system theoretically derived in [30]. The shear and normal stress (τ and σ) distributions at the interface are:

$$\tau(x) = T\cos(\theta)\frac{\alpha}{w}\mathrm{e}^{\alpha x} \tag{6.11}$$

and

$$\sigma(x) = T\sin(\theta)\frac{2\beta}{w(1-K)}\mathrm{e}^{\beta x}(\cos(\beta x) + K\sin(\beta x)) \tag{6.12}$$

where x is the distance from the peeling line, α, β and κ are parameters that depend on geometry and mechanical properties:

$$\begin{gathered}\alpha = \sqrt{\frac{G_I}{Y_t ab}};\quad \beta = \sqrt[4]{\frac{3Y_I}{Y_t ab^3}};\\ K = 1 - \frac{\sin(\theta)}{\sqrt{2Y_t I_t(1-\cos(\theta))/T - \beta b/2\cos(\theta) + \sin(\theta)}}\end{gathered} \tag{6.13}$$

where G_I and Y_I are the interface shear and elastic modulus, G_t and Y_t the tape interface shear and elastic modulus, I_t its momentum of inertia, and a the interface thickness. Considering an interface of setae with fixed mechanical parameters, the shear stress distribution can be derived approximately using (6.11). Gradually reducing the interface stiffness allows to distribute the stress over a larger area and therefore to increase the adhesive strength of the attachment. This is shown in Fig. 6.6a, where results are calculated using the same mechanical properties as in the first Section and the geometry shown in the inset. For the sake of simplicity, we only plot the shear stress distribution considering $\theta = 0$ and applying a force $T = 0.1$ µN. We consider both the tape and interface to be isotropic, so that $G = Y/2(1 + \nu)$, with $\nu = 0.3$. Shear stress along the length of the adhesive area for various stiffness values of the soft interface Y_I compared to that of the elastic layer Y_t are plotted in Fig. 6.6a. Results show how layer softening can have a considerable beneficial effect in reducing stress concentrations towards the tip of the pad, redistributing them more uniformly along the whole length of the interface area. For the same applied external force, in the considered case a tenfold reduction can be

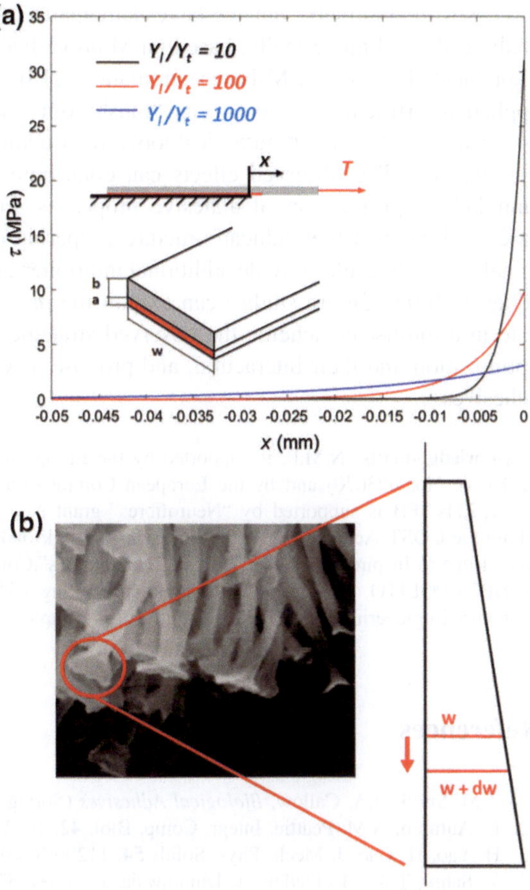

Fig. 6.6 **a** Geometry of the considered thin elastic layer with a "ventral" soft interface and an applied horizontal force *T*. Shear stress distributions are plotted along the length of the attached region for varying ratios between contact layer stiffness Y_l and tape stiffness Y_t. Softer contact layers are shown to reduce stress concentrations with respect to stiffer layers. **b** Tapered geometry of terminal contact element width to increase peeling line ($w \rightarrow w + dw$) during delamination (image from [25])

obtained in the normal stresses, with an approximately equivalent increase in the adhesive strength of the layer. Another simple strategy to tune the adhesive strength is to generate tapered geometries. Indeed, most of the observed setal elements in insects and geckos display width and thickness variations over the length of the setae [9]. As predicted by Kendall's model, these structures would increase the adhesive strength as the delamination proceeds, thanks to the increasing peeling line length (Fig. 6.6b) [7]. These broadened contact units are characteristic features observed not only in gecko adhesion systems [31], but also in those of insects and spiders [32].

6.5 Conclusions

In summary, we have reviewed how structural, geometric and mechanical features contribute on different scale levels to optimizing adhesion in biological adhesives. Many of these effects have been observed experimentally in the literature and discussed qualitatively, but we have demonstrated how they can also be derived analytically and numerically, based on Multiple Peeling Theory [16], its numerical implementation and FEM-based simulations [18]. These features can readily be applied to artificial adhesives, e.g. "mushroom"-like structured surfaces [12, 33, 34], using the developed numerical tools for structural, geometrical and mechanical optimization. The observed effects can contribute on different scale levels, with cumulative optimization of adhesive properties. For example, in the considered cases of branched hierarchical structures, tapering or contact softening in the terminal elements could provide additional multiplier effects in the enhancement of the adhesion force. Future studies can exploit the proposed approach to further elucidate in a multiscale scheme the observed strategies found in Nature for adhesion optimization and their interaction, and propose new structural designs for artificial adhesives.

Acknowledgements N.M.P. is supported by the European Research Council (ERC PoC 2015 SILKENE No. 693670) and by the European Commission FET Proactive "Neurofibres" grant No. 732344. FB is supported by "Neurofibres" grant No. 732344. This work was carried out within the COST Action CA15216 "European Network of Bioadhesion Expertise: Fundamental Knowledge to Inspire Advanced Bonding Technologies".Computational resources were provided by HPC@POLITO, a project of Academic Computing within the Department of Control and Computer Engineering at the Politecnico di Torino (http://www.hpc.polito.it).

References

1. A.M. Smith, J.A. Callow, *Biological Adhesives* (Springer, Berlin, 2006)
2. K. Autumn, A.M. Peattie, Integr. Comp. Biol. **42**, 1081 (2002)
3. H. Yao, H. Gao, J. Mech. Phys. Solids **54**, 1120 (2006)
4. V. Sahni, T.A. Blackledge, A. Dhinojwala, J. Adhes. **87**, 595 (2011)
5. V. Sahni, J. Harris, T.A. Blackledge, A. Dhinojwala, Nat. Commun. **3**, 1106 (2012)
6. I. Grawe, J.O. Wolff, S.N. Gorb, J. R. Soc. Interface **11**, 20140477 (2014)
7. M. Varenberg, N.M. Pugno, S.N. Gorb, Soft Matter **6**, 3269 (2010)
8. M. Kamperman, E. Kroner, A. Del Campo, R.M. McMeekin, Adv. Eng. Mater. **12**, 335 (2010)
9. A. Pantano, N.M. Pugno, S. Gorb, Int. J. Fract. **171**, 169 (2011)
10. H. Peisker, J. Michels, S.N. Gorb, Nat Commun. **4**,1661 (2013)
11. A.K. Geim, S.V. Dubonos, I.V. Grigorieva, K.S. Novoselov, A.A. Zhukov, S.Y. Shapoval, Nat. Mater. **2**, 461 (2003)
12. G. Carbone, E. Pierro, S.N. Gorb, Soft Matter **7**, 5545 (2011)
13. M. Piccardo, A. Chateauminois, C. Fretigny, N.M. Pugno, M. Sitti, J. R. Soc. Interface **10** (2013)
14. K.L. Johnson, K. Kendall, A.D. Roberts, Proc. R. Soc. A: Math. Phys. **324**, 1558 (1971)
15. K. Kendall, J. Phys. D Appl. Phys. **8**, 1449 (1975)
16. N.M. Pugno, Int. J. Fract. **171**, 185 (2011)
17. L. Afferrante, G. Carbone, G. Demelio, N.M. Pugno, Tribol. Lett. **52**, 439 (2013)
18. L. Brely, F. Bosia, N.M. Pugno, Interface Focus **5**, 20140051 (2015)
19. L. Qu, L. Dai, M. Stone, Z. Xia, Z.L. Wang, Science **322**, 238 (2008)
20. M.P. Murphy, B. Aksak, M. Sitti, Small **5**, 170 (2009)

21. H.E. Jeong, J.K. Lee, H.N. Kim, S.H. Moon, K.Y. Suh, Proc. Natl. Acad. Sci. **106**, 5639 (2009)
22. M.P. Murphy, S. Kim, M. Sitti, A.C.S. Appl, Mater. Interfaces **1**, 849 (2009)
23. R.A. Sauer, M. Holl, Comput. Methods Biomech. **16**, 577 (2013)
24. F. Bosia, S. Colella, V. Mattoli, B. Mazzolai, N.M. Pugno, RSC Adv. **4**, 25447 (2014)
25. C. Guo, J. Sun, Y. Ge, W. Wang, D. Wang, Z. Dai, Sci. China Life Sci. **55**, 181 (2012)
26. Y. Hsia, E. Gnesa, F. Jeffery, S. Tang, C. Vierra, in *Metal, Ceramic and Polymeric Composites for Various Uses* (J. Cuppoletti), Intech, Rijeka, Croatia, p. 303 (2011)
27. S.W. Cranford, A. Tarakanova, N.M. Pugno, M. Buehler, Nature **482**, 72 (2012)
28. J.O. Wolff, I. Grawe, M. Wirth, A. Karstedt, S. Gorb, Soft Matter **11**, 2394 (2015)
29. M. Elices, G.V. Guinea, J. Gomez, J. Planas, Eng. Fract. Mech. **69**, 137 (2002)
30. D.H. Kaelble, Trans. Soc. Rheol. **4**, 45 (1960)
31. N.W. Rizzo, K.H. Gardner, D.J. Walls, N.M. Keiper-Hrynko, T.S. Ganzke, D.L. Hallahan, J. R. Soc. Interface **3**, 441 (2006)
32. A.B. Kesel, A. Martin, T. Seidl, Smart Mater. Struct. **13**, 512 (2004)
33. A.V. Spuskanyuk, R.M. McMeeking, V.S. Deshpande, E. Arzt, Acta Biomater. **4**, 1669 (2008)
34. L. Heepe, A.E. Kovalev, A.E. Filippov, S. Gorb, Phys. Rev. Lett. **111**, 104301 (2013)

Chapter 7
Double Peeling Mechanism Inspired by Biological Adhesive Systems: An Experimental Study

Lars Heepe, Saverio Raguseo and Stanislav N. Gorb

Abstract In this study, we have performed, for the first time, a detailed experimental test of the double peeling configuration, including the role of pre-strain and different initial peeling angles. Double peeling systems consist of two adhesive tapes in contact with a substrate and are loaded at a common hinge. Such systems are widely observed in the hairy attachment pads of insects, arachnids, and reptiles. It was previously shown that the antagonistic work of such opposing tape-like contacts lead to enhanced, stable, and robust attachment, if compared to the single peeling configuration. The obtained results are in very good agreement with recent theoretical models and may help explaining the functional mechanisms of biological attachment systems.

7.1 Introduction

Hairy attachment pads of insects (e.g. [1–3]), arachnids (e.g. [4]) and reptiles (e.g. [5–9]) were previously imaged using scanning electron microscopy (SEM) to analyze this particular contact geometry in great detail [10] (Fig. 7.1). This study revealed that in the majority of studied species the smallest hierarchical level of setae, responsible for the formation of intimate contact with the substrate, appears to be a two-dimensional tape-like spatula [10]. Based on these images, obtained for representatives of different animal groups, an interesting correlation between the geometrical properties of setal tips and animal weight was found by applying Kendall's peeling model [11] and the principle of contact splitting [12], in order to

L. Heepe (✉) · S. Raguseo · S.N. Gorb
Department of Functional Morphology and Biomechanics, Zoological Institute, Kiel University, Am Botanischen Garten 1-9, 24118 Kiel, Germany
e-mail: lheepe@zoologie.uni-kiel.de

Present Address:
L. Heepe
Mads Clausen Institute, University of Southern Denmark, NanoSYD Alsion 2, 6400 Sønderborg, Denmark

L. Heepe et al. (eds.), *Bio-inspired Structured Adhesives*, Biologically-Inspired Systems 9, DOI 10.1007/978-3-319-59114-8_7

explain, why most biological hairy adhesive systems, involved in locomotion, rely on spatula-shaped terminal elements [10]. However, even if the peeling geometry is appropriate to describe the behavior of a single spatula, animal locomotion involves simultaneous peeling of many spatulae, at the level of adhesive pads (Fig. 7.1), at the level of contra-lateral legs (Fig. 7.2a, b), and sometimes also at the level of contra-lateral toe pads within one foot (Fig. 7.2c). Thus, in climbing animals, the situation is more complex and resembles a hierarchical multiple peeling configuration. On the macroscale, however, the peeling configuration of climbing animals may be reduced, for simplicity, to the peeling of two elastic tapes loaded at common hinge (Fig. 7.2d), further called double peeling configuration. Motivated by this observation, preliminary double peeling experiments, according to the configuration shown in Fig. 7.2d, were performed using a flexible (Young's modulus of order 1 MPa) and a rigid tape (Scotch®, Young's modulus of order 1 GPa). Here, the hinge was loaded quasi-statically with different masses starting from two distinct initial peeling angles 0° and 90°. For the flexible tape, clearly an optimal angle with maximal peeling force emerged, which for the given elasticity and adhesive energy was approximately 65° (Fig. 7.3a). For the rigid tape no such optimal angle could be observed: the highest mass was kept at the smallest angle (Fig. 7.3b). These results were later published in [13], accompanied by a first version of a multiple peeling theory, see [14].

Actually, the double, or V-shape, peeling configuration was already both experimentally [15] and theoretically [15–17] investigated. Peeling models, both of single and double peeling configuration, with further features that are important for the field of animal adhesion were implemented, including the role of

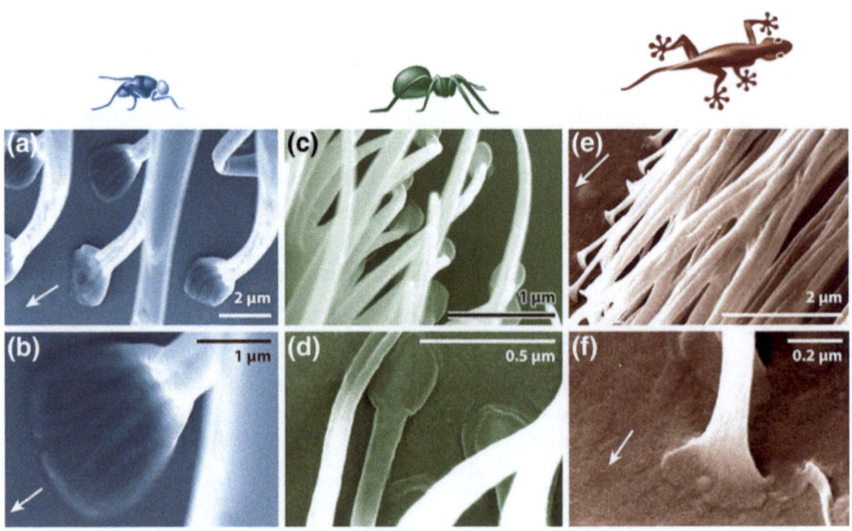

Fig. 7.1 Spatulate tips of tarsal adhesive setae in climbing animals. Reproduced with kind permission from Annual Reviews [31]

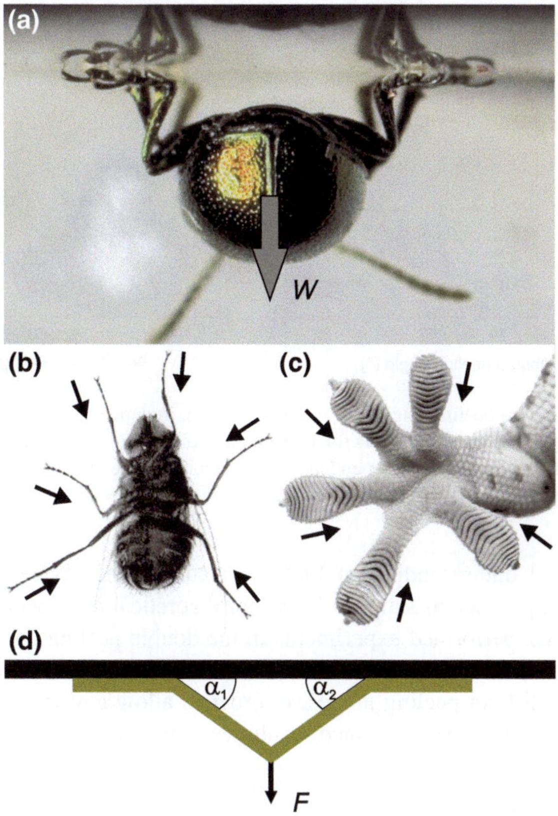

Fig. 7.2 **a** A beetle *Gastrophysa viridula* in contact with a glass ceiling (lateral view from behind). The *arrow* indicate weight force *W* acting at the center of mass. **b** Fly *Calliphora vicina* in contact with a glass ceiling (view from above through the glass ceiling). *Arrows* indicate contra-lateral leg movement. **c** Foot of a gecko *Gekko gecko* in contact with a glass ceiling (view from above through the glass ceiling). *Arrows* indicate contra-lateral toe pad movement. **d** Simplified mechanical model of the ceiling situation (lateral view from behind, similar to (**a**)). Two elastic tapes are in contact with a rigid ceiling and having a common hinge, where a load *F* is applied. The tapes make an angle with respect to the substrate of α_1 and α_2

large-deformations (e.g. [14, 18, 19]), pre-tension (e.g. [14, 20–22]), frictional sliding (e.g. [19, 22]), self-healing (e.g. [22, 23]), hierarchy (e.g. [14, 24, 25]), geometry (e.g. [15, 20, 26]), elastic and adhesive heterogeneities (e.g. [27]). For a current review see [28] and citation therein. Moreover, recent studies focused on the stability of the double peeling configuration [29, 30].

In spite of numerous biological observations on the peeling geometry at the macro-, micro- and nanoscales, and elaborate theoretical models, the detailed experimental proof of double-peeling models has previously not been carried out. Such an experiment is, however, of a great importance, because it would aid in

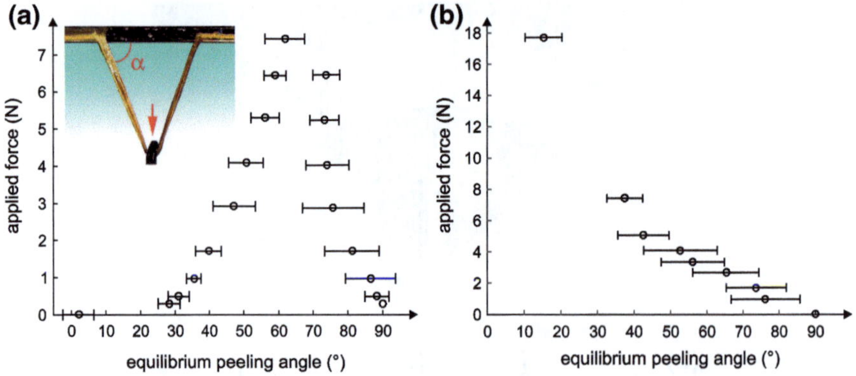

Fig. 7.3 **a** Equilibrium peeling angles for the flexible tape, in dependence of the applied force at the common hinge (see *arrow* in *inset*) for two initial peeling angles 0° and 90°. **b** Equilibrium peeling angles for the rigid tape, in dependence of the applied force at the common hinge for two initial peeling angles 0° and 90°

better conceptual understanding of biological attachment devices and providing potentially an experimental support for recent theoretical considerations.

We have, thus, performed experiments in the double peeling configuration with an elastic model adhesive tape with different levels of pre-strain. Peeling was initiated at two distinct peeling angles, in order to allow covering whole potential regimes of detachment. The obtained results were then compared with recent theoretical models.

7.2 Experimental

In order to understand the double peeling behavior of hinged tapes, we have carried out experiments using a well-defined, elastic model adhesive tape, described previously [31–33]. This biologically inspired adhesive tape made from polyvinylsiloxane (PVS) had a Young's modulus E = 1.1–1.5 MPa [34] and a backing layer thickness t of approximately 250 μm. One side of the tape is covered by about 29,000 hexagonally distributed mushroom-shaped adhesive elements per square centimeter. Samples used for the experiments were 250 mm long and had a width w of 20 mm. Figure 7.4a shows the principal configuration of the double peeling experiments. An adhesive tape (dark gray) is in contact with smooth glass plates (light gray). A load F is applied at the hinge (black circle), in the middle of the tape, which makes peeling angles α_1 and α_2 with the substrate. All experiments were started at two distinct initial peeling angles 0° and 90°. In order to reliably start from an initial peeling angle of 0°, two glass plates were brought together leaving a small gap for the hinge (Fig. 7.4b). The tape was then adhered to both glass plates. In order to achieve an initial angle of 90°, a "pull strap" was formed with the tape (Fig. 7.4c).

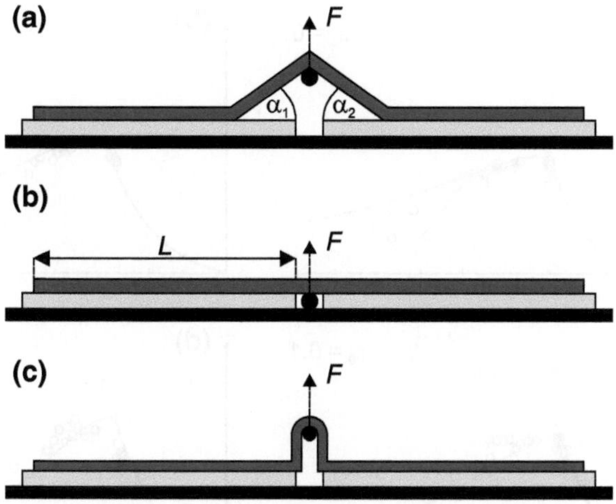

Fig. 7.4 Schematic of the experimental setup. **a** Principal double peeling configuration: the tape (*dark gray*) is in contact with glass substrates (*light gray*) and loaded with a force F at the hinge in the middle of the tape. The tape makes peeling angles with the substrate of α_1 and α_2. **b** Initial double peeling configuration starting with a peeling angle of 0°. **c** Initial double peeling configuration starting with a peeling angle of 90°. L is the length of one side of the tape in contact with the substrate without any pre-strain

Prior each experiment the tape was pressed onto the glass by rolling an aluminum cylinder of 16 g over the entire tape. Subsequently, the hinge was loaded using a peeling tester DS4 (TETRA GmbH, Ilmenau, Germany) [35]. Both types of experiments were performed quasi-statically. The tape was displaced by the hinge in a discrete way with increments of order 5 mm. Equilibrium peeling angles were recorded after each step and a waiting time of 1 min. Corresponding angles α_1 and α_2 were determined graphically by images taken with a photo camera. For data analysis, always the mean value of both angles was taken. The same experiments were then repeated, but now with the tape being stretched before adhering it to the glass plate by defined pre-strains of $\varepsilon_0 = \{0.05; 0.1; 0.15; 0.2; 0.25\}$. As indicated in Fig. 7.4b an initial length L of the tape was stretched to length L', thus yielding a pre-strain of $\varepsilon_0 = (L' - L)/L$. Practically, the tape was fixed at one edge of the glass plate, stretched to the corresponding length L' and then carefully attached to the glass plate.

7.3 Results and Discussion

7.3.1 Double Peeling Experiments

Figure 7.5 shows the peeling forces obtained for the different pre-strains ε_0 of the adhesive tapes (open circles). Similar to the preliminary experiments (Fig. 7.3a) an

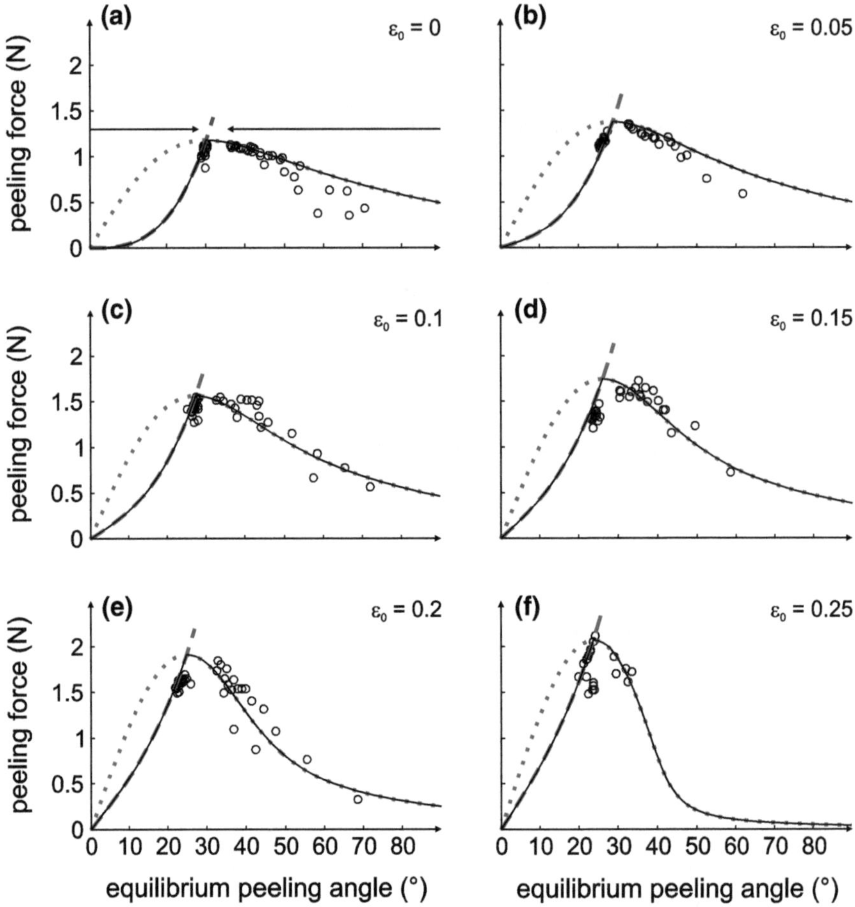

Fig. 7.5 Experimental results (*open circles*) of double peeling a soft microstructured elastic tape with different pre-strains $\varepsilon_0 = \{0.05; 0.1; 0.15; 0.2; 0.25\}$, (**a**–**f**) respectively, in comparison with recent theories. *Dashed lines* were plotted according to (7.1) and *dotted lines* were plotted according to (7.2). *Thin black lines* indicate the area of validity of the peeling process. *Arrows* in (**a**) indicate peeling started at 0° and 90° respectively

optimum angle was observed with a maximum in peeling force. This maximum peeling force was reached irrespective the initial peeling angles 0° and 90° (indicated by arrows in Fig. 7.5a). For experiments starting from 0°, the tape first elastically deformed without peeling up to a critical angle. When this critical angle was reached, a steady-state peeling force was reached at constant peeling angle. This is in agreement with early experimental tests of the double peeling configuration reported by Gent et al. [15] in 1986.

For experiments starting from 90°, the peeling angle decreased with increasing load and also reached a steady-state behavior at a critical angle. These critical angles, for both initial peeling angles 0° and 90°, should coincide [29, 30] and refer to the experimentally observed optimum peeling angle α_{opt}. However, a certain hysteresis of the optimum angle was observed, i.e. the maximum peeling force was obtained at slightly different equilibrium peeling angles, for peeling initiated at 0° and 90°. To the moment, we do not have a conclusive hypothesis for this observed hysteresis, but potential mechanisms may involve different frictional sliding [19], causing energy dissipation, and bending behavior [17] when starting at this two extreme initial angles. Moreover, the discrete nature (the presence of many individual adhesive contact elements instead of one, see Sect. 7.2) of the adhesive interface and the discrete way of hinge displacement during the load application may also play a role.

We now compare the results obtained with recent double peeling theories reported in [14, 15, 29, 30], which assume a no slip boundary condition at the interface of the tape and the substrate. Depending on the initial peeling angle α_i, we used the following relations of the applied load F and the peeling angle α:

$$F = \begin{cases} 2E^{*}A\sin\alpha\left(\frac{1}{\cos\alpha} - 1 + \varepsilon_0\right), & \alpha_i \le \alpha_{opt} \\ 2E^{*}A\sin\alpha \cdot \varepsilon_c, & \alpha_i \ge \alpha_{opt} \end{cases} \qquad (7.1, 7.2)$$

where $\varepsilon_c = \cos\alpha - 1 + \varepsilon_0 + \sqrt{(1-\cos\alpha-\varepsilon_0)^2 + 4\lambda - \varepsilon_0^2}$, $E^{*} = E/(1-\nu^2)$ is the reduced elastic modulus with ν being Poisson's ratio (here assumed to be 0.5), $A = wt$ is the cross-sectional area of the tape, ε_0 is a pre-strain, and $\lambda = \gamma/E^{*}t$ with γ being the work of adhesion.

For initial peeling angles $\alpha_i \le \alpha_{opt}$, (7.1) (dashed lines in Fig. 7.5) predicts the pure elastic deformation of the tape in the double peeling configuration without detachment (peeling) [15], in accordance to our experimental observations. Here, for the general case, we have added a pre-strain ε_0. For initial peeling angles $\alpha_i \ge \alpha_{opt}$ we have used (7.2) (dotted lines in Fig. 7.5) derived in [14].

With E^{*} = 1.5 MPa, $\gamma = 6.5$ J/m^2, and prescribed pre-strains, (7.1) and (7.2) show very good agreement with the experimental data using only a single set of parameter (Fig. 7.5). This indicates the general validity of double peeling theories. We note that $\gamma = 6.5$ J/m^2 is in the same order of magnitude as the one obtained for a similar tape in the single peeling configuration [32], but it is approximately larger by a factor of100 than the experimentally obtained work of adhesion for the detachment of individual mushroom-shaped adhesive elements covering this tape (see Sect. 7.2), which involves crack velocities during detachment of order cm/s [34], much larger than in the present study.

In the present study, we did not obtain data at peeling angles significantly lower than the optimum angle (Fig. 7.5), simply because already the first applied displacement (of order 5 mm) was sufficient to initiate peeling.

Furthermore, the experimental data confirm the prediction that with an increasing pre-strain the maximum peeling force increases and shifts to lower α_{opt} [21, 30]. Interestingly, a critical pre-strain $\varepsilon_{0,cr}$ can be predicted [21, 30], for which the adhered tape should spontaneously detach from the substrate without external load. For the given elasticity and work of adhesion we find a critical pre-strain

$\varepsilon_{0,cr} \approx 0.18$. Thus, at least the tapes with pre-strains $\varepsilon_0 = 0.2$ and $\varepsilon_0 = 0.259$ in the present study should have detached spontaneously. This, however, was not observed and may be a peculiarity of the adhesive tape used in the present study. As described in Sect. 7.2, one side of the tape is covered with mushroom-shaped adhesive elements, thus the adhesive interface is not smooth and continuous, but rather microstructured and discrete. Since any tension of the tape is received by the backing layer, the adhesive interface is only marginally affected by the applied pre-strains. Indeed, the stretched backing layer applies certain shear forces to individual mushroom-shaped adhesive elements, but these adhesive elements were shown to be very robust against applied pull-off loads, even for high tilt angles, with only a marginal decrease in load capacity [36].

7.3.2 Biological Relevance

When insects walk on the ceiling, almost no normal load to the ceiling surface is detected, but rather high shear forces directed toward the center of the body mass [37]. In this orientation, setae are recruited by bending in their tilted direction and it seems, that by such a proximal pull, the adhesive system, in particular spatulae, of the insect is activated by sliding-induced pre-strain [21, 38] and/or by sliding-induced increase in the real contact area of spatulate elements with the substrate [39]. Anisotropy of contact forces is a result of the structural anisotropy at different levels (macro-level: orientation of pads, microlevel: slope of the setal shafts, and orientation of the spatulae). Such anisotropy of attachment forces and shear activation was previously demonstrated for geckos [40–44], flies [45], and spiders [46].

Observations of the antagonistic work of legs in the beetle *Chrysolina fastuosa* (Fig. 7.6a–c) revealed, that if five of the maximum of six legs are ablated, the animal is no longer able to adhere to the ceiling, even if the average adhesive force of the single leg (measured for all six legs acting in concert and then normalized by the number of legs) should be sufficient to resist the body weight on the ceiling [37]. However, as soon as two intact legs remain, the beetle can stay perfectly well on the ceiling (Fig. 7.6b) [37]. Similar results were obtained in vertical pulling experiments with the hunting spider *Cupiennius salei*. The overall attachment ability was shown to strongly depend on the antagonistic work of opposing legs [47]. Moreover, for beetle with only three intact legs remaining on one side of the body, the animal could reorient one leg to the other side of its body and stayed on the ceiling (Fig. 7.6c) [37].

In light of above results, the explanation of this phenomenon is rather simple. The situation of the single leg in contact can be simulated by a piece of sticky tape adhering to the ceiling and holding a certain mass. The peeling angle in such a geometry would be 90°, which according to the Kendall model [11] generates rather low peeling force. When two or multiple tapes are combined, the peeling angle will be generally lower than 90°, due to the self-stabilization effect of such geometry.

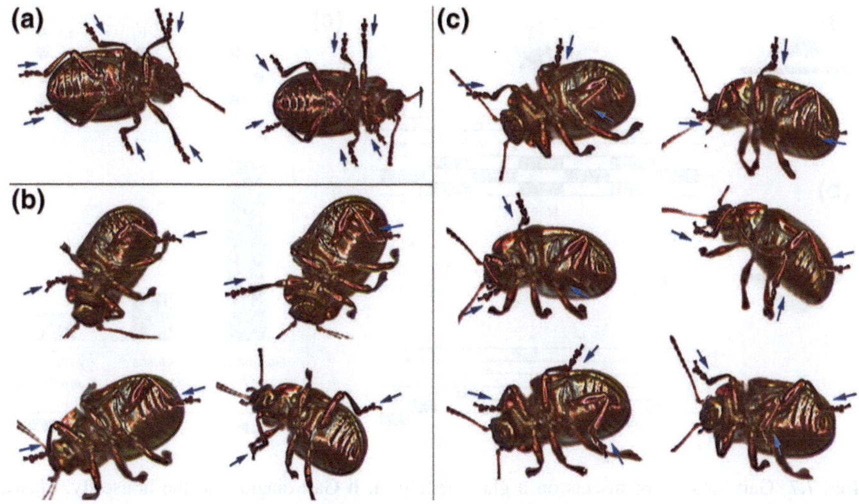

Fig. 7.6 **a**, **c** Posture of the chrysomelid beetle *Chrysolina fastuosa* on the ceiling. **a** Intact beetles. **b** Two legged beetles (four legs are removed). *Arrows* indicate orientation of leg tarsomeres. **c** Three legged beetles (three legs on one body side are removed). Please note the rearrangement of leg positions, in order to generate shear towards the center of the body mass (*arrows*). Reproduced from [37] with permission of Springer

Observations of the gait pattern in flies walking on a horizontal surface revealed, that three opposing legs were in the swing phase (moving), whereas the other three legs remained in the stance phase (motionless) (Fig. 7.7a). For the flies walking on a vertical surface, it was shown that their gait pattern is not different from the pattern on the horizontal surface [37]. However, for flies walking on the ceiling, they moved only two legs and four legs remained in a stance phase (Fig. 7.7b) [3]. This result led us to assume that the gait pattern may contribute to the flys ability to adhere dynamically on the ceiling, by providing an optimal relationship between the body weight and the supporting contact configuration.

Experiments with insect species that had different safety factors (attachment force divided by the body weight) showed that insects with a safety factor above 10 could employ a tripod gait similar to the one observed during locomotion on the floor. Insects with safety factors in the range of 4–5 changed their gait, similar to that reported above for the fly. Insects with safety factors lower than 3 changed their gait to where only one leg is in the swing phase and the other five remain in the stance phase (Fig. 7.7c). These results show that dynamical adhesion in insects may also strongly rely on a multiple peeling configuration especially in ceiling locomotion [37].

Thus, it is plausible to assume that biological adhesive systems try to actively use changes in their configuration of their tape-like contact geometry, in order to control and/or maintain adhesion. In order to securely stay on the ceiling, animals tend to keep their macroscopic peeling angle below the optimum. In this regime no

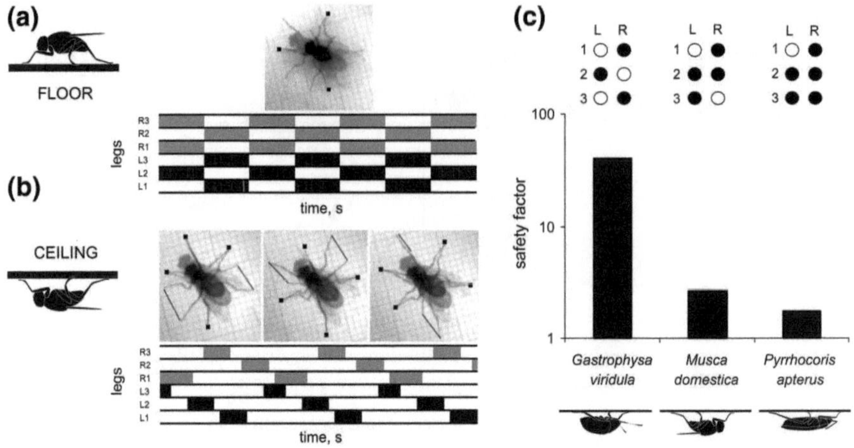

Fig. 7.7 Gait patterns of insects on a glass ceiling. **a**, **b** Gait diagram of the house fly, *Musca domestica*, on the floor (**a**) and ceiling (**b**). Two pictures were superimposed to show two succeeding steps. *Black squares* indicate the feet that are in the stance phase. **c** Relationship between the safety factor (attachment force divided by the body weight) and the gait pattern on the ceiling in different insects. *Insets* show feet (*black dots*) that are typically in contact while walking on a smooth ceiling. Images **a, b** are reproduced from [3] by permission of Oxford University Press. Image **c** is reproduced from [37] with permission of Springer

peeling occurs. Interestingly, in flies clinging to a glass ceiling, it was observed that their feet were in constant slow motion towards the body center of mass [48], potentially due to the tarsal secretions produced in insect adhesive systems [1]. After a certain sliding motion of the feet they were quickly extended and re-attached, by this maintaining a position with a low macroscopic peeling angle. Quick and easy detachment may be achieved by a simple active change of the macroscopic peeling angle above α_{opt} by e.g. proximo-distal peeling in case of flies [45] and disto-proximal peeling or digital hyperextension in case of geckos [49]. This may be possible, since their available contact area is restricted to the pad area and, thus, only a short peeling distance is necessary before complete detachment of a pad or a foot occurs.

7.4 Conclusions

The present study demonstrates, for the first time, a detailed experimental test of the double peeling configuration, including the role of pre-strain and different initial peeling angles. A comparison between experimental results and recent theories showed excellent agreement. However, a unique property of the double peeling configuration is not in the existence of an optimum peeling angle with a maximum in peeling force, as previously stated by other authors. An optimum angle does also exist for the single peeling configuration, i.e. for the vertical peeling force component [11]. It is rather the self-stabilization of the double peeling configuration, which always, irrespective of the initial peeling angles, tends to its optimum configuration. In its optimum, peeling in this configuration is always stable. Moreover, the double peeling configuration seems to have two different regimes: for $\alpha_i < \alpha_{opt}$, there is pure elastic deformation of the tapes without detachment (peeling) until $\alpha_i = \alpha_{opt}$; for $\alpha_i > \alpha_{opt}$, there is a peeling-dominated regime with decreasing peeling angle with increasing applied load. Although it seems that climbing animals make use of the characteristics of this double (multiple) peeling configuration, it would be worth observing the actual ceiling situation for various insects, spiders, and geckos, including the arrangement of peeling angles at the various hierarchical levels.

Acknowledgements This book chapter is based on the original publication L. Heepe, S. Raguseo, and S.N. Gorb, An experimental study of double peeling mechanism inspired by biological adhesive systems, Appl. Phys. A, DOI:10.1007/s00339-016-0753-9, (2017), with permission of Springer. Extensive work of V. Kastner on the preliminary experiments is greatly acknowledged. We would like to thank E. Appel for assistance with Fig. 7.2a. We would like to thank A. Kovalev for helpful comments on the manuscript. This work was partially supported by CARTRIB Project of The Leverhulme Trust (S.N. Gorb) by projects CP 1550 and 1623 by a grant of the Cluster of Excellence 80 The Future Ocean (L. Heepe and S.N. Gorb). The Future Ocean is funded within the framework of the Excellence Initiative by the Deutsche Forschungsgemeinschaft (DFG) on behalf of the German federal and state governments. S. Raguseo greatly acknowledges support of the Erasmus + programme of the European Union.

References

1. S. Gorb, *Attachment Devices of Insect Cuticle* (Kluwer Academic Publishers, London, 2001)
2. S.N. Gorb, R.G. Beutel, Naturwissenschaften **88**, 530 (2001)
3. S.N. Gorb, Am. Entomol. **51**, 31 (2005)
4. J.O. Wolff, S.N. Gorb, *Attachment Structures and Adhesive Secretions in Arachnids* (Springer, Berlin, 2016)
5. A. Haase, Arch. Naturgesch. **66**, 321 (1900)
6. R. Ruibal, V. Ernst, J. Morphol. **117**, 271 (1965)
7. U. Hiller, Z. Morphol, Tiere **62**, 307 (1968)
8. D.J. Irschick, C.C. Austin, K. Petren, R.N. Fisher, J.B. Losos, O. Ellers, Biol. J. Linn. Soc. **59**, 21 (1996)
9. K. Autumn, P.H. Niewiarowski, J.B. Puthoff, Annu. Rev. Ecol. Evol. Syst. **45**, 445 (2014)
10. M. Varenberg, N.M. Pugno, S.N. Gorb, Soft Matter **6**, 3269 (2010)
11. K. Kendall, J. Phys. D Appl. Phys. **8**, 1449 (1975)
12. E. Arzt, S. Gorb, R. Spolenak, Proc. Natl. Acad. Sci. U.S.A. **100**, 10603 (2003)
13. N. Pugno, S. Gorb, in *Proceedings of the 12th international conference on fracture, July 1217, Ottawa, Canada, USA* (2009)
14. N. Pugno, Int. J. Fract. **171**, 185 (2011)
15. A.N. Gent, S. Kaang, J. Appl. Polym. Sci. **32**, 4689 (1986)
16. J.G. Williams, Int. J. Fract. **87**, 265 (1997)
17. K.T. Wan, J. Adhes. **70**, 197 (1999)
18. A. Molinari, G. Ravichandran, J. Adhes. **84**, 961 (2008)

19. M.R. Begley, R.R. Collino, J.N. Israelachvili, R.M. McMeeking, J. Mech. Phys. Solids **61**, 1265 (2013)
20. Z. Sun, K.T. Wan, D.A. Dillard, Int. J. Solids Struct. **41**, 717 (2004)
21. B. Chen, P. Wu, H. Gao, J. R. Soc. Interface **6**, 529 (2009)
22. D. Labonte, W. Federle, J. R. Soc. Interface **13**, 20160373 (2016)
23. K. Kendall, J. Phys. D Appl. Phys. **11**, 1519 (1978)
24. B. Chen, P.D. Wu, H. Gao, Proc. R. Soc. A **464**, 1639 (2008)
25. F. Bosia, S. Colella, V. Mattoli, B. Mazzolai, N.M. Pugno, RSC Adv. **4**, 25447 (2014)
26. A. Pantano, N.M. Pugno, S.N. Gorb, Int. J. Fract. **171**, 169 (2011)
27. S. Xia, L. Ponson, Phys. Rev. Lett. **108**, 196101 (2012)
28. Z. Gu, S. Li, F. Zhang, S. Wang, Adv. Sci. **3**, 1500327 (2016)
29. L. Afferrante, G. Carbone, G. Demelio, N. Pugno, Tribol. Lett. **52**, 439 (2013)
30. C. Putignano, L. Afferrante, L. Mangialardi, G. Carbone, Beilstein J. Nanotechnol. **5**, 1725 (2014)
31. L. Heepe, S.N. Gorb, Annu. Rev. Mater. Res. **44**, 173 (2014)
32. S. Gorb, M. Varenberg, A. Peressadko, J. Tuma, J. R. Soc. Interface **4**, 271 (2007)
33. K. Dening, L. Heepe, L. Afferrante, G. Carbone, S.N. Gorb, Appl. Phys. A **116**, 567 (2014)
34. L. Heepe, A.E. Kovalev, A.E. Filippov, S.N. Gorb, Phys. Rev. Lett. **111**, 104301 (2013)
35. X. Jin, J. Strueben, L. Heepe, A. Kovalev, Y.K. Mishra, R. Adelung, S.N. Gorb, A. Staubitz, Adv. Mater. **24**, 5676 (2012)
36. L. Heepe, G. Carbone, E. Pierro, A.E. Kovalev, S.N. Gorb, Appl. Phys. Lett. **104**, 011906 (2014)
37. S.N. Gorb, in *Handbook of Adhesion Technology*, ed. by L.F.M. da Silva, A. Öchsner, R.D. Adams (Springer, Berlin, 2011)
38. Q.H. Cheng, B. Chen, H.J. Gao, Y.W. Zhang, J. R. Soc. Interface **9**, 283 (2012)
39. A. Filippov, V.L. Popov, S.N. Gorb, J. Theor. Biol. **276**, 126 (2011)
40. K. Autumn, Y.A. Liang, S.T. Hsieh, W. Zesch, W.P. Chan, T.W. Kenny, R. Fearing, R. J. Full, Nature **405**, 681 (2002)
41. H. Gao, X. Wang, H. Yao, S.N. Gorb, E. Arzt, Mech. Mater. **37**, 275 (2005)
42. G. Huber, S.N. Gorb, R. Spolenak, E. Arzt, Biol. Lett. **1**, 2 (2005)
43. K. Autumn, A. Dittmore, D. Santos, M. Spenko, M. Cutkosky, J. Exp. Biol. **209**, 3569 (2006)
44. Y. Tian, N. Pesika, H. Zeng, K. Rosenberg, B. Zhao, P. McGuiggan, K. Autumn, J. Isralachvili, Proc. Natl. Acad. Sci. U.S.A. **103**, 19230 (2006)
45. S. Niederegger, S. Gorb, Y. Jiao, J. Comp. Physiol. A **187**, 961 (2002)
46. S. Niederegger, S.N. Gorb, J. Comp. Physiol. A **192**, 1223 (2006)
47. E. Wohlfart, J.O. Wolff, E. Arzt, S.N. Gorb, J. Exp. Biol. **217**, 222 (2014)
48. V.B. Wigglesworth, J. Exp. Biol. **129**, 373 (1987)
49. A.P. Russell, J. Zool. Lond. **176**, 437 (1975)

Chapter 8
The Role of Effective Elastic Modulus in the Performance of Structured Adhesives

Di Tan, Yingjie Zheng and Longjian Xue

Abstract Structured adhesives have been found to play a key role in the locomotion of hundreds of species of animals. Fibrillar and smooth adhesive pads are the two basic structures, which can greatly reduce the effective elastic modulus of the adhesive pads. Inspired by the natural structured adhesives, various artificial structures possessing adhesive abilities have been developed. In this contribution, after the introduction of concepts of elastic modulus and effective elastic modulus, both the natural and artificial structured adhesives are introduced. The strategies to reduce the effective elastic modulus of bioinspired structured adhesives are then summarized and discussed.

8.1 Introduction

Billions of years' evolution, the living beings on the earth have greatly developed. The number of natural materials in living beings is limited, however, a wealth of structures endow these materials various mechanical properties in an extremely wide range. For example, though both a tree trunk and a cotton fiber are mainly composed of cellulose, an important polysaccharide, their mechanical properties are quite different. While the cotton fiber is flexible, the tree trunk is rigid enough to support the weights of branches, leaves and itself; additionally, the tree trunk should allow the delivery of nutrition and water to the highest top of the tree in order to efficiently compete for light. The assembly of cellulose into porous structures with a gradient in the radical direction of a trunk meets the requirement of multi-functions [1, 2]. Similar gradient porous structure has also been found in bamboos which can possess a very large aspect ratio [3]. Another example is calcium carbonate which is very fragile. The combination of layers of calcium carbonate nanocrystals and an acidic matrix protein between the layers makes the mollusk shell a hard tissue [4].

D. Tan · Y. Zheng · L. Xue (✉)
School of Power and Mechanical Engineering, Wuhan University, South Donghu Road 8, Wuhan 430072, China
e-mail: xuelongjian@whu.edu.cn

L. Heepe et al. (eds.), *Bio-inspired Structured Adhesives*,
Biologically-Inspired Systems 9, DOI 10.1007/978-3-319-59114-8_8

Also composed of ~97% calcium carbonate, the egg-shape provides the eggshell a strong mechanical protection to the embryo [5]. All these examples suggest the importance of the structures for various functions.

Another very interesting example is the adhesion abilities of some animals bearing structured adhesive pad. Gecko is one of the most well-known species. It has been revealed that the hierarchical setae on gecko's toe pad play the key role in achieving the ability of reversible adhesion. The seta is mainly composed of β-keratin, a rigid material with rather high elastic modulus, E, of 2–4 GPa [6, 7]. Bulk β-keratin is too stiff to be able to form a proper contact with nature surfaces possessing unpredictable roughness. However, in the form of nanohairs, β-keratin can raise a great contact area, generating van der Waals forces large enough to support the body weight of geckos [8].

Even though E is an inherent property of a material, structuring the material can greatly change the apparent elastic modulus, which is also called effective elastic modulus, E_{eff}, and therefore the final mechanical properties of the object. The structured β-keratin has a much lower E_{eff}, which in turn contributes to a high adhesion performance. This kind of fine structure in animal's toe pad has a critical influence on the adhesion and friction properties. Following this idea, this chapter is arranged as following: the concepts of elastic modulus and effective elastic modulus are introduced first, followed by the introduction of natural structured adhesives. The methods to fabricate bioinspired structured adhesives and the parameters influencing adhesion performances are then discussed.

8.2 Elastic Modulus

Elastic modulus, E, which is also called Young's modulus, indicates the resistance of a material to be deformed elastically (non-permanently) when a loading force is applied. As shown in Fig. 8.1, the E is defined as the ratio between stress δ (force F per unit area A, $\delta = F/A$) and strain ε (ratio of deformation Δl over initial length L, $\varepsilon = \Delta l/L$) of a material as long as the hook's law applies:

$$E = \delta/\varepsilon = (F/A)/(\Delta l/L) \tag{8.1}$$

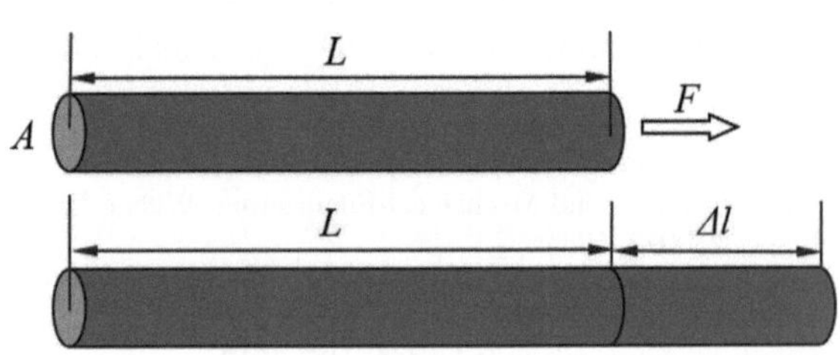

Fig. 8.1 An object with length of L, and cross-section area of A. Under a tensile force F, the length of the object extends Δl

E is one of the most important properties of a material. As long as the applied load is less than the yield strength of the material, E can be used to predict the elongation or compression, which allows engineers to find reasonable designs for controllable deformation by using appropriate materials.

There are two qualitative words to express E: stiffness and softness. Comparing two rods, the one with a smaller E shows a larger strain under the same stress. In other words, the rod is easier to be stretched, which is called softness. If E of a rod is very large, the rod is difficult to be elongated. The rod is stiff. The axial stiffness, k, can be calculated as following:

$$k = AE/L \tag{8.2}$$

From (8.2) we can find that stiffness of an object is linearly proportional to the elastic modulus of the material and depends on the geometry parameters. While elastic modulus is an inherent property of a material, stiffness is the rigidity of an object, a property of the structure.

8.3 Effective Elastic Modulus

Though E of a material is a very important parameter for various applications, it may not applicable in some cases. For instance, the properties of a material differ quite a lot from its bulk state when the size of the material reduces down to nanometers exposing much more atoms to the surface. Similarly, the mechanical properties of a composite are determined by the content and the arrangement of reinforcements in matrix. Taking composite as the example: each component in the composite has its own intrinsic E; and it is also clear that the E of every component cannot represent that of the composite. Considering the composite as a new homogeneous material, it has an apparent elastic modulus, E_{eff}.

Ideally, E_{eff} of a composite material can be estimated from the mixing law. In the following, we will take the unidirectional fiber reinforced composite as the example (Fig. 8.2). In the composite, continuous fibers are arranged parallel in the matrix. The volume fraction of fibers is V_{f} and the volume fraction of the matrix is $V_{\mathrm{m}} = 1 - V_{\mathrm{f}}$. Supposing strong interactions between the fibers and the matrix, the loading force along the long axis of the fibers causes the same strain in different components:

$$\varepsilon_c = \varepsilon_f = \varepsilon_m \tag{8.3}$$

where $\varepsilon_c, \varepsilon_f, \varepsilon_m$ are the strains of the composite, the fibers and the matrix, respectively. The E_{eff} of the unidirectional fiber reinforced composite ($E_{\mathrm{eff,c}}$) can be calculated by the following equation:

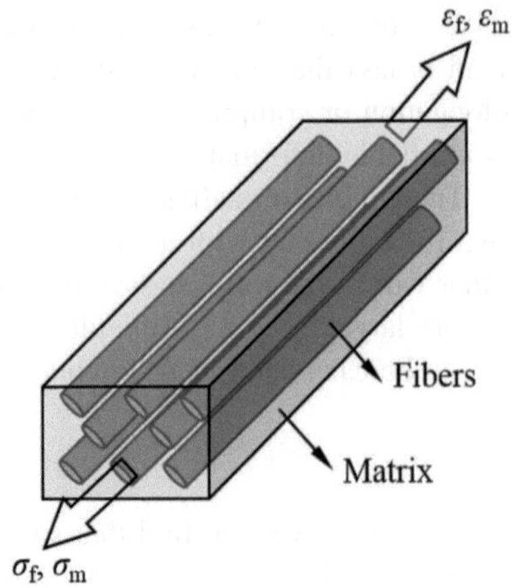

Fig. 8.2 Schematic drawing of unidirectional fibers reinforced composite under the stress (σ) along the long axis of the fibers and the strain is ε. The subscript f and m denotes fibers and matrix, respectively

$$E_{\text{eff,c}} = E_f V_f + E_m (1 - V_f) \tag{8.4}$$

where E_f and E_m are the moduli of the embedded fibers and the matrix, respectively. The real $E_{\text{eff,c}}$ could be different from the calculated value as the irregularity of the fibers and their alignments. A correction factor α is proposed to make the theoretical value closer to the real one:

$$E_{\text{eff,c}} = \alpha [E_f V_f + E_m (1 - V_f)] \tag{8.5}$$

More generally, an ideal composite with components A and B, whose volume fractions are V_A and $V_B = 1 - V_A$, and the elastic moduli are E_A and E_B, respectively, the effective elastic modulus can be written as:

$$E_{\text{eff}} = \alpha (E_A V_A + E_B V_B) \tag{8.6}$$

For the structured materials, like the porous materials, they can be considered as the composites composed of one kind of material and air. For instance, component B can be considered as air. Therefore, the effective elastic modulus of the porous material, $E_{\text{eff,p}}$, depends mainly on the volume fraction of the material A:

$$E_{\text{eff,p}} \rightarrow \alpha E_A V_A \tag{8.7}$$

It offers a great opportunity to continuously tune E_{eff} of a structured material. Moreover, the structure (or the assembly) of the composite components has a strong influence on the final E_{eff}. For example, in the unidirectional fiber reinforced composite, E_{eff}s in different directions, e.g. along and perpendicular to the long axis of fibers, differ quite a lot. It infers that structured materials with a same volume fraction but different structures could possess E_{eff}s with huge difference, which have a strong influence on various functional properties like adhesion and friction.

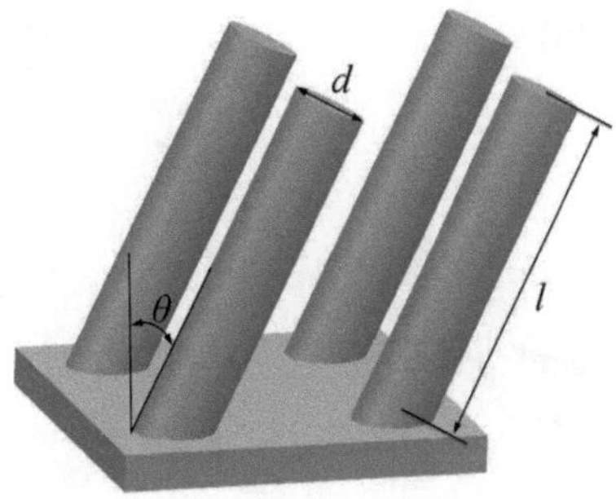

Fig. 8.3 Schematic illustration of pillar array with a pillar density ρ. The pillar has a diameter of d, a length of l and a tilting angle of θ

8.4 Natural Structured Adhesives Regulated by E_{eff}

In nature, all the surfaces have fractal roughness with multiple overlapping wavelengths. The existence of roughness has a tremendous influence on the contact formation between different surfaces; therefore macroscopic solids normally do not adhere to each other. On the other hand, the attractive forces, like van der Waals forces, are always present between two objects when they are close to each other in a distance in the order of a few atomic distances. In order to shorten the distance between two solid surfaces down to atomic scale and to form large area of real contact, at least one moiety of the contacting pair should be elastically very soft. In other words, the E (or E_{eff}) of the contacting-pair moiety should be greatly reduced. This principle has already been adopted in the structured adhesives in biology.

For instance, the materials of adhesive setae in several species of animals possess large E, i.e. 2–4 GPa of keratin in lizards [6, 7] and 7.2 GPa of resilin in ladybird beetle's tarsal setae [9]. The E_{eff} of the adhesive setae, which could be considered as pillar arrays composed of a material with elastic modulus of E (Fig. 8.3), can be calculated from the pillar diameter d, pillar length l, aspect ratio (AR = l/d), tilting angle θ ($0° < \theta < 90°$), and pillar density ρ [7]:

$$E_{eff} = \frac{3\pi E \rho d^2 \cos(\theta)}{64(\mathrm{AR})^2 \sin^2(\theta)} \tag{8.8}$$

From (8.8), the E_{eff} of a structured surface composed of rigid material could be greatly decreased by increasing AR (by increasing l or decreasing d) or increasing tilting angle θ, which are exactly the strategies for those animals to gain strong adhesion abilities.

8.4.1 Structured Adhesives in Nature

Gecko is a representative animal bearing structured adhesives in nature. The toe pad of gecko has several structural levels (Fig. 8.4): Thousands of branched setae with

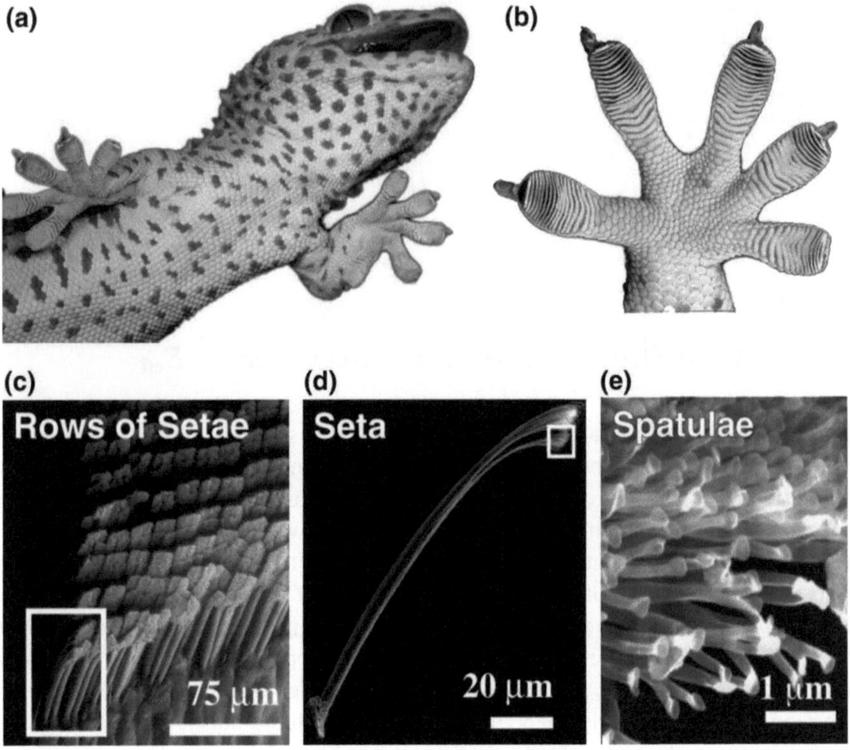

Fig. 8.4 Gecko adhesive structures. **a** Ventral view of a Tokay gecko (*Gekko gecko*) on glass surface. **b** Photo image of the foot of a Tokay gecko, showing the structure of tape-like lamellae. **c** SEM image of rows of setae on the tape-like lamellae. The zoomed-in single gecko seta in *white box* in (**c**) is shown in (**d**). The zoomed-in single spatula at the end of gecko seta in *white box* in (**d**) is shown in (**e**) [10]

setal stem of 30–130 μm long, 5–10 μm in diameter and density of ~14 400 setae/mm^2 originate from thin tape-like lamellae [10, 11]. Each seta is branched into secondary seta with length of 2–3 μm and diameter of 0.2–0.3 μm. At the end of each secondary seta, it is a spatular structure with width of 100–200 nm and thickness of 5–20 nm [12]. Moreover, the setae are tilted at certain angle towards the tarsal end of the toe pad (Fig. 8.5). It has been theoretically demonstrated that the tilted configuration can reduce the E_{eff} of fiber arrays [13]. The multileveled hairy structure together with the slanted configuration greatly reduces E_{eff} of the keratin setae on gecko's toe pad, facilitating the formation of numerous direct contacts with the counterpart surface. Millions of these direct contacts therefore insert great van der Waals forces between the surfaces, making gecko toe pad one of the most effective reversible adhesives [8].

Based on (8.8) and the requirement to form reliable contacts with rough surface, one may consider a hairy structure with a finest diameter and a length as long as

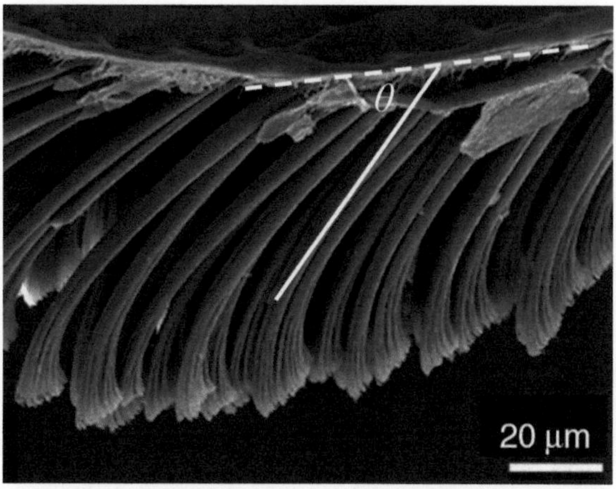

Fig. 8.5 The slanted arrangement of setae (*Straight line*) against the tape-like lamellae (*dashed line*) at an angle of θ on Tokey gecko's toe pad [14]

possible should be the best option. Then the question raises, why geckos use such a complex structure? Before answering this question, it should be noticed that strong adhesion is just one of the most important functions that gecko's toe pad should have [10, 14]. Gecko's toe pad should also possess the following properties: high structure stability, high tolerance to contamination and high switching efficiency between strong adhesion and easy detachment, in order to remain the ability of forming effective contacts on unpredictable substrates at a high frequency of steps (step intervals of $\sim$20 ms was reported [15]). As a nature surface is usually covered with stone fragments, dust particles, pollens etc., which will contaminate the hairy structure during the steps, the setae therefore require the ability of self-cleaning [16]. The setae should have enough stability during the self-cleaning that a thick setal stem is preferred, otherwise the thin setae may form cluster as shown in Fig. 8.6a. A thicker setal stem is also helpful to keep the setae in a slanted position (Fig. 8.6b), which is thought to be responsible for the asymmetric adhesion/friction forces or the ability to switch between strong adhesion and easy detachment. Though the thick seta roots guarantee the stability of setae and allow a larger length of setae to reach the valleys on a rough surface (Fig. 8.7), the density of setae is much lower. By splitting the thick setae into several levels of smaller branches (Fig. 8.6c), both requirements of high stability and large density of contact points (strong adhesion ability) can be realized [17].

Using the same strategy, several groups of animals, like spiders, flies and beetles in independent evolutionary lineages have the hairy adhesive organs (Fig. 8.8a). Recently, the careful examination of 225 climbing animal species, whose weights cover more than seven orders of magnitude, suggested the area of adhesive pad A is almost linearly proportional to the body mass m, $A \propto m^{1.02}$ [18]. Divided by the body mass, all these animals bearing the same area of adhesive pad! On the other hand, however, the adhesive setae with finer diameter and larger density are found in heavier animals. Since lizards are known as the heaviest animal using reversible adhesion for their locomotion, they may have the smallest, the most condensed hierarchical adhesive hairs (Fig. 8.8b) [19, 20]. As the surface area of an object

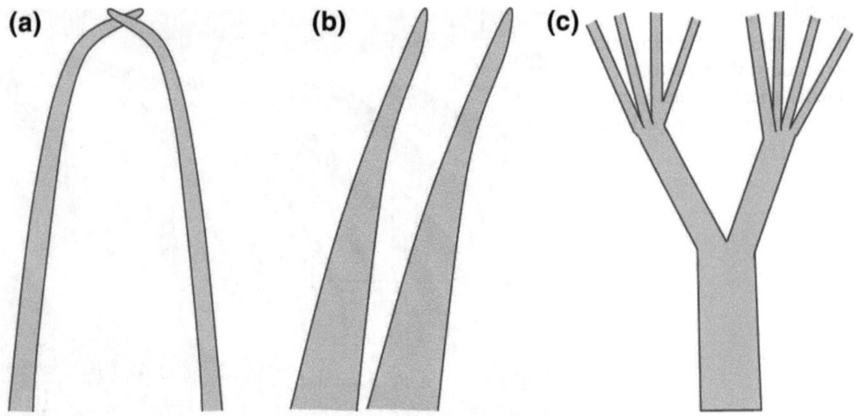

Fig. 8.6 Schematic drawing of the stabilities of pillar arrays with same height but different configurations. **a** Clustering happens in the pillar array with thin stems; **b** stable pillars with thick stem close to the root of pillars; **c** stable hierarchical pillar with thick stem and branched thin ends

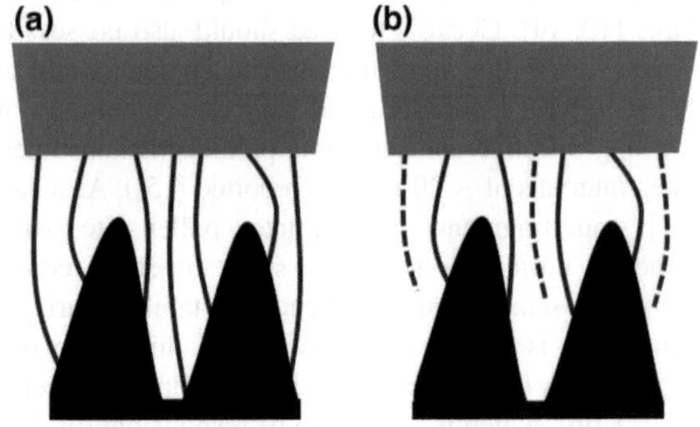

Fig. 8.7 Schematic drawing of the influence of the length of setae on the formation of contacts on a rough surface. **a** Long setae (*solid lines*) have a high possibility to get into contact with the rough surface (*black*); **b** short setae have a low possibility to form contacts with the rough surface that part in contact (*solid lines*) and part out of contact (*dashed lines*)

increases slower than the volume (or mass) with an increase in the linear size of the object, the adhesive system in large living beings, such as geckos, must be more effective (adhesion per unit attachment area) than in smaller animals, such as flies and beetles. That is to say, the hierarchical setae on gecko's toe pad is the most efficient structured adhesive system.

Another very important feature of gecko seta is the spatular terminal, as well as in animals like spiders, flies and beetles. Generally, the spatulas are quite thin as compared to their width or length. For example, the thickness of gecko spatula is only around 1/10 of its width [12]. One very interesting feature is that the spatulas on different animals have a larger width at the free end of spatula. This phenomenon can be very well explained by peeling theory [20]. If we consider spatula as an elastic film with thickness h and width $w(l)$, where $w(l)$ changes along the length l, as illustrated in Fig. 8.9. If it is peeled off from a rigid surface at a constant angle θ, the force F needed to initiate the detachment can be estimated by the following equation:

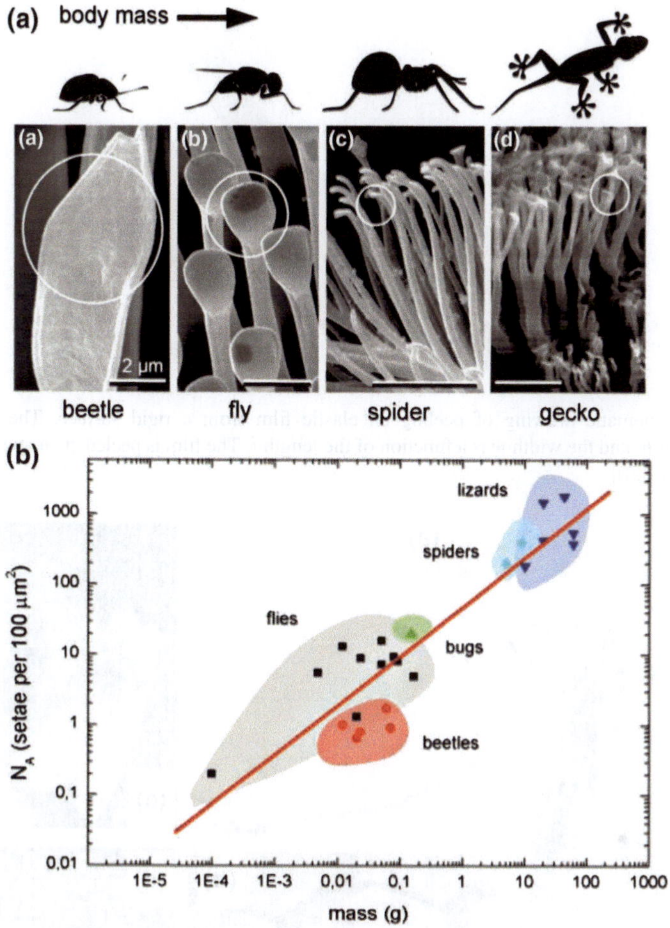

Fig. 8.8 **a** Seta array in animals bearing hairy adhesive pads. The heavier the animal, the finer the seta. **b** Dependence of the seta density (N_A) on the body mass (m) of animals bearing hairy adhesive pads of diverse animal groups. The *red line* is the linear fit: log N_A (m^{-2}) = 13.8 + 0.699 log m (kg), R = 0.919 [19]

$$F = \frac{R}{1 - \cos\theta} w(l) \tag{8.9}$$

where R is the energy required to fracture a unit area of interface at certain crack speed. From the equation, the peeling force is proportional to the width $w(l)$. At the spatular end, a larger peeling force is required to finally detach the whole spatula, which could avoid the "slipping" during the animal's locomotion (Fig. 8.9).

Different from the hairy structures in animals like geckos, smooth adhesive pads have been found in some animals like tree frogs and grasshoppers, as shown in Fig. 8.10 [21, 22]. The so-called smooth pad does not mean a flat solid surface. For

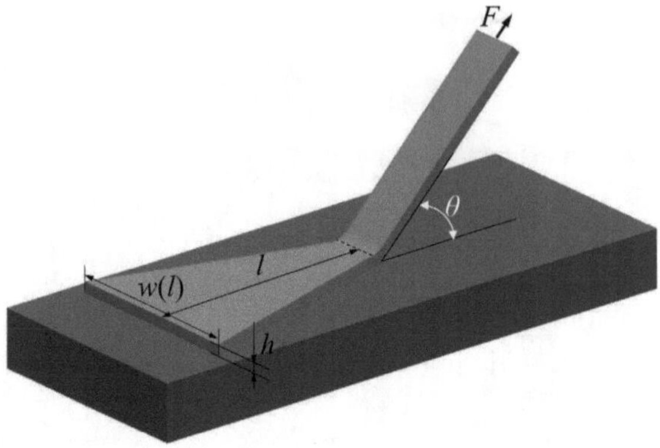

Fig. 8.9 Schematic drawing of peeling an elastic film from a rigid surface. The film has a thickness of h, and the width w is a function of the length l. The film is peeled at an angle of θ with a force of F [20]

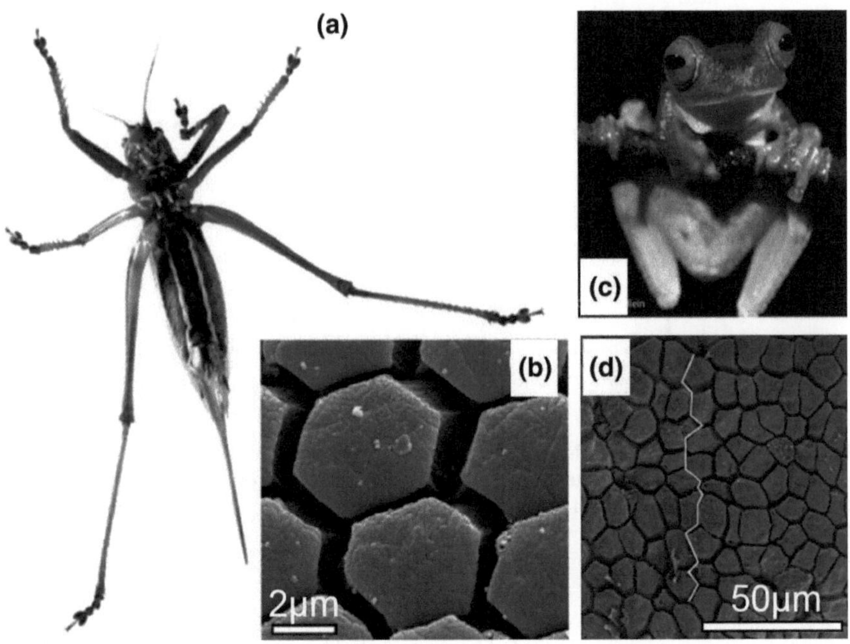

Fig. 8.10 **a** Bush cricket *Tettigonia viridissima* on a vertical glass surface and **b** the SEM image of its euplantulae surface [21]; **c** Tree frog *Rhacophorus pardalis* and **d** the SEM image of its toe pad topography [22]

example, the toe pad of tree frog *Rhacophorus pardalis* is composed of array of hexagonal epithelial cells with diameters in the range of 10 μm separated by microchannels with width of 2–3 μm and depth of around 5 μm. Recently, dense array of keratin nanopillars were also found in the toe pad epithelium cells of rock frog *Staurois parvus*. The height profile provided by AFM image showed concave tip geometry of the keratin nanopillars [23]. These special structures could be developed for the holding of mucus on top of the toe pad. Another very important feature is that the ends of these nanopillars are almost in the same plane, while that on gecko's toe pad are not. The same height of nanopillars is helpful to form capillary bridges on all of the nanopillars at the same time and therefore to generate high adhesion forces (capillary forces). On the other hand, the arrangement of nanohairs on gecko's toe pad indicates the importance of shearing for geckos to gain high adhesion forces, which is supported by the fact that the normal loading and unloading of a probe on gecko's toe pad cannot detect any adhesion forces [14].

8.4.2 Natural Adhesives with Material Softening

It is noticed gecko adhesion increases following the relative humidity within the tested range between 30 and 80% at a constant temperature (Fig. 8.11) [24]. The

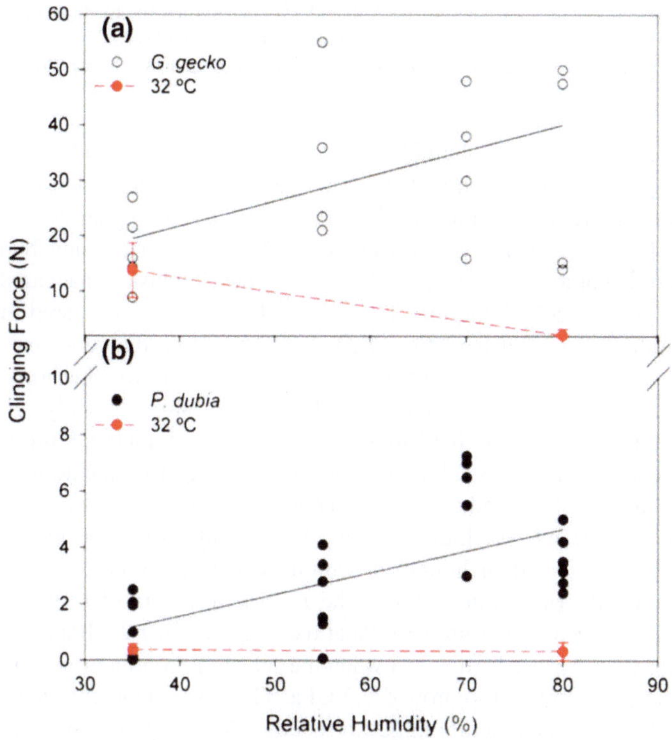

Fig. 8.11 The adhesion of the toe pads of (**a**) *Gekko gecko* (*G. gecko*) and (**b**) *Phelsum dubia* (*P. dubia*) at constant temperature of 12 and 32 °C under variable relative humidity. Red lines and symbols show results of 35% and 80% trials at 32 °C designed to test for an interaction of temperature and humidity; error bars represent ±2 SE [24]

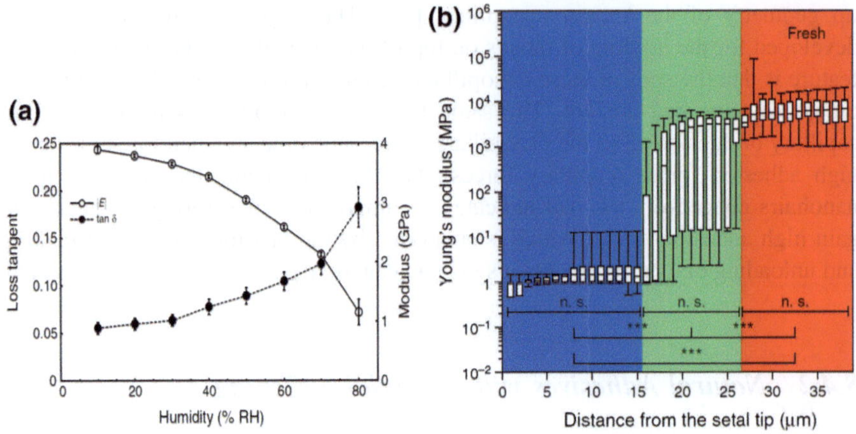

Fig. 8.12 **a** The viscoelastic parameters of *Gekko gecko*'s seta at different humidity tested by the dynamic mechanical analysis. Each data point is the mean of 9 trials, and the error bars are standard deviation [27]. **b** Young's modulus of fresh adhesive pad on the foreleg of female *Coccinella septempunctata* obtained by AFM nanoindentations (1 μm spacing) along the long axis of seta. The borders of the *boxes* define the 25th and 75th percentiles, the median is indicated by a *horizontal line*, and the *error bars* define the 10th and 90th percentiles (*n. s.* not significant, *** highly significant). The *background colours* indicate the different seta sections [9]

dependence of adhesion on humidity was also found in the spider *Philodromus dispar*. Slightly different from that found on gecko, the highest adhesion found on the spider is at a relative humidity of around 70% [25]. As higher adhesion forces were detected by the using of hydrophilic AFM probes than using hydrophobic ones at different humidity, Sun et al. [26] came to the conclusion that capillary force contributes to the gecko adhesion. They also calculated and compared the magnitudes of both van der Waals and capillary forces based on the geometry of sphere on plane and claimed that capillary force dominates the gecko adhesion as long as the relative humidity is higher than 16%. However, Autumn's study rejected this theory and proposed a mechanism of the changing in material property [27]. Mechanical properties of setal keratin in gecko's toe pad, including elasticity, tensile strength, fracture, and dynamic response were found to possess a significant dependence on the relative humidity (Fig. 8.12a), supporting the hypothesis that setal keratin is softened in humid environment and contributes to the adhesion enhancement [28]. More interestingly, the E in a fresh seta of beetle *Coccinella septempunctata* decreases from 6.8 GPa at the setal root to 1.2 MPa at the setal tip (Fig. 8.12b) [9]. This gradient is lost when the seta underwent a drying process and the E of the seta increased to around 7.2 GPa. The elastic gradient plays the same role as the hierarchical structure in gecko seta that the root part of seta provides the structure stability and the soft or finer setal ends help to gain higher adhesion. If we consider the water softened seta as a composite, it represents the second mechanism to reduce E (E_{eff}) of the material and to enhance the adhesion performances.

8.5 Artificial Structured Adhesives

The comparative studies have revealed two highly specialized structures for biological adhesion: hairy and smooth adhesive pads. Though these two structures are distinctly different, both of them have greatly reduced E_{eff}, which in turn maximize the possibility to form effective contacts with the counterpart surface. From the structure point of view, the smooth pad is also kind of pillar-like structure, but with a very small AR. Moreover, it is found that the structure rather than the surface chemistry contributes mainly to the fabulous adhesion abilities of geckos. This is supported by the finding that van der Waals forces are the primary adhesive mechanism [8, 29] in dry adhesion (gecko adhesion), though several other forces, like electrostatic forces [30] and vacuum pressure [31–33] could also contribute. Anyway, it suggests a minimum requirement for the material selection when fabricating the structured adhesives. For the reality, the cheap gain and easy manipulation are the main criteria to select materials that the materials like thermoplastic polystyrene (PS) [34], UV-curable polyurethane acrylate (PUA) [35], polyimide [36], polydimethylsiloxane (PDMS) [37, 38], carbon nanotubes (CNT) [39, 40], and various other materials [41–48] are used. The performances of bioinspired structured adhesives made of various materials are summarized in Table 8.1.

Guided by the idea to gain small E_{eff} mimicking the nature, the basic configuration for structured adhesive is pillar array [49]. Generally, there are two main manufacturing methods to get micro- and nanopillars. One is bottom-up method including micro/nanocasting, gas phase growth and so on, while the second one, top-down method includes photolithography, electron beam lithography, etc. Moreover, the structural features like tip geometries, slanted configuration, necking, porosity and hierarchy can further tune the E_{eff} of the structure and therefore the adhesion performances.

8.5.1 Basic Pillar/Fiber Arrays

8.5.1.1 Manufacture Methods

Within the bottom-up methods, micro/nanocasting is the widely used method to fabricate bioinspired structured adhesives [50]. The templates, including anodic aluminum oxide (AAO), silicon and photoresist molds prepared by lithography and other microfabrication techniques, play the central role. For the preparation of nanoscale pillar arrays, AAO templates are widely used because AAO is easy to get and the pore parameters can be tuned continuously [34]. The diameter (40–400 nm) and the depth (few hundreds of nanometers to hundreds of micrometers) in AAO

Table 8.1 Summary of recently published bioinspired hairy adhesives

Materials	Fiber shape	Diameter (μm) height (μm)	E (MPa)	Adheison force	Shear force	Preload	References
Polyurethane	Tilted spatula	380; 1000	0.3	–	3 kPa	0.6 kPa	[43]
Polyurethane	Spatula	4.5; 20	3	180 kPa	–	120 kPa	[69]
Polyurethane	Mushroom	5 (stalk), 9 (tip); 20	69–690	180 kPa	–	–	[47]
Polyurethane	Flared tips	50 (stalk), 100 (tip); 100	2.9, 9.8	64 kPa (stiff) 24 kPa (soft)	–	50 mN	[41]
Polyvinylsiloxane	Mushroom	40; 100	3	60.6 kPa	2 kPa	2 kPa	[11]
Polyvinylsiloxane	Hexagon	10, 20, 50, 95; 1	0.94–2.86	–	41 kPa (dry) 16 kPa (wet)	60 mN	[21]
Polyvinylsiloxane	Mushroom	48; 100	0.94–2.85	116 kPa	–	90 mN	[33]
Carbon nanotube	–	0.08; 400	10^6	50 kPa	360 kPa	500 kPa	[40]
Carbon nanotube	Tilted	0.01–0.015; 0.1	10^6	200 kPa	1000 kPa	–	[39]
Polypropylene	Tilted	0.6; 1.8	≈1500	45 kPa	298 nN/fiber	1 mN	[48]
Polypropylene	–	0.6; 18	1000	10nN/fiber	20 kPa	<1 kPa	[42]
Polyurethane acrylate	Tilted hierarchal spatula	Nano: 0.6; 2.8 Micro: 5; 25	19.8	–	90 kPa	3 kPa	[35]
Polyurethane acrylate	Tilted	0.1; 1	19.8–320	110 kPa	–	3 kPa	[87]
Polyurethane acrylate	Tilted	0.1; 1	19.8	–	95 kPa	3 kPa	[88]
Polydimethylsiloxane	Mushroom	10; 20	1–10	90 kPa	–	–	[53]
Polydimethylsiloxane	Wedge	50 (base); 200	0.44, 1.75, 2.63	5.1 kPa	17 kPa	–	[38]
Polydimethylsiloxane	Mushroom	8; 30	1–10	219 kPa	–	750 g	[52]
Ge/parylene	–	0.2; 3	6200 ± 200	~40 μN	–	300 μN	[96]
Parylene	–	0.25; 2.5	2800	7 kPa	–	10 kPa	[44]
Polythiophene	Tilted	0.2; 8–20	1500	800 ± 80 kPa	1740 ± 100 kPa	–	[45]

(continued)

Table 8.1 (continued)

Materials	Fiber shape	Diameter (μm) height (μm)	E (MPa)	Adheison force	Shear force	Preload	References
Polystyrene	–	0.27; 0.5	3000	100 μN	–	500 μN	[34]
	Spatula		3000	200 μN			
	Mushroom		3000	700 μN			
Poly (glycerolco-sebacateacrylate)	Cone	0.1–1; 0.8–3	0.38–10.68	–	48 kPa	–	[46]
Polystyrene-*block*-poly (2-vinyl pyridine)	Spatula	0.3; 1.5	28	20 μN	–	300 μN	[91]

template can be very well controlled by the selection of proper electrolyte, applied voltage, temperature, and electrolytic time [51]. Micropillars are normally replicated from silicon [52] and photoresist [53] molds. Silicon molds are fabricated by dry or wet etching techniques, of which the process is either quite slow or expensive. Compared to silicon molds, photoresist templates are much cheaper and relatively easier to fabricate. The selective exposure of photosensitive materials through a photo mask can decompose or cross-link the irradiated areas depending on the photoresist used. The subsequent wet-etching of the undesired areas converts the photoresist film into 3D structures. Once the mold is ready, it could be used for many times. The fluidic materials are poured onto the molds with micro- or nano-hole arrays. After the solidification, the negative replica of the mold (thus the pillars/fibers) can be released from the mold by simple peeling off the sample or selective etching away the mold. The geometrical parameters of the replicated pillars/fibers, such as diameter, depth and period are determined by the templates.

Using the catalyst iron (Fe) on silicon wafer precoated with an adhesion layer of aluminum (Al) as the seeds (templates), arrays of carbon nanotube (CNT) either single-walled or multi-walled could be easily obtained. The pyrolysis of certain hydrocarbons in different conditions (carbon source, buffer gas type and pressure, temperature, etc.) could result in vertical aligned single-walled carbon nanotube (SWCNT) or multi-walled carbon nanotube (MWCNT) [39, 40, 54, 55]. For example, when all the conditions are identical, the addition of 1% thiophene to the carbon source favored SWCNT with diameters of 1.1–1.9 nm, while the using of 10% thiophene resulted in MWCNT. While the diameter of SWCNT ranges from several Å up to around 6 nm, the diameter of MWCNT could reach 100 nm [56]. On the other hand, both of them can reach a height of several hundred of micrometers, thus a super-high AR!

Top-down methods, opposite to the bottom-up methods, including technologies like photolithography, electron beam lithography, micro/nano printing technology and so on, directly generate pillar arrays without the using of molds but masks. Fibrils with diameter between sub-100 nm to micrometers can be obtained by using corresponding masks. Through the photomask, the exposed area is either decomposed/etched (in case the photoresist, positive ones) or cross-linked (negative photoresist). For example, SU-8® has become the favorite negative photoresist for high-aspect-ratio and three-dimensional lithographic patterning due to its excellent coating, planarization and processing properties as well as its mechanical and chemical stabilities [57]. The micro/nano printing technologies need templates with micro/nano hole arrays. The templates are put inversely to smash down into semifluid organic materials. The micro/nano pillar arrays complementary to the templates are then obtained by taking off the templates.

8.5.1.2 Influence of Pillar Size on Adhesion

It is well accepted that a material with low stiffness (elastic modulus) is easier to conform to rough surfaces under a given applied load. However, the disadvantages associated with the use of a low-modulus material is also quite obvious. It tends to creep and foul by particulate contamination [58]. Employing stiff materials, a fibrillar interface can deliver a compliant response as already demonstrated in natural adhesives. Inspired by this fact, pillar arrays made of a variety of materials have been developed to investigate the influence of various strcutural parameters of pillar arrays on the adhesion performances.

PDMS, a relative soft material, has been widely used as the model material, as it is simple to use, easy to control, and commercially available. Compared with the flat controls, the PDMS pillar arrays showed greatly enhanced adhesions. In the pillar arrays, discrete contact points form between the pillar top and the counterpart surface that the detachment on each pillar is an independent case. The detachment on each pillar involves the crack initiation and propagation. Therefore, numerous simultaneous detachments result in a high pull-off force. Increasing the pillar height from 5 to 100 μm with a fixed diameter, or in another word the increase in AR, significantly decreased E_{eff} of PDMS pillar array from around 1.4 to 0.6 MPa

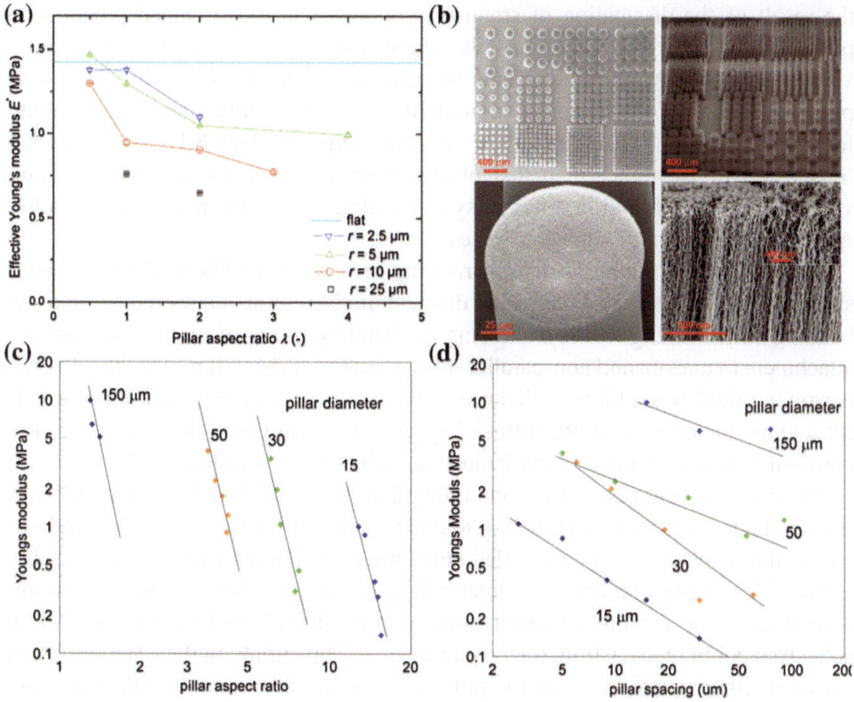

Fig. 8.13 **a** The dependence of effective elastic modulus of PDMS micropillars on the aspect ratio of pillars [37]. **b** SEM images of top view (*top left*), three-dimensional view (*top right*) and detailed structure of single micropillar (*bottom*) of vertically aligned carbon nanotube (VACNT) with different pillar diameters (15–150 μm) and spaces between pillars. **c**, **d** The dependence of effective elastic modulus of VACNT on **c** the aspect ratio of VACNT micropillars and **d** the space between VACNT micropillars [59]

(Fig. 8.13a) [37]. It should be noted that the PDMS pillar arrays with a low AR of 0.5 showed a similar E_{eff} as that of the flat control, which is considered as the result of the contribution of backing layer during the indentation test. Moreover, the drop in E_{eff} is more pronounced for AR smaller than 2 and is less significant at larger values. This kind of E_{eff} dropping was also demonstrated in the micropatterned vertically aligned CNT (VACNT) forests (Fig. 8.13b). With an AR of ~15 and pillar diameter of 15 μm E_{eff} of VACNT is less than 0.2 MPa (Fig. 8.13c) [59]. It should be mentioned here that the elastic modulus of CNT is around 1000 GPa. It means a reduction of 7 orders of magnitude! Moreover, the increase in the space between micropillars also reduced the E_{eff} of VACNT (Fig. 8.13d). Accordingly, the increase in AR increased the pull-off strength. In the PDMS pillar array, the pull-off strength was found to increase with $AR^{0.3}$, and to decrease with pillar radius with an exponential index of −0.4 [37]. It suggests that a pillar array with larger AR or smaller diameter can gain high compliance and thus a higher adhesion performance. In contrast, a higher pull-off strength is clearly seen for VACNT micropillars with larger diameter. However, the dependence on AR in the VACNT micropillars is the same as that found in PDMS pillars, which is considered to be the result of the dissipation of strain energy stored in the stretched pillars at the pull-off point where an adhesive fails. The examination on macroscopic arrays of vertically-aligned MWCNT detected an adhesion force of 11.7 N/cm^2 [60], comparable to the gecko adhesion of around 10 N/cm^2, though the independent atomic force microscopic measurements on vertically-aligned MWCNTs found adhesion forces 200 times higher than that observed for gecko foot-hairs [61]. Although variable scenarios found in different systems, all point out the importance of a high AR for obtaining high adhesion forces.

In some animal's adhesive hair, narrowing neck was found beneath the tip end, as shown in Fig. 8.14a [62]. The smaller diameter in the region of neck reduces the local bending stiffness (Fig. 8.14b), allowing the bending/tilting of the tip to form better attachment to uneven and non-parallel substrates (Fig. 8.14c). It means the pillar with narrowing neck has a higher tolerance to the rough surface, which in turn provides tilt-tolerant attachment. In the artificial system, it is also found the advantage of the narrowing neck to enhance the tolerance to high compressive loads [63].

Moreover, recent researches found the pillar radius of dry adhesive systems has great influence on the contamination tolerance and self-clean ability [64, 65]. Using nano-, microspheres as the modelled contamination, the relative size of the sphere to the bioinspired pillar arrays (diameter, length and space between pillars) strongly determines how far the adhesion abilities are affected and to what extent the adhesives could recover from the contamination. Depending on the relative size, the particles either retain on top of the pillar array, or partially sink into the blank area between pillars, or totally trapped among the pillars (Fig. 8.15) [63, 64]. For the first two cases, the contaminations may hinder the direct contact between two surfaces, but they could be easily removed by simple shearing. In the last case, the particles can get into the space between pillars. Therefore, it is possible for the pillars to form direct contact with the target surface generating a relative high adhesion force, but the simple shearing will not remove the particles away. It

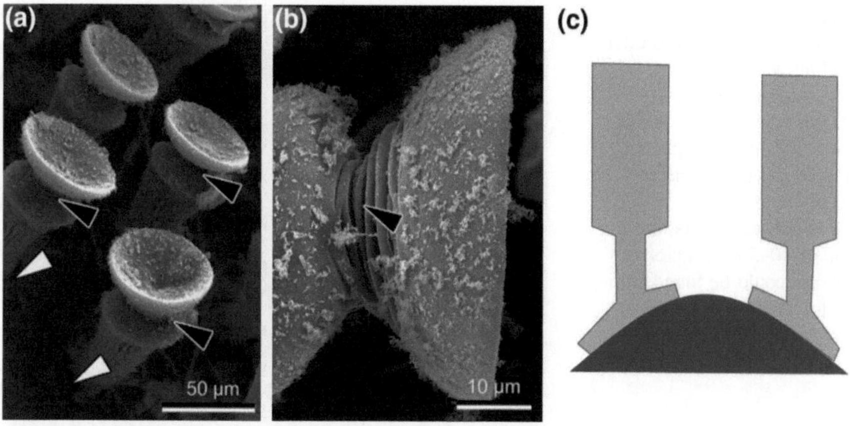

Fig. 8.14 **a**, **b** SEM images of the tarsal adhesive microstructures in dytiscid beetle *Dytiscus marginalis. Black arrowheads* indicate the narrowing neck structure below the adhesive disc; *White arrowheads* indicate the base of the stalk [62]. **c** Diagram of the proper contact forming between micropillars (*light gray*) with narrowing neck beneath the functional tips and a rough surface (*dark gray*)

suggests the self-cleaning by physical shearing in nanopillars is much easier than microscaled ones, as the smaller the pillar radius is, the harder can the particles get into pillar arrays, especially for large particles. For sure, the stiffness of the material also plays a role. For example, nanopillar array composed of a hard thermoplastic polypropylene (PP) was able to recover 96–115% of its shear adhesion after fouling with relative small and large but not medium particles. In contrast, the adhesion of a micropillar array made of PDMS can only recover 40–55% after the contaminations by medium and large particles, while small particles could even be uptaken by the soft PDMS [64].

8.5.2 Tip Geometry

8.5.2.1 Manufacture Methods

It has been demonstrated in many animals that the tip geometries of setae determine their adhesion abilities. For example, the mushroom-like setae on male ladybugs generate much stronger force on smooth surface than the female ones which do not bear mushroom-tipped setae. As discussed above, the overhang in mushroom-like tip is very thin that it could adapt to surfaces, enhancing the contact formation and ultimately the adhesion forces. The understanding of the importance of tip geometries in hairy adhesives has inspired many fabrication techniques. Because of the nature of different techniques, the templates for pillars normally have flat [66] or hemispherical [34] pore bottoms. Pillars replicated from these templates are

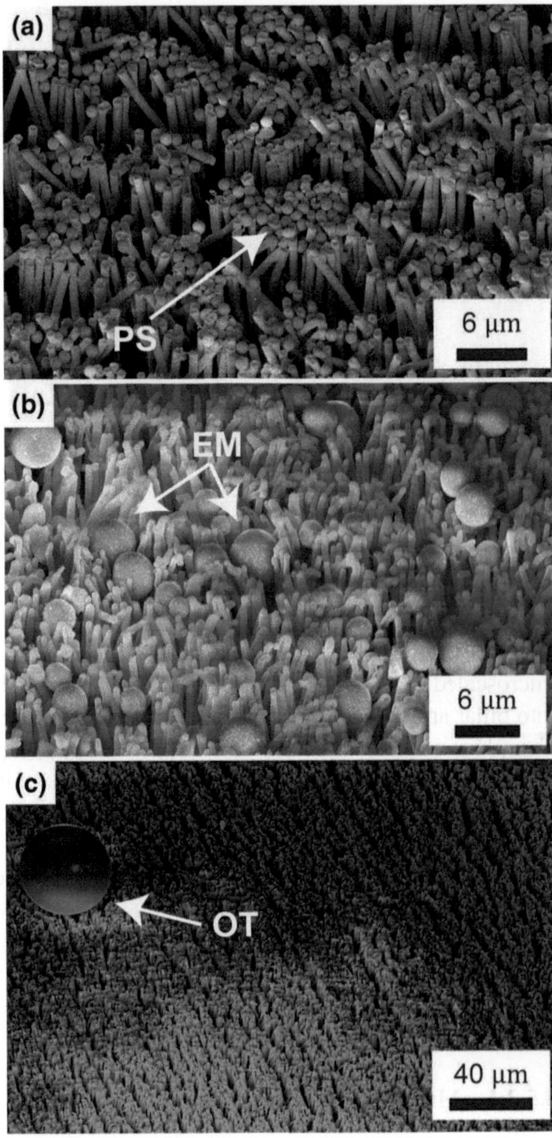

Fig. 8.15 Polypropylene fibrillar arrays 40 steps after fouling with **a** 1 μm polystyrene spheres, **b** 3–10 μm glass spheres and **c** 40–50 μm glass spheres. **b** 3–10 μm glass spheres were deeply embedded (EM) between fibers; **c** only one of 40–50 μm glass spheres could be found sitting on top (OT) of the fibrillar array [64]

therefore have flat or hemispherical tips. The partial wetting of pores in template offers the possibility to fabricate pillars with concave tips or even with tubular tips [66, 67]. In order to get bioinspired tip geometries in micro- and nano-pillars, either specially designed template or postshaping of the pillars is required [35, 68–71].

The studies on biological hairy adhesives have demonstrated the importance of mushroom-like (in some cases also called T-shaped, or pancake-like) tips on the normal adhesions. Several techniques have been developed to fabricate pillar arrays

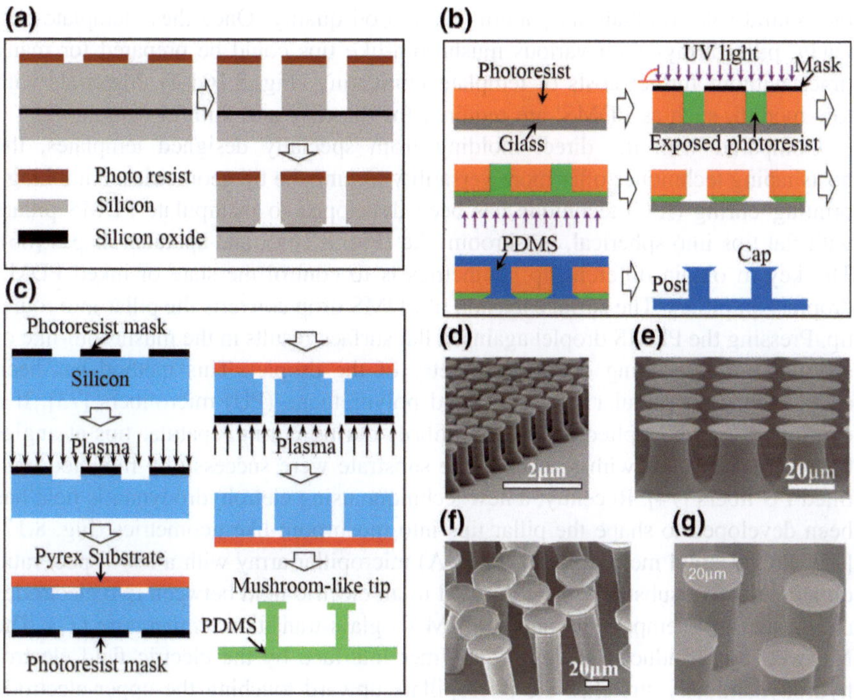

Fig. 8.16 **a–c** Schematic drawing of three procedures to prepare molds for micropillar arrays with mushroom-like tips; and SEM images of resulted structures are shown in (**d–f**), respectively. **g** SEM image of micropillars terminated with mushroom-like tips prepared by IPC technique [14, 52, 66, 69, 72]

with mushroom-like tips: (1) Making use of the effect of etching stop, where the vertical etching in silicon wafer cannot proceed vertically anymore but expand laterally when the etching front comes into a layer of silicon oxide, a negative mushroom-like profile can be shaped (Fig. 8.16a). (2) Similarly, the lateral wet etching in the exposed back-side layer (without any mask, Side A) in positive photoresist can also add a negative mushroom-like profile to the negative pillar profile which is formed by exposing the top side of the photoresist (Side B) with a mask (Fig. 8.16b) [72]. The back-side exposure should be short enough that the resulted mushroom tips can have thin overhangs. In both techniques, the time for lateral etching should be controlled within a proper time window: too-short time will not result in overhangs, long-time etching will detach the whole photoresist layer and fail the whole preparation. (3) In a more controlled way, the exposures of photoresist with masks from both sides convert the exposed areas precisely into negative profile of pillar with mushroom-like tip [52]. In this technique, the mask on one side (Side A) with relative larger feature size is for mushroom-like profile, while the mask on the other side (Side B) is for the pillars, as depicted in Fig. 8.16c. Therefore, the alignment of masks on both sides and the exposure from Side A play the central role in obtaining patterns with good quality. Once these templates are ready, pillar arrays with various mushroom-like tips could be prepared for many times without further costs of template fabricating (Fig. 8.16d–f). Materials with low moduli, such as PDMS, are required for the easy and faithful demolding.

Compared with the direct molding from specially designed templates, the postshaping techniques offer more versatility to tune the tip geometries. An inking–printing–curing (IPC) technique has been developed to manipulate PDMS pillars with flat tips into spherical, mushroom-like (Fig. 8.16g), and spatular shapes [66]. The key to obtain different tip geometries is to control the state of inked PDMS droplets on pillars. The surface tension of PDMS drop converts the pillar into round tip. Pressing the PDMS droplet against a flat surface results in the mushroom-like or spatular tips depending on the symmetry of the droplet. This method has been adapted to graft spatular tips onto tilted polyurethane (PU) microfibers [73]. If a controlled load is applied to bend the fibrils during curing, spatular tips at angles between 0° and 90° with respect to the substrate were successfully mounted onto tilted PU fibers [74]. Recently, a new technique using eletrohydrodynamic field has been developed to shape the pillar tips into mushroom-like geometries (Fig. 8.17) [75]. Poly(methyl methacrylate) (PMMA) micropillar array with a low aspect ratio on a conductive substrate was subjected to an electric field between two electrodes and heated to a temperature above PMMA's glass transition temperature (T_g). The Maxwell force induced at the air–polymer interface by the electric field electro-hydrodynamically stretches the micropillars upward touching the upper electrode (Fig. 8.17a, b). PMMA is then spread onto the upper electrode's surface forming the mushroom-like tips because of the electrowetting effects under a continuous electric field (Fig. 8.17c, d). The diameter of the mushroom tip can be controlled by the voltage applied. The simulation and experimental results showing the whole evolving procedure of a pillar are shown in Fig. 8.17e, f, respectively. During this procedure, AR of the pillar is greatly enlarged that the corresponding E_{eff} is sharply reduced.

Compared to the micropillars, the shaping of nanopillar tips are much more difficult that only few technique has been developed. Recently, we succeeded to shape polystyrene (PS) nanopillar arrays with dimensions comparable to the gecko setae from hemispherical into flat, mushroom-like, and spatular tips by cold-pressing, hot-pressing, and shear-pressing (Fig. 8.18) [34]. The selective opening of pore bottoms in AAO template offering strong support to the PS nanorod array during the manipulation process turns out to be the key step in this technique. While it has been widely demonstrated that the reducing in pillar diameter can increase the adhesion performance, the tip geometries at nanoscale is the critical feature for further enhancement of adhesion.

8.5.2.2 Influence of Tip Geometry on Adhesion

A strong influence of the tip geometry of fibrils on adhesion performance has been demonstrated in various artificial fibrillar adhesives at micro- and nanoscales.

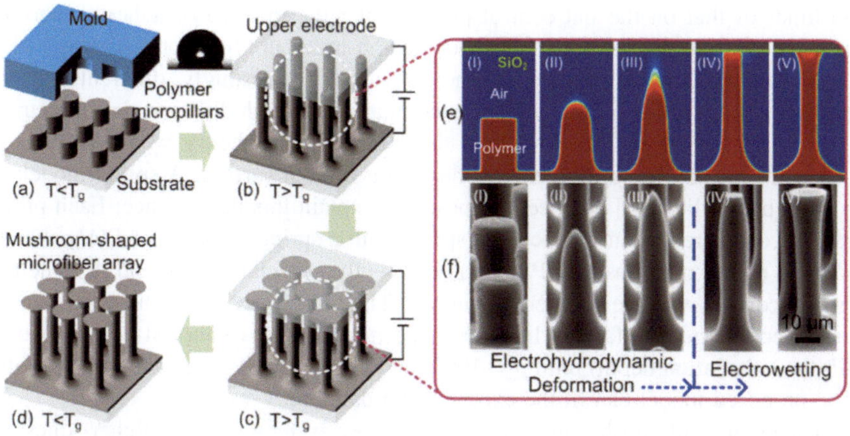

Fig. 8.17 Schematic illustration of the procedure to fabricate micropillar array terminated with mushroom-like tips with a high aspect ratio. **a** Micropillar array with a low aspect ratio on conductive substrate prepared by hot embossing. **b** A voltage is applied between two electrodes while the polymer micropillar array is sandwiched between two electrodes and heated to the glass transition temperature of the polymer. Micropillars are electrohydrodynamically brought upward into contact with the upper electrode and **c** spread onto the upper electrode by electrowetting. **d** Micropillar array terminated with mushroom-like tips are ready after the removal of the upper electrode. *Insets* **e** and **f** show the evolutions of micropillar array induced by an external electric field numerically and experimentally [75]

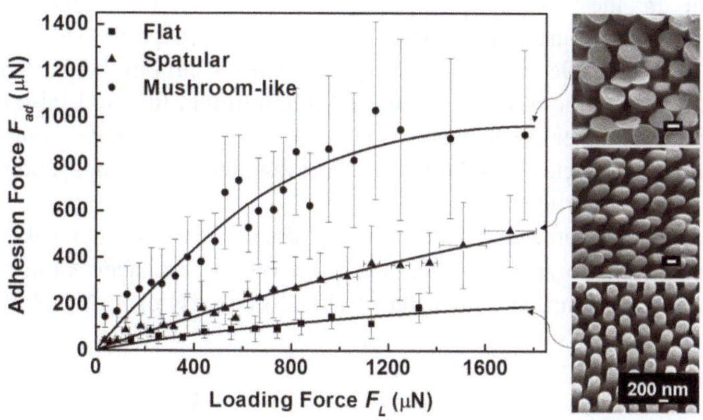

Fig. 8.18 The adhesion dependences of PS nanorod arrays terminated with flat, spatular and mushroom-like tips on the loading forces. The loading dependence is fitted with a spring model of contact possibility. SEM images of the corresponding structures are shown on the *right side* [34]

Comparing to the pillar arrays with flat, spherical, and concave tips, mushroom-tipped (T-shape) PDMS micropillars showed the best performance of normal adhesion with and without the presence of water at the contacting interface [66, 71, 76]. The highest pull-off force found in pillars with mushroom-like tips is

30 times of that on the flat control [71]. Similar dependence of adhesion performance on the tip geometry has also been demonstrated in PS nanopillar arrays with the dimension close to the gecko setae [34]. More interestingly, different loading dependences have been found on the nanopillar arrays with different tip geometries (Fig. 8.18). However, this is against the classical JKR theory which predicts a constant adhesion force for a rigid sphere on a smooth surface. A spring model of contact possibility [34] has been proposed to explain this dependence: Each pillar can be considered as an independent spring with a spring constant of E/Al_0, where E is the elastic modulus of a PS nanorod, l_0 is the original length, and A is the tip area in contact with the counterpart surface. The probability p of contact formation to the counterpart surface is linearly proportional to the "nanorod compression" Δl (the indentation depth) until $p = 100\%$ and the contact persists until the pull-off. The proposed model can fit the experimental data pretty well.

The pillars with mushroom-like tips in all those systems showed higher adhesion forces than their reference pillars with a same pillar diameter d and period λ, as shown in Fig. 8.19. Considering the main contribution to dry adhesion is van der Waals forces, mushroom-like pillars take the advantage to possess a larger contact diameter, $D > d$ (Fig. 8.19a, b). With this in mind, we may come to the result that the pillar with diameter D will have a higher adhesion force than that with a diameter of d. Obviously, this is against the experimental finding that pillar arrays with smaller diameter have relatively larger adhesion forces! If we consider the pillar with mushroom-like tip as a composite pillar with apparent diameter of D (dashed box in Fig. 8.19b) that the blank part beneath the overhang is the second component in the "composite" pillar. According to (8.6), the pillar with mushroom-like tip thus has a lower E_{eff} that the solid pillar with diameter of D shown in Fig. 8.19c. Therefore, the pillar with mushroom-like tip possess the highest adhesion in these three structures. The mushroom-like tip is also suggested to contribute in other mechanisms: (1) high compliance to the rough surface because of its nanometer scaled thickness, arresting the crack propagation; (2) high tolerance to small particles.

More detailed parameters of the overhang structure in mushroom-like tips have been demonstrated to strongly affect the adhesion performances. Considering the overhang as a thin film with diameter of D and thickness of h (Fig. 8.20a), it has

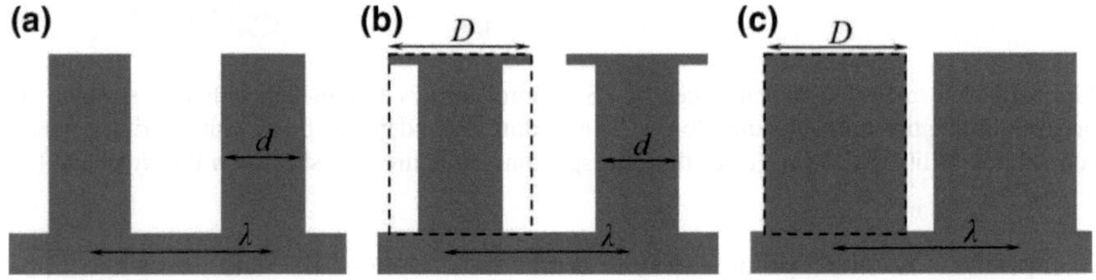

Fig. 8.19 Diagram of pillar array with **a** pillar diameter of d, and period of λ; **b** pillar diameter of d, diameter of mushroom-like tip of D, and period of λ; **c** pillar diameter of D, and period of λ. The *dashed boxes* in **b** and **c** have the same volume

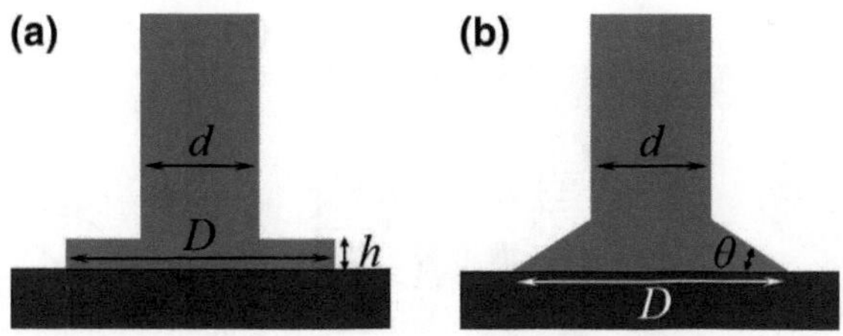

Fig. 8.20 Schematic illustration of pillar (diameter of *d*) terminated with mushroom-like tips with different overhang structures. **a** A film-like overhang with diameter of *D* and thickness of *h*; **b** a wedge-shaped overhang with diameter of *D* and angle of θ

been suggested that the optimal values of D/d and h/d are within 2–3 and 0.2–0.3, respectively [77]. With the optimized parameter, the stress at the contact interface is homogeneously distributed [70, 78]. Therefore, the detachment of a mushroom-like tip from a surface initiates at the pillar center, and propagates outwards until the complete detachment. The procedure has been captured and demonstrated by a superfast camera [31]. The cavitation under the mushroom tip would then contribute a vacuum pressure to the pull-off force during the detachment from a smooth surface, which is around 10% of the total adhesion force [79]. However, all the surfaces in reality, especially the natural ones, have certain roughness that the vacuum pressure would contribute much less than the value estimated or the contribution is ignorable. In many techniques, the fabricated overhang in mushroom-like tip has wedge-like cross-sections with an edge angle of θ as depicted in Fig. 8.20b [34, 65, 66, 73]. The optimal values of D/d and θ were estimated to be 1.1 and 45°, respectively, based on the Dugdale–Barenblatt cohesive zone model [80].

Though it has been demonstrated in several systems that the spatular tips show less normal adhesion than the mushroom-like tips [34, 66, 71], most of these animals bear spatular tips [20]. It infers a high adhesion force is not the only criteria for the evolution of structured adhesives. The asymmetric nature of spatular tip offers directional adhesion and friction, which is of critical importance in animals' locomotion. Replicated from specially designed template, squared micropillars having rectangular overhangs facing one direction have been fabricated (Fig. 8.21a, b) [81]. The highest adhesion was detected in the retracting direction normal to the surface. Along the direction overhangs pointing, the retractions at ±30° with respect to the normal direction showed the largest difference in the normal pull-off forces with a ratio of 2.3 (Fig. 8.21c). With IPC technique, round micropillars with overhangs in crescent shape have been developed (Fig. 8.21d, e). Enhanced kinetic friction was found in the shearing direction against spatular end, while static friction was stronger in the opposite direction (Fig. 8.21f). The contributions of the spatular terminal to the frictions in opposite directions are different, providing a new mechanism for asymmetric friction [82].

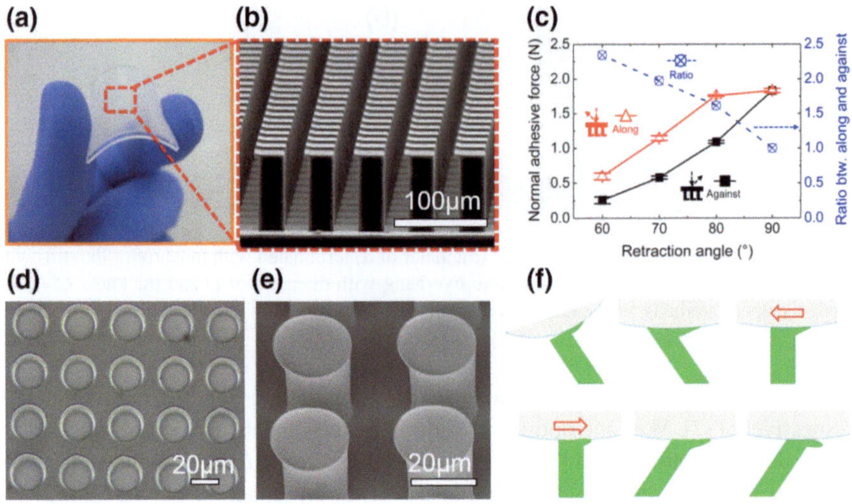

Fig. 8.21 **a** Optical and **b** SEM images of *squared* micropillars with *rectangular* overhangs. **c** The dependence of normal adhesion forces on the retraction angles along and against the orientation of overhangs and the corresponding ratio between the normal adhesion forces at these two retraction angles. Each data point represents 20 measurements [81]. **d** Optical and **e** SEM images of the spatular tips on micropillars with an aspect ratio of 1.5. **f** Schematic drawing of the sliding of a round probe (*gray*) on the micropillar terminated with spatular tip (*green*) in the directions against (*up row*) and toward (*bottom row*) spatular end. *Red arrow* indicates the moving direction of the probe. When sliding against the spatular end, the overhang contributes mainly to the kinetic friction while in the opposite direction contributes to the static friction [82]

8.5.3 Tilted Configuration

While the vertical exposure of a surface to UV light/electron beam generates structures perpendicular to the surface, the natural thought is that the tilted-placing of an exposing surface to the incident light/electron beam may form slanted structures. Indeed, tilted holes and pillars have been successfully realized by exposing tilted-placed positive photoresist PMGI SF-15 and negative photoresist SU-8 to the incident UV lights, respectively [83, 84]. However, the simple tilting of polySi in SF6/Ar plasma will not lead to an angled etching. This is because the incident ions are accelerated through the plasma sheath and will etch normally to the substrate surface no matter how the substrate is placed. The installation of a Faraday cage can change the situation and guide the ions to the etching of slanted holes on the substrate. Replicated from such templates, tilted PUA pillars with diameter of 350 nm, length of 2.8 μm, and angles with respect to the substrate surface varying from 30° to 60° have been successfully replicated [35].

Vertical pillars can be turned into slanted or bent states by post-treatments, avoiding the preparation of special templates. Polymers become soft when they are heated to a temperature above their *Tg*s, and turn back to glassy states fixing the structure of the polymeric pillars by decreasing the temperature below their *Tg*s. For

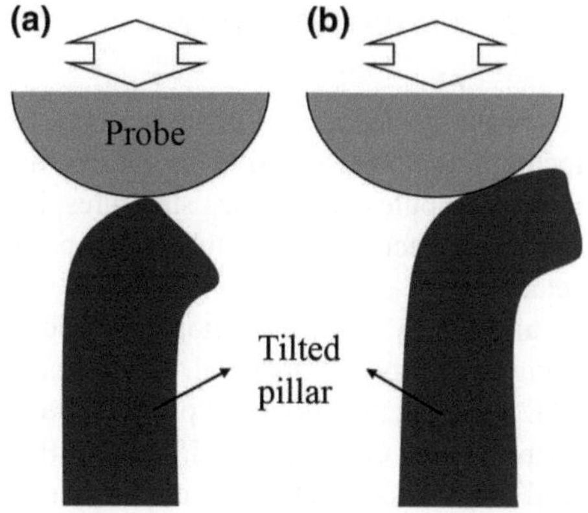

Fig. 8.22 Schematic illustration of tilted pillar into contact with a probe with **a** one contact point and **b** certain area. Please note that the pillars are not bent under the pressure but just tilted

example, vertically aligned polypropylene (PP, $E = 1.5$ GPa) micropillars with diameter of 600 nm, length of 18–20 μm were turned into slanted pillars with a tilting angle of 45° by applying an angled pressure through two heated rollers. Subjecting polymeric pillars to asymmetric treatments (e.g. electron- or ion-beam irradiation from certain direction) can arise shrinking or expanding of the polymer pillar on one side, resulting in the bending of pillars towards the designed direction [85–88].

It is obvious that the effort (force) for bending a pillar to a defined deflection is significantly smaller than that needed for axial compression. According to (8.8), E_{eff} of the slanted structure is much smaller than the vertical ones. A large tilting angle θ thus allows the expectation for a high adhesion. In actual experiments, however, both lower [84] and higher [87] adhesions were found on tilted pillars as compared to the vertically aligned pillars, which are related to the actual contact areas. For example, the tilted pillar with flat tip may offer a sharp edge on pillar top or the side wall of the pillar for the formation of contact (Fig. 8.22). While the former case (Fig. 8.22a) results in a reduced adhesion, the second case (Fig. 8.22b) could form a larger contact area and therefore a strong adhesion. It suggests once again the importance of tip geometries. Moreover, the tilted pillar also offers a great chance to gain directional adhesion and friction [85–87], similar to the spatular ends. The shear adhesion in the tilting/bending direction of the pillar is remarkably higher than that in the opposite direction. For example, shear adhesion of ~110 kPa was detected in the tilting direction of stooped PUA nanopillars and only ~22 kPa adhesion force was found in the opposite direction [87]. The highest ratio of adhesion forces in opposite directions was reported to be ~7.6 in the tilted Janus PUA nanopillars (90 nm in diameter and 900 nm in height) with one side coated with several-nanometer-thick platinum (Pt) by oblique metal deposition [86].

8.5.4 Porous Structure

It should be noted that many insects bearing setae use wet adhesion for their locomotion. The adhesive secretions are delivered to the contacting interface through tubules or porous structures inside the setae. Besides the function of delivering secretions, the tubular or porous structure also reduces the E_{eff} of the setae.

Inspired by this fact, nanopores inside pillar arrays have been developed by the using of a special material, diblock copolymer poly(styrene-*b*-2-vinylpyridine) (PS2VP). The two blocks of polystyrene (PS) and poly(2-vinylpyridine) (P2VP) are connected via chemical bond. Though PS and P2VP blocks are not compatible and tend to phase separate from each other forming two distinct phases, the chemical linking between them hinders the phase separation that domains in the range of nanometers form. This kind of demixing is therefore called microphase separation. The molecular weight, block ratio, Flory–Huggins segment–segment interaction parameter and the confinements control the microphase-separated structures [89]. PS2VP nanopillar arrays with diameter of 300 nm, length of 1.5 μm and lattice period of 500 nm supported by ~110 μm thick layer were replicated from AAO template. The resulted PS2VP nanopillar array has an E_{eff} of ~117 MPa. The annealing in 60 °C ethanol for 4 h and the subsequent removal of ethanol converted the P2VP domains into nanopores [90]. The generation of nanopores greatly reduced the E_{eff} to around 25 MPa, ~78% reduced. As P2VP is hydrophilic and can absorb water, the exposure of porous PS2VP nanopillar array to humid

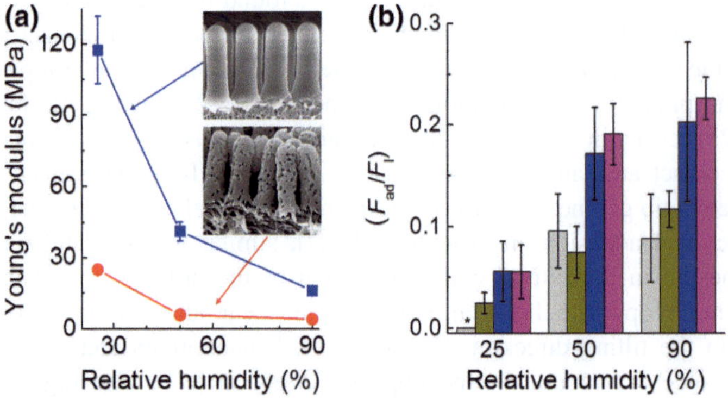

Fig. 8.23 **a** Effective elastic moduli of solid-hemispherical and porous-hemispherical nanorod arrays at different relative humidities. *Insets* show the corresponding structures. **b** Normalized adhesion (F_{ad}) to loading forces (F_l) (F_l ~ 300 μN) on porous smooth PS-*b*-P2VP films (*gray*), solid-hemispherical (*dark green*), porous-hemispherical (*blue*) and porous-footlike (*magenta*) adhesive pads at different relative humidities. Each data point in panel (**a**) or bar in panel (**b**) represents at least five measurements at different positions on two samples. The *error bars* indicate standard deviations [91]

environment with relative humidity of around 90% further reduced the E_{eff} to around 4 MPa. Comparing to the original solid nanopillar array, the E_{eff} at high humidity is more than 96% reduced, which accordingly enhanced the adhesion 3 times (Fig. 8.23) [91]. The switching between dry and wet environments is therefore able to reversibly turn on and off the adhesion. Mimicking the secretion function of insects, like beetles, liquid was delivered to the contacting interface through the nanopores inside PS2VP nanopillars changing the dry adhesion into wet adhesion [92]. The close look at the adhesion performances at different humidity revealed that the high humidity contributes to a high adhesion in two ways: reducing the E_{eff} and enhancing the formation of direct contacts.

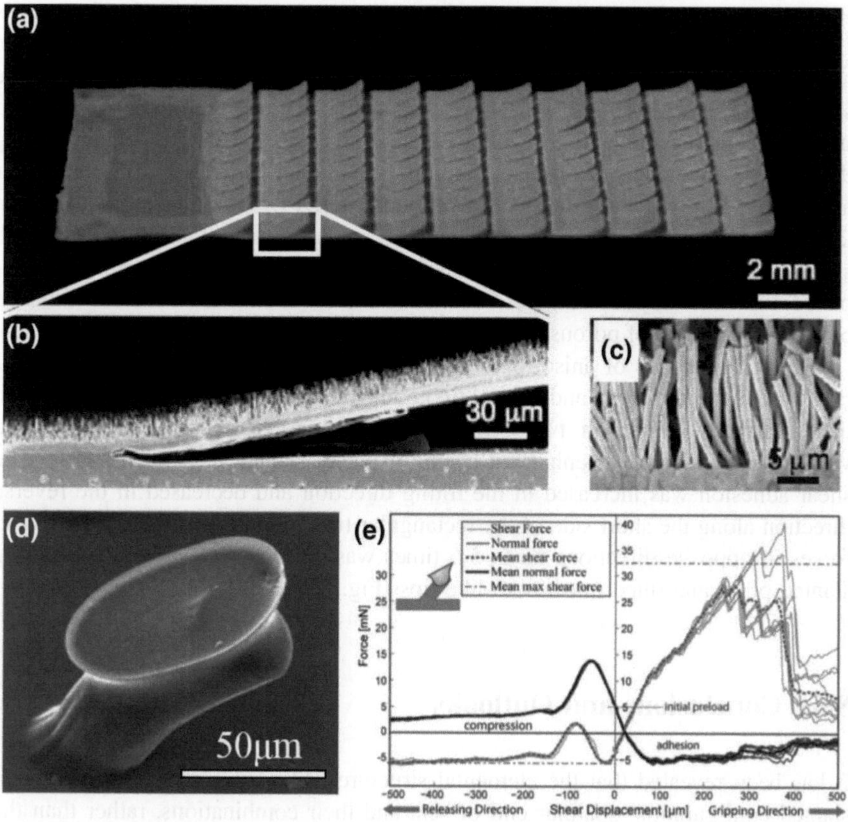

Fig. 8.24 **a–c** HDPE nanofiber arrays supported by thin lamellar flaps over 2 cm^2 flat backing layer. **a** An overview of the HDPE adhesive pad and enlarged images of **b** the single flap and **d** the nanofiber arrays[95]. **d** SEM images of angled polyurethane microfiber terminated with angled mushroom-like tip. **e** The shear forces on the pillars shown in **d** in the directions of gripping and releasing, and the normal adhesion forces after the shearing in both directions [74]

8.5.5 Combination of Several Structural Features

In nature, the aforementioned elemental features are not isolated from each other but contribute to the functionality of biological systems in a combinational way. Gecko is a typical example that uses pillar structures combining several features: the dimension across several levels (hierarchical pillars), special tip, slanted configuration and water-softened material.

The combination of different levels of pillar sizes along the long axis of pillars, in other words the hierarchical structure, sharply reduces E_{eff} of the structure. Greater reduction in E_{eff} can be expected by constructing more levels of hierarchy [13]. As the simplest hierarchical system, the 2-level pillar array made of PDMS or PUA, however, showed lower adhesion strength than the single-level controls [35, 93]. In contrast, enhanced adhesion on two-level PU pillars with each level terminated with flat mushroom-like tips was detected as compared to the single-level pillars [94]. The adhesion enhancement is considered to be the result of increased surface contact as well as increased extension of the pillars during detachment. Interestingly, the integration of lamellar structure with nanopillar arrays [95, 96], i.e. high-density polyethylene (HDPE) nanofiber arrays on thin lamellar flaps (Fig. 8.24a–c), showed greatly enhanced shear adhesion on rough surface as compared to the HDPE nanofiber array, though the inverse case happened on a smooth glass surface. Great adhesion enhancements were also realized by the combination of porous structure, spatular tip and material softening [91, 92].

The performances of anisotropic adhesion could be enhanced by the combination of slanted pillar stem and special tip geometries. For example, the anisotropy in the shearing adhesion in two orthogonal directions on rectangular tip-capped vertical-micropillars was enhanced by the using of slanted micropillars [97]. The shear adhesion was increased in the tilting direction and decreased in the reverse direction along the short side of the rectangular tip. Similarly, a difference in shear forces in opposite directions up to 5.6 times was detected in a system combining slanted pillar and tilted mushroom-like tips (Fig. 8.24d–e) [74].

8.6 Conclusions and Outlooks

It has been revealed that the elemental structures, like hairy structure, hierarchy, slanted configuration, spatular end of seta and their combinations, rather than the chemical component of seta control the adhesion abilities of geckos. Though the gecko seta is composed of a stiff material β-keratin, those elemental structures sharply reduce the effective elastic modulus of the gecko setae. Therefore, we propose to use effective elastic modulus as an important parameter to describe and to tune the adhesion abilities of structured adhesives. In this way, by constructing a very well-designed structure, one could turn a material into adhesive or non-adhesive state. And surely, the chemical properties of the material should also be carefully considered. Besides the inherent properties of the material, the environment in which the structured adhesive will be used, the ease handling and availability should also be considered in the material selection.

Various structures mimicking biological structured adhesives have been successfully designed and fabricated. Great adhesion forces, even surpass the biological models have been achieved in some materials, like in stiff materials of PS and CNT. On the other hand, however, the performances of bioinspired structured adhesives are still far behind the biological systems which have strong adhesion on unpredictable/dirty surfaces, easy detachment, self-cleaning ability and long durability. To fully mimic the adhesion abilities of these animals, there is still a long way to go.

Acknowledgements L. Xue thanks "Young 1000 Talents Plan", National Natural Science Foundation of China (51503156, 51611530546) for support.

References

1. G.A. Simons, in *Paper Presented at the 19th Symposium (International) on Combustion, Technion-Israel Intitute of Technology Haifa, Israel*, 8–13 August (1982)
2. J.E. Stone, A.M. Scallan, J. Polym, Sci. C **11**, 13 (1965)
3. X. OuYang, R.N. Jin, L.P. Yang, Y.G. Wang, L.Y. Yang, RSC Adv. **4**, 28699 (2014)
4. S. Michio, S. Kazuko, K. Toshihiro, Y. Yuya, N. Tatsuya, K. Takashi, N. Hiromichi, Science **325**, 1388 (2009)
5. P. Hunton, Revista Brasileira De Ciência Avícola **7**, 67 (2005)
6. G. Huber, S. Orso, R. Spolenak, U.G. Wegst, S. Enders, S.N. Gorb, E. Arzt, Int. J. Mater. Res. **99**, 1113 (2008)
7. K. Autumn, C. Majidi, R.E. Groff, A. Dittmore, R. Fearing, J. Exp. Biol. **209**, 3558 (2006)
8. K. Autumn, M. Sitti, Y.A. Liang, A.M. Peattie, W.R. Hansen, S. Sponberg, T.W. Kenny, R. Fearing, J.N. Israelachvili, R.J. Full, Proc. Nat. Acad. Sci. USA **99**, 12252 (2002)
9. H. Peisker, J. Michels, S.N. Gorb, Nat. Commun. **4**, 1661 (2013)
10. K. Autumn, A.M. Peattie, Integr. Comp. Biol. **42**, 1081 (2002)
11. R. Ruibal, V. Ernst, J. Morphol. **117**, 271 (1965)
12. B.N.J. Persson, S. Gorb, J. Chem. Phys. **119**, 11437 (2003)
13. M. Schargott, Bioinspir. Biomim. **4**, 026002 (2009)
14. L.J. Xue, S. Martin, S.N. Gorb, in *Biomaterials Surface Science*, 1st edn., ed. by A. Taubert, J.F. Mano, J.C. Rodríguez-Cabello (Wiley, New York, 2013)
15. Y. Tian, N. Pesika, H. Zeng, K. Rosenberg, B. Zhao, P. McGuiggan, K. Autumn, J. Israelachvili, Proc. Nat. Acad. Sci. USA **103**, 19320 (2006)
16. W.R. Hansen, K. Autumn, Proc. Nat. Acad. Sci. USA **102**, 385 (2005)
17. S.N. Gorb, A.E. Filippov, Beilstein J. Nanotechnol. **5**, 837 (2013)
18. D. Labonte, C.J. Clemente, A. Dittrich, C.Y. Kuo, A.J. Crosby, D.J. Irschick, W. Federle, Proc. Nat. Acad. Sci. USA **113**, 1297 (2016)
19. E. Arzt, S. Gorb, R. Spolenak, Proc. Nat. Acad. Sci. USA **100**, 10603 (2003)
20. M. Varenberg, N.M. Pugno, S.N. Gorb, Soft Matter **6**, 3269 (2010)
21. M. Varenberg, S.N. Gorb, Adv. Mater. **21**, 483 (2008)
22. J. Iturri, L.J. Xue, M. Kappl, L. García-Fernández, W.J. Barnes, H.J. Butt, A. del Campo, Adv. Funct. Mater. **25**, 1499 (2015)

23. D.M. Drotlef, E. Appel, H. Peisker, K. Dening, A. del Campo, S.N. Gorb, W.J. Barnes, Interface Focus **5**, 20140036 (2015)
24. P.H. Niewiarowski, L. Stephanie, L.H. Ge, E. Hagan, A. Dhinojwala, PLoS ONE **3**, e2192 (2008)
25. J.O. Wolff, S.N. Gorb, Proc. Boil. Sci. **279**, 139 (2012)
26. W. Sun, P. Neuzil, T.S. Kustandi, S. Oh, V.D. Samper, Biophys. J. **89**, 14 (2005)
27. J.B. Puthoff, M.S. Prowse, M. Wilkinson, K. Autumn, J. Exp. Biol. **213**, 3699 (2010)
28. M.S. Prowse, M. Wilkinson, J.B. Puthoff, G. Mayer, K. Autumn, Acta Biomater. **7**, 733 (2010)
29. K. Autumn, Y.A. Liang, S.T. Hsieh, W. Zesch, W.P. Chan, T.W. Kenny, R. Fearing, R. J. Full, Nature **405**, 681 (2000)
30. A.E. Kovalev, S.N. Gorb, Tribol. Lett. **48**, 103 (2012)
31. L. Heepe, A.E. Kovalev, A.E. Filippov, S.N. Gorb, Phys. Rev. Lett. **111**, 104301 (2013)
32. L. Heepe, A.E. Kovalev, S.N. Gorb, Beilstein J. Nanotechnol. **5**, 903 (2013)
33. M. Varenberg, S. Gorb, J.R. Soc, Interface **5**, 383 (2008)
34. L.J. Xue, A. Kovalev, F. Thöle, G.T. Rengarajan, M. Steinhart, S.N. Gorb, Langmuir **28**, 10781 (2012)
35. H.E. Jeong, J.K. Lee, H.N. Kim, S.H. Moon, K.Y. Suh, Proc. Nat. Acad. Sci. USA **106**, 5639 (2009)
36. A.K. Geim, S.V. Dubonos, I.V. Grigorieva, K.S. Novoselov, A.A. Zhukov, S.Y. Shapoval, Nat. Mater. **2**, 461 (2003)
37. C. Greiner, A. del Campo, E. Arzt, Langmuir **23**, 3495 (2007)
38. A. Parness, D. Soto, N. Esparza, N. Gravish, M. Wilkinson, K. Autumn, M. Cutkosky, J.R. Soc, Interface **6**, 1223 (2009)
39. L. Qu, L. Dai, M. Stone, Z. Xia, Z.L. Wang, Science **322**, 238 (2008)
40. L. Ge, S. Sethi, L. Ci, P.M. Ajayan, A. Dhinojwala, Proc. Nat. Acad. Sci. USA **104**, 10792 (2007)
41. E. Cheung, M. Sitti, Langmuir **25**, 6613 (2009)
42. J. Lee, C. Majidi, B. Schubert, R.S. Fearing, J.R. Soc, Interface **5**, 835 (2008)
43. D. Santos, S. Kim, M. Spenko, A. Parness, in *IEEE International conference on robotics and automation, Roma, Italy*, 10–14 April (2007)
44. T.S. Kustandi, V.D. Samper, D.K. Yi, W.S. Ng, P. Neuzil, W. Sun, Adv. Funct. Mater. **17**, 2211 (2007)
45. G. Lu, W. Hong, L. Tong, H. Bai, Y. Wei, G. Shi, ACS Nano **2**, 2342 (2008)
46. A. Mahdavi et al., Proc. Nat. Acad. Sci. USA **105**, 2307 (2008)
47. S. Kim, E. Cheung, M. Sitti, Langmuir **25**, 7196 (2009)
48. J. Lee, R.S. Fearing, K. Komvopoulos, Appl. Phys. Lett. **93**, 191910 (2008)
49. R. O'Rorke, T. Steele, H. Taylor, J. Adhes. Sci. Technol. **30**, 362 (2016)
50. D. Sameoto, C. Menon, Smart Mater. Struct. **19**, 103001 (2010)
51. H. Masuda, K. Yada, A. Osaka, Jpn. J. Appl. Phys. **37**, L1340 (1998)
52. J. Davies, S. Haq, T. Hawke, J. Sargent, Int. J. Adhes. Adhes. **29**, 380 (2009)
53. D. Sameoto, C. Menon, J. Micromech. Microeng. **19**, 115002 (2009)
54. C.T. Kingston, B. Simard, Anal. Lett. **36**, 3119 (2003)
55. L. Qu, L. Dai, Adv. Mater. **19**, 3844 (2007)
56. M. Kumar, Y. Ando, J. Nanosci. Nanotechnol. **10**, 3739 (2010)
57. A. del Campo, C. Greiner, J. Micromech. Microeng. **17**, R81 (2007)
58. A. Jagota, S.J. Bennison, Integr. Comp. Biol. **42**, 1140 (2002)
59. B. Chen, P. Goldberg Oppenheimer, T.A. Shean, C.T. Wirth, S. Hofmann, J. Robertson, J. Phys. Chem. C **116**, 20047 (2012)
60. Y. Zhao, T. Tong, L. Delzeit, A. Kashani, M. Meyyappan, A. Majumdar, J. Vac. Sci. Technol. B **24**, 331 (2006)
61. B. Yurdumakan, N.R. Raravikar, P.M. Ajayan, A. Dhinojwala, Chem. Commun. **30**, 3799 (2005)

62. L. Heepe, G. Carbone, E. Pierro, A.E. Kovalev, S.N. Gorb, Appl. Phys. Lett. **104**, 011906 (2014)
63. S. Gorb, M. Varenberg, A. Peressadko, J. Tuma, J.R. Soc, Interface **4**, 271 (2007)
64. A.G. Gillies, J. Puthoff, M.J. Cohen, K. Autumn, R.S. Fearing, ACS Appl. Mater. Interfaces **5**, 6081 (2013)
65. Y. Mengüç, M. Röhrig, U. Abusomwan, H. Hölscher, M. Sitti, J. R. Soc. Interface **11**, 20131205 (2014)
66. A. del Campo, C. Greiner, I. Álvarez, E. Arzt, Adv. Mater. **19**, 1973 (2007)
67. Z. Cheng, J. Gao, L. Jiang, Langmuir **26**, 8233 (2010)
68. D. Santos, M. Spenko, A. Parness, S. Kim, M. Cutkosky, J. Adhes. Sci. Technol. **21**, 1317 (2007)
69. S. Kim, M. Sitti, Appl. Phys. Lett. **89**, 261911 (2006)
70. G. Carbone, E. Pierro, S.N. Gorb, Soft Matter **7**, 5545 (2011)
71. A. Del Campo, C. Greiner, E. Arzt, Langmuir **23**, 10235 (2007)
72. Y. Wang, H. Hu, J. Shao, Y. Ding, ACS. Appl. Mater. Interfaces **6**, 2213 (2014)
73. M.P. Murphy, B. Aksak, M. Sitti, J. Adhes. Sci. Technol. **21**, 1281 (2007)
74. M.P. Murphy, B. Aksak, M. Sitti, Small **5**, 170 (2009)
75. H. Hu, H. Tian, X. Li, J. Shao, Y. Ding, H. Liu, N. An, ACS Appl. Mater. Interfaces **6**, 14167 (2014)
76. D.M. Drotlef, L. Stepien, M. Kappl, W.J.P. Barnes, H.J. Butt, A. del Campo, Adv. Funct. Mater. **23**, 1137 (2013)
77. G. Carbone, E. Pierro, Small **8**, 1449 (2012)
78. Z. Peng, S. Chen, Soft Matter **8**, 9864 (2012)
79. L. Heepe, M. Varenberg, I. Yan, S.N. Gorb, J. R. Soc. Interface **8**, 585 (2011)
80. B. Aksak, K. Sahin, M. Sitti, Beilstein J. Nanotechnol. **5**, 630 (2014)
81. S. Seo, J. Lee, K.S. Kim, K.H. Ko, J.H. Lee, J. Lee, ACS Appl. Mater. Interfaces **6**, 1345 (2014)
82. L. Xue, J. Iturri, M. Kappl, H.J. Butt, A. del Campo, Langmuir **30**, 11175 (2014)
83. J. Yu, S. Chary, S. Das, J. Tamelier, N.S. Pesika, K.L. Turner, J.N. Israelachvili, Adv. Funct. Mater. **21**, 3010 (2011)
84. B. Aksak, M.P. Murphy, M. Sitti, Langmuir **23**, 3322 (2007)
85. M.W. Moon, T.G. Cha, K.R. Lee, A. Vaziri, H.Y. Kim, Soft Matter **6**, 3924 (2010)
86. H. Yoon, H.E. Jeong, T.I. Kim, T.J. Kang, D. Tahk, K. Char, K.Y. Suh, Nano Today **4**, 385 (2009)
87. T.I. Kim, H.E. Jeong, K.Y. Suh, H.L. Hong, Adv. Mater. **21**, 2276 (2009)
88. T.I. Kim, C. Pang, K.Y. Suh, Langmuir **25**, 8879 (2009)
89. C. Park, J. Yoon, E.L. Thomas, Polymer **44**, 6725 (2003)
90. Y. Wang, L. Tong, M. Steinhart, ACS Nano **5**, 1928 (2011)
91. L. Xue, A. Kovalev, K. Dening, A. Eichler-Volf, H. Eickmeier, M. Haase, D. Enke, M. Steinhart, S.N. Gorb, Nano Lett. **13**, 5541 (2013)
92. L. Xue, A. Kovalev, A. Eichlervolf, M. Steinhart, S.N. Gorb, Nat. Commun. **6**, 6621 (2011)
93. C. Greiner, E. Arzt, A. del Campo, Adv. Mater. **21**, 479 (2009)
94. M.P. Murphy, S. Kim, M. Sitti, ACS Appl. Mater. Interfaces **1**, 849 (2009)
95. J. Lee, B. Bush, R. Maboudian, R.S. Fearing, Langmuir **25**, 12449 (2009)
96. M.T. Northen, C. Greiner, E. Arzt, K.L. Turner, Adv. Mater. **20**, 3905 (2008)
97. Y. Wang, X. Li, H. Tian, H. Hu, Y. Tian, J. Shao, Y. Ding, J. R. Soc. Interface **12**, 106 (2015)

Chapter 9
Biological Microstructures with Enhanced Adhesion and Friction: A Numerical Approach

Valentin L. Popov, Alexander E. Filippov and Stanislav N. Gorb

Abstract Numerical approach to study of the biological structures with enhanced adhesion and friction is presented. The presentation is mainly devoted to the study of natural and artificial structures of a plate with elastic fibers interacting with rough fractal surface, but some closely related systems are discussed also.

1. To optimal elasticity of adhesives mimicking gecko foot-hairs.
2. Flexible tissue with adhesive fibers interacting with surface.
3. Fibrillar adhesion with no clusterisation: Functional significance of material gradient along adhesive setae of insects.
4. Spatial model of the Gecko foot hair: Functional significance of highly-specialized non-uniform geometry.
5. Shear induced adhesion: contact mechanics of biological spatula-like attachment devices.

9.1 To Optimal Elasticity of Adhesives Mimicking Gecko Foot-Hairs

Recently a great attention has been paid to dry adhesion which takes place at nano-scales and is relevant for some biological objects [1–7]. Very important theoretical and experimental studies have been performed in order to gain a deeper

V.L. Popov (✉)
Institut für Mechanik, FG Systemdynamik und Reibungsphysik, Technische Universität Berlin, Sekr. C8-4, Raum M 122, Straße des 17. Juni 135, 10623 Berlin, Germany
e-mail: v.popov@tu-berlin.de

A.E. Filippov
Donetsk Institute for Physics and Engineering, National Academy of Science, Donetsk 83114, Ukraine
e-mail: filippov_ae@yahoo.com

S.N. Gorb
Department of Functional Morphology and Biomechanics, Zoological Institute of the University of Kiel, Am Botanischen Garten 1–9, 24098 Kiel, Germany
e-mail: sgorb@zoologie.uni-kiel.de

L. Heepe et al. (eds.), *Bio-inspired Structured Adhesives*,
Biologically-Inspired Systems 9, DOI 10.1007/978-3-319-59114-8_9

insight into these questions. It was found in particular, that a foot of the Gecko is covered by a layer of hair (fibers). Each of these fibers branches out into about 10^3 thinner ones. These smaller fibers end with a thin (5–10 nm) leaf-like plates. The latter structures are already small enough to be able to follow the surface roughness profile at almost molecular scale.

The existence of so powerful natural system has stimulated nano-technological efforts in this direction. Some success has been achieved to fabricate surface patterns with polymers to mimic the structure of setae and spatulas in gecko foot-hair [6–8]. In some cases these synthetic systems have already properties comparable to natural ones. Very recent novel artificial structures are based on relatively hard materials like carbon nano-tubes or micro-electromechanically produced 'organorods' [9, 10]. These new structures show very good adhesive properties.

Natural Gecko foot has very complex structure and its adhesion is combined effect of many impacts on different scales [4, 11–13]. This foot is a result of development driven by the principle of natural selection and very likely it is highly optimized. In its turn, being a result of the natural selection it certainly conserves incidental fingerprints of some "fossil foot" which gradually evolved into present-day structure. It is not necessary to copy it absolutely, but just use the main idea of fibers to make artificial molecular structures (like nano-tubes or mushroom-like structures) as technological and as efficient as possible. So, an original goal of the numerical study is to simulate artificial structure of elastic structures contacting with a rough fractal surface by Van der Waals forces to find an optimal relation between the system parameters.

Besides of the practical efficiency novel nano-structures have common theoretical advantage. It is easier to simulate them numerically than complex mesoscopic ones. But, even artificially made construction is normally complex many-body system moving in all 3D directions.

In is convenient to construct first very simple model in which one can restrict the motion by vertical z-direction (which is orthogonal to the mean positions of two contacting plates) only. An effective elasticity of the fibers K_{eff} in this case can be treated as an integral combination of bending of the hairs with their extension under Van der Waals force.

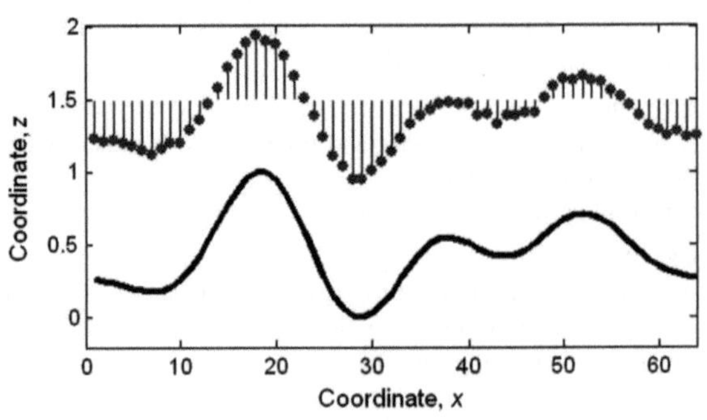

Fig. 9.1 Conceptual structure of the model. *Black line* presents a fragment of rough surface. The *lines* with *gray circles* schematically show elastic bonds

Conceptual structure of such a simplified model takes a form shown in Fig. 9.1. Black line in the plot presents a *small fragment* of fractal rough surface.

Despite of its simplicity such a model structure reminds many of the recently fabricated polymer surfaces. See for example system with carbon hairs [10]. These structures form dense nano-scale brushes of almost parallel bundles providing extremely good contact to surface on nano-scale. They have strong nanometer-level adhesion force/area relation $1.6 \pm 0.5 \times 10^{-2}$ nN nm^{-2} that is 200 times higher than those observed for gecko foot-hairs. Typical stress-displacement curves during a step loading–unloading cycle show a high hysteresis loop with adhesion region up to 20 nm.

From theoretical point of view these structures differ from the natural Gecko hair system. In natural system a finest hair does not reach the molecular scale and ends in a thin spatula. However, even oversimplified initial model can be restricted by simple pull-off contact force, but must include an adjustment of the fiber terminals (which substitute spatulas) to the rough surface. With account of this model was build in [14] as conceptually shown in Fig. 9.1.

It combines elastic interaction and chemical potential connecting each fiber with the surface. Standard attracting potential is Van der Waals interaction:

$$U_{\mathrm{VdW}} = U\left[2\left(\frac{z}{z_{\mathrm{VdW}}}\right)^{-6} - \left(\frac{z}{z_{\mathrm{VdW}}}\right)^{-12}\right]/12, \tag{9.1}$$

where U_{VdW} is a characteristic energy of the adhesion potential and z_{VdW} is a position of the potential minimum.

Each flexible fiber can be elastically extended with an elastic energy:

$$U_{elastic} = K_{eff}(z - Z)^2/2, \tag{9.2}$$

where Z is an equilibrium position of the end of fiber. As usually, corresponding forces in the equations of motion are determined by the following derivatives:

$$F_{\mathrm{VdW}} = \partial U_{\mathrm{VdW}}/\partial z, \quad F_{elastic} = \partial U_{elastic}/\partial z \tag{9.3}$$

For the numerical studies it is normally convenient to normalize all the energies, noise intensity and space scales to the units of Van der Waals potential F_{VdW} and ζ_{VdW}. Typical values for the values are as follows: $F_{\mathrm{VdW}} = 10$–70 nN; $\sigma = 0.1$–1.0 nm (in particular, for nano-tubes $\sigma = 0.34$ nm and $\varepsilon = 12$ meV). The elastic constant in the units $K = K_{eff}\,\sigma^2$ belongs to the interval $0.1 < K < 10$, stiffness $S = 0.1$ K.

In the dimensionless units one has:

$$U_{\mathrm{VdW}} = U\left[2r^{-6} - r^{-12}\right]/12 \quad \text{and} \quad U_{elastic} = K(z - Z)^2/2; \tag{9.4}$$

The randomness in the microscopic system can be caused by a number of sources: by the fractal surface $z = w(x)$, by complex dynamics of the fibers and by temperature fluctuations which are important at molecular scales.

In a numerical model [14] fractal surface can be generated as x-depending data array according to the standard definition:

$$w(x) = \frac{1}{2\pi} \int_{q_{\min}}^{q_{\max}} c(q) \cos(qx + \xi), \tag{9.5}$$

where $c(q) = c_0 q^{\beta}$ and ξ is random δ-correlated phase:

$$\langle \zeta(x) \zeta(x') \rangle = \delta(x - x') \tag{9.6}$$

Actual surface $w(x)$ never can be fractal starting from zero to infinite distances. It is so-called semi-fractal, having some limited spectrum of wave vectors $q_{\min} < q < q_{\max}$ in the definition (9.5). The amplitude of its maximums and minimums (in other words its standard deviation) is limited also:

$$\left\langle (w(x) - \langle w(x) \rangle)^2 \right\rangle^{1/2} \simeq A. \tag{9.7}$$

Here factor A describes a characteristic physical "roughness" of the surface.

At some regions of the parameters total potential combined by adhesion and elastic forces (9.4) has two valleys with comparable depth. At fluctuating parameters these comparable energy valleys provoke dynamic jumps of the fiber ends between two alternative (attached-detached) states. The time averaged dynamic exchange leads to an attraction of the tissue to the surface [14].

Total potential is shown in Fig. 9.2 in the form of families of the curves calculated at different values of the distance and elastic constant. It is seen directly that in both cases there are some regions of the parameters at which the potential has two valleys with comparable depth. Due to general reasons of physical kinetics one can expect that at fluctuating parameters two comparable energy valleys can cause jumps between alternative states of the system.

In zero approximation, one can neglect an interaction between the bonds. The equilibrium in this case is given by the balance of the elastic and Van der Waals forces $F_{VdW} = F_{elastic}$. For the fractal surface one has to solve this equation numerically. This procedure has been performed for different distances between the surfaces and different roughness A. Each value of A determines a family of the relations between the sticking force and elastic constant. However, there are many natural surfaces which have (at least) the same fractal structure at levels close to the molecular scale.

So, it seems possible to choose the structure and elasticity of artificial structure quite close to the optimal one. In normalized units it was found [14] that the optimal elasticity is close to unit when $A \approx 1$. It leads to the highest sticking force of the

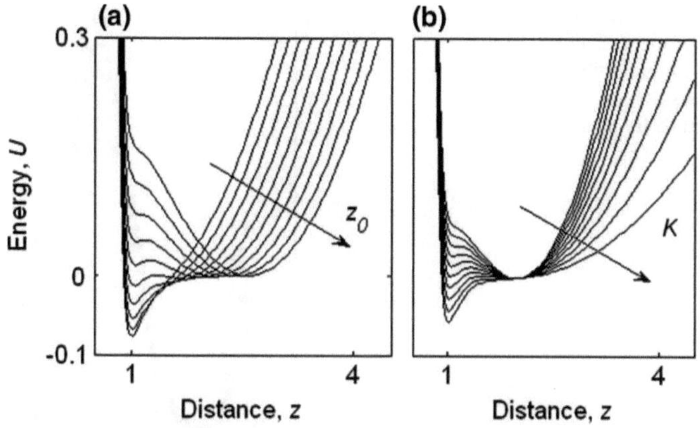

Fig. 9.2 Combined potential of the elastic and Van der Waals forces shown at varied distance *A* and elastic constant *K* (**a** and **b** respectively)

same order. In the units of problem it corresponds to a characteristic distance and energy of the Van der Waals forces. In other words, to create strong adhesion, it is good to have stiff fibers. Certainly, the ideal case $A \approx 1$ corresponds to extremely small roughness, which is comparable to the atomic scales and may not exist in the reality. Very likely, artificial system has to combine both features: soft basic tissues on relatively high scales (to adapt preliminary large-scale part of the surface structure) and hard short fibers to fit very last nano-scales.

To complete the study, the system was simulated in dynamical approach. The equations of motion can be written as follows:

$$\frac{\partial^2 z_k}{\partial t^2} = -\gamma \frac{\partial z_k}{\partial t} + F_{elastic,k} + F_{\mathrm{VdW},k} + \zeta(z_k; t); \tag{9.8}$$

The random source of fluctuations and dissipation $\gamma \frac{\partial z}{\partial t}$ are also included to simulate both thermal and dynamic impacts to the system with combined effective temperature T_{eff}:

$$\langle \zeta(z_k; t')\zeta(z_k; t) \rangle = D\delta(z_{k'} - z_k)\delta(t' - t); \quad D = 2k_B \gamma T_{eff} \tag{9.9}$$

When the parameters are close to the optimum total potential has two close minimums. Dynamic chaos and randomness cause chaotic exchange of the bond states between the minimums. This effective "exchange interaction" leads to a better adjusting of the flexible "surface" formed by the ends of the fibers to the contact surface and results in stronger attraction between two surfaces. This effect is also known in friction physics: sliding or vibrating of the plate (which is transferred down to molecular scales as a random noise, or a form of "liquid layer") may increase the contact area and adhesion force. Time averaged probability $P(z)$ to find

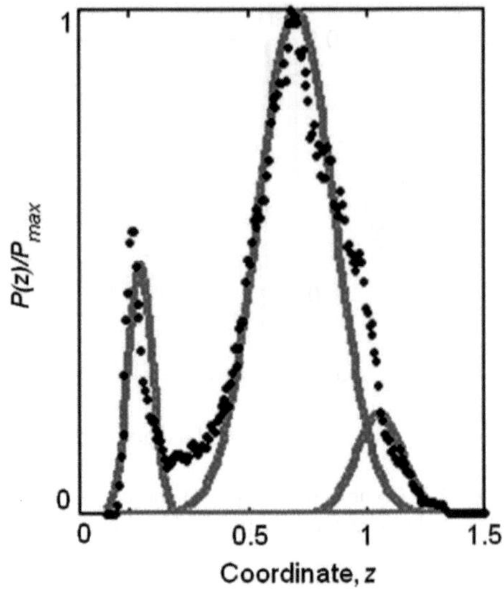

Fig. 9.3 Time-averaged histogram for the probability distribution to find a fiber end point in a position z normalized to its maximum value $P(z)/P_{\max}(z)$

a fiber end point in a position z at optimal relation between the interactions is shown in Fig. 9.3.

The numerical experiment has been repeated at different roughness of the surface (resulting in particular in different mean distances between the plates, and for three different elastic constants: $K < K_{opt}$, $K \approx K_{opt}$ and $K > K_{opt}$.

The results of the simulations are summarized in Fig. 9.4. Each vertical cross-section in the gray scale maps corresponds to a particular histogram $P(z)/P_{\max}(z)$. Dark gray depicts higher probability $P(z)$. The subplots (a), (b) and (c) correspond to $K < K_{opt}, K \approx K_{opt}$ and $K > K_{opt}$ respectively. It is seen from the plots, that if the elastic constant is too large, all the bonds are mainly attracted to the upper plate. They can not fit properly the rough surface and total force is too small. In opposite limit the ends of the fibers fit the surface "ideally". But, in this case the "springs" are too weak to return them well to the upper surface. Very long extension of the bonds and big hysteresis are necessary to detach all the pad from the adhesive surface.

Presented results favour to a conclusion that some optimal elastic properties of the fibers do exist. It is found that the fibers must be stiff enough to provide a compromise between good adhesion and attachment-detachment properties of the system. This compromise is possible because even already made artificial systems are "overbuild". A small part of the fibers in good contact (sometimes less then 1%) produces strong total force. It is shown that stiffness of the fibers give some advantage in dynamics. Such hairs fluctuate mainly in intermediate position, randomly jumping from attached to detached states. As result, relatively rough hair system ignores short-range peculiarities of the surface, but provides sufficient force

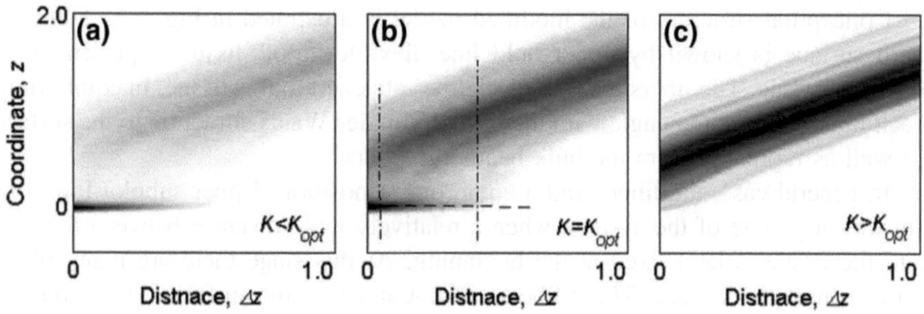

Fig. 9.4 Gray scale maps for the histograms of z-coordinate distribution shown for different deviations of the elastic plate from the rough surface. *Dark gray color* corresponds to high probability $P(z)$. Subplots **a**, **b** and **c** correspond to $K < K_{opt}, K \approx K_{opt}$ and $K > K_{opt}$ respectively

and smooth attachment-detachment process for all the system. In the optimum the structure made with such fibers is expected to adhere to wide variety of the shapes of fractal surfaces with enough adhesion forces.

9.2 Flexible Tissue with Fibers Interacting with Adhesive Surface

In previous part of this chapter we simulated numerically artificial structure of elastic fibers contacting with a rough fractal surface by Van der Waals forces. The main goal was to find an optimal relation between the system parameters. The force balance equations were solved in 1 + 1 dimensional system for different values of elastic constant and variable surface roughness and optimal elasticity has been calculated to provide maximum cohesion force between the plate and the surface. In agreement with experimental observations it was found, that high flexibility of the fibers is not always good for efficiency of the system and artificial adhesives must be made from stiff enough polymers.

Despite of its efficiency for the numerical simulations the model with the only motions in z-direction allowed needs in a further development. Besides theoretical understanding of this there are recent experimental observations which favor to a strong importance of the orthogonal motions in the system under consideration. In particular, sliding or vibrating in orthogonal direction of the plate which is transferred down to molecular scales may increase the contact area and change a relation between the adhesion and friction impacts to the interaction between the surfaces depending on the contact roughness or angle of the force applied [11, 13, 15].

Below the model is extended to incorporate additional degree of freedom to allow bending and rotation of the fibers. We study how an intensity of the interaction as well as attachment and detachment scenarios depend on different starting distances, surface roughness and pulling angles.

Conceptual structure of the modified model is illustrated in Fig. 9.5. Adhesive hard surface is shown by upper bold line, flexible "foot" tissue is presented by down grey line. The fibers are thin black lines attached to this tissue. In equilibrium each fiber forms some angle with the tissue. Van der Waals attraction to the surface as well as repulsion from the bulk bends the fibers.

In general case, the fibers find a compromise position. Upper subplot Fig. 9.1a shows early state of the system, when a relatively large distance between the foot and the tissue is kept fixed at the beginning. At this stage there are many fibers remote from the surface. These fibers can not attach to the surface at the moment. Later stage of the attraction is shown in second subplot Fig. 9.1b, where relatively short part of the system is shown to make visible the fibers, which are bended and/or extended and attached to different points of the upper surface.

The model equations of modified version are basically close to the equations used above. However, 1D potential that connects each fiber with the surface must be changed by 2D Van der Waals potential:

$$U_{\mathrm{VdW}} = U\left[2\left(\frac{r}{r_{\mathrm{VdW}}}\right)^{-6} - \left(\frac{r}{r_{\mathrm{VdW}}}\right)^{12}\right]/12. \tag{9.10}$$

Here U is as before characteristic energy of the adhesion potential, but $r = \{x, z\}$ is a 2-dimensional vector and r_{VdW} is a position of the potential minimum in 2D space.

More complex changes appear for the elastic part of the interaction. The flexible fibers should be constructed now from relatively large number of segments $j = 1, 2, \ldots n_{\max}$ with the coordinates $\{r_i, r_j\}$. To reduce the calculations one can suppose that only the final segment of every fiber is attached to the surface via Van der Waals potential. It means that above term $U_{\mathrm{VdW}} = U_{\mathrm{VdW}}^{n_{\max}}$ in energy valids for the

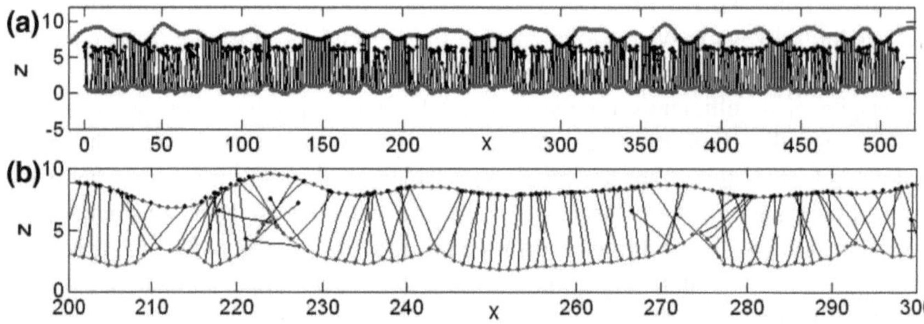

Fig. 9.5 Conceptual structure of the modified model. *Bold grey upper* and *down lines* present the fragments of rough surface flexible tissue respectively. The *lines* with *black circles* show instant positions of the elastic fibers. The subplot **a** presents an early stage of the contact with many disconnected bonds, a closer view of the system at later stage, when majority of the fibers have distorted to attach the surface is shown in the subplot (**b**)

$j = n_{max}$ only. In zero approximation, one can neglect also an interaction between the bonds.

Each flexible segment can be elastically extended with an elastic energy:

$$U^{ij}_{elastic} = K_{eff}(r_i - r_j)^2/2. \tag{9.11}$$

New term of interaction here is a bending energy. Mathematically it is constructed to prevent too strong mutual deviations between the directions in every pair of the nearest segments (which is associated to the bending of the fibers):

$$U^{j}_{bending} = K_{bending}\left[1 - (r_j - r_{j+1}) \cdot (r_{j-1} - r_j)/\left|(r_j - r_{j+1})\right| \cdot \left|(r_{j-1} - r_j)\right|\right]. \tag{9.12}$$

Corresponding forces in the equations are equal to the derivatives in two direction. Now complete set of the forces is: $F^z_{\mathrm{VdW}} = \partial U_{\mathrm{VdW}}/\partial z$; $F^x_{\mathrm{VdW}} = \partial U_{\mathrm{VdW}}/\partial x$, $F^{z,ij}_{elastic} = \partial U^{ij}_{elastic}/\partial z$; $F^{x,ij}_{elastic} = \partial U^{ij}_{elastic}/\partial x$ and $F^{z,j}_{bending} = \partial U^{j}_{bending}/\partial z$; $F^{x,j}_{bending} = \partial U^{j}_{bending}/\partial x$ respectively.

As it was already done in the previous section, it is convenient to normalize all the energies, noise intensity and space scales to the units of Van der Waals potential F_{VdW} and $\zeta_{\mathrm{VdW.}} < 10$, stiffness $S = 0.1$ K.

In dimensionless units one finally has:

$$\begin{aligned} U_{\mathrm{VdW}} &= U\left[2r^{-6} - r^{-12}\right]/12; \quad U^{ij}_{elastic} = K\left(r_i - r_j\right)^2/2; \\ U^{j}_{bending} &= S\left[1 - \left[r_j - r_{j+1}\right] \cdot \left(r_{j-1} - r_j\right)/\left|\left(r_j - r_{j+1}\right)\right| \cdot \left|\left(r_{j-1} - r_j\right)\right|\right]. \end{aligned} \tag{9.13}$$

Fractal surface, is generated according to the same definition as in (9.5)–(9.7) $z = w(x) = \frac{1}{2\pi}\int_{q_{\min}}^{q_{\max}} c(q)\cos(qx + \xi)$ with the same limited spectrum of wave vectors $q_{\min} < q < q_{\max}$.

At some regions of the parameters total potential has two valleys with comparable depth. At fluctuating parameters these comparable energy valleys provoke dynamic jumps of the ending fiber segments between two alternative (attached-detached) states. The time averaged dynamic exchange leads to an attraction of the tissue to the surface. Below in numerical experiments we measure all distances in units of fixed interval between two nearest fibers $\Delta x_k = const \equiv 1\,\mathrm{nm}$ and vary the roughness amplitude up to $A_{\max} = 10\Delta x_k$. In the same units the elastic constant is kept fixed $K = 1$ and the stiffness is changed up to $S_{\max} = 10\,\mathrm{K}$.

Last addition which completes the equation of motion relates to the transformations of the flexible tissue. We suppose that last segment of each fiber is rigidly attached to the tissue. It means that the coordinates of the tissue segments coincide with the positions of the first elements of the fibers: $\{x_k^{j=1}, z_k^{j=1}\}$. It is supposed also that model tissue has its own elasticity K^z_1 in z-direction $U^1_{elastic} =$

$K_1^z\left[(z_k^1 - z_{k+1}^1)^2 + (z_k^1 - z_{k-1}^1)^2\right]/2$ while a distance between the segments in x-direction (defining an array of the fibers) is kept fixed $\Delta x_k = const \equiv 1$.

The tissue is supported also in z-direction by an external force $F^z_{external}$. As result, one has additional interactions for the coordinates $\{z_k^{j=1}\}$:

$$F^{z,1}_{elastic} = \partial U^1_{elastic}/\partial z; \quad \text{and}\, F_k^{z,1} = F^z_{external} \tag{9.14}$$

The complete set of the dynamic equations can be written now as follows:

$$\begin{aligned}\frac{\partial^2 z_k^j}{\partial t^2} &= -\gamma\frac{\partial z_k^j}{\partial t} + F^{z,ji}_{elastic,k} + F^{z,j}_{bending,k} + F^{z,j=n_{\max}}_{\mathrm{VdW},k} + \delta_{j1}(F^{z,1}_{elastic,k} + F^z_{external}) + \zeta(x_k^j, z_k^j; t);\\ \frac{\partial^2 x_k^j}{\partial t^2} &= -\gamma\frac{\partial x_k^j}{\partial t} + F^{x,ji}_{elastic,k} + F^{x,j}_{bending,k} + F^{x,j=n_{\max}}_{\mathrm{VdW},k} + \zeta(x_k^j, z_k^j; t);\end{aligned} \tag{9.15}$$

Here δ_{j1} is a Kronecker symbol. The random source $\xi(z_j; t)$ and dissipation $\gamma\ \partial z_j/\partial t$ are also included to simulate both thermal and dynamic impacts to the system with an effective temperature T_{eff}:

$$\left\langle \zeta(x_k^j, z_k^j; t)\zeta(x_{k'}^{j'}, z_{k'}^{j'}; t')\right\rangle = D\delta(x_{k'}^{j'} - x_k^j)\delta(z_{k'}^{j'} - z_k^j)\delta(t' - t) \tag{9.16}$$

$$D = 2k_B\gamma T_{eff}$$

A set of numerical experiments has been performed with the above system. Typical procedure is organized as follows. One generates a fragment of random rough surface (each time new to accumulate a statistics) and place a flexible tissue at some initial distance z_0.

Initially the tissue is flat with the fibers arranged as a regular array of lines and oriented orthogonally to the tissue in direction of the surface. After this, the system is allowed to move naturally according to (9.15)–(9.16) and to attract to a its dynamic equilibrium. Time run to get the equilibrium is determined by a characteristic relaxation rate $t_{\max} \propto \gamma^{-1}$. The coefficient here should be 'a-posteriori' estimated from a time behavior in real simulations, while systems goes to stationary asymptotics. Normally the run for this model takes approximately: $t_{\max} \cong 200\gamma^{-1}$.

Typical observation shows that at intermediate stage of the process terminal ends of the revolving flexible fibers intensively move along surface in x-direction. This ability to move along x-axis allows them much better adapt to a fractal surface than it was possible for a hard "bristle" (with the fibers extendable in z-direction only) [15].

As a result, transversal motion leads to a strongly nonuniform redistribution of the contact density along the surface. To some extend, one can say that fractal **redistribution of the fibers along transversal direction compensates fractal surface structure in z-direction.**

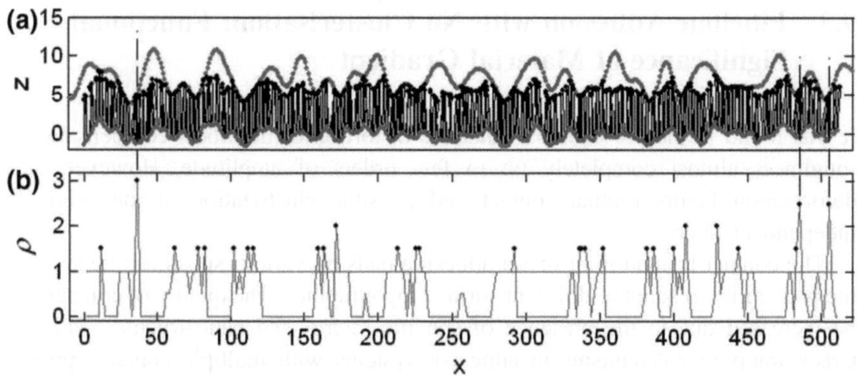

Fig. 9.6 Comparison between the instant places of good fiber attachment (**a**) and nonuniform density distribution (**b**) of the attachment positions along x-axis (normalized to the mean density). The places with the highest densities are marked by the *dash-dot lines*

Figure 9.6 illustrates a typical redistribution of the fibers at intermediate time moment of their adaptation to the surface. The instant places of good fiber attachment are directly seen in subplot (a). They are compared to a nonuniform density distribution of the attachment positions along x-axis in subplot (b). The density is normalized to a mean fiber density over the system equal to $\rho_0 = N_{fibers}/L_{tissue}$. The places with the highest densities are marked by the dash-dot lines for a better comparison.

The mechanical properties of the system depend on the parameters: surface roughness, fiber stiffness and elasticity. In [16] the numerical experiments have been performed at wide interval of the parameters. For each parameter the system starts from a trial state with completely disconnected bonds and tends to a dynamic equilibrium. It was found that when the surface roughness A increases it provokes stronger distortion of the fibers. **This distortion compensates the roughness almost completely up to few orders of amplitude.**

The results favour to a conclusion that transversal motion leads to a strong redistribution of the fibers along the surface. It compensates fractal surface structure in z-direction. As result, the tissue is much better attracted by the fibers to the surface and matches it finally with extremely large contact area for a wide variation of the surface roughness A. Additionally it was found in [16] that transversal distortions of the fibers strongly affect scenario of their detachment from the surface when the tissue is pulled off by an external force. The detachment process in every local point depends on a random orientation of the fibers caused by the surface randomness and almost do not depend on a direction of the external force F_{ext} in a wide interval of the pulling angles.

9.3 Fibrillar Adhesion with No Clusterisation: Functional Significance of Material Gradient

It was found in the previous part that the distortion of the fibers compensates the roughness almost completely up to few orders of amplitude. However, such deformation favors mutual contact and possible clustrization of the originally independent fibers.

The contact formation of insect adhesive pads on various substrates depends on the pad ability to adapt to different surface topographies. The quality of contact may be increased due to the presence of specific micro- and nanostructures [17–21]. Crack trapping mechanisms in adhesive systems with multiple contacts provide advantages in attachment on rough substrates [22]. Also hierarchical organization of insect pad structures enables formation of multiple contacts that contribute to an enhancement of overall length of the total peeling line [23].

We have recently shown that thin tape-like contact tips of hairs (setae) in combination with applied shear force lead to the formation of maximal real contact area without slippage within the contact [24]. Due to this reason, the material flexibility is important for contact formation of adhesive pads. Flexible materials may generate large contact area between the pad and substrate at minimal normal load. On the other hand, elongated structures, made of too flexible materials, have low mechanical stability [25]: insect setae made of too soft material can buckle and collapse resulting in so called clusterisation/condensation [26, 27]. Due to such clusterisation, functional advantage from multiple adhesive contacts may strongly decrease. That is why, material properties of insect adhesive setae represent an optimization problem, which is solved in the course of biological evolution by the presence of gradients of thickness and mechanical properties. Thickness gradients of insect setae are well known in various adhesive setae due to numerous scanning electron microscopy studies [17]. Recently, we presented the combined study on the material structure and local mechanical properties in tarsal setae of the beetle *Coccinella septempunctata* and demonstrated the presence of a material gradient at the level of each single seta [28].

Setal elasticity modulus, probed by AFM, ranges from 1.2 MPa at the tip [28] to 6.8 GPa at the base. At the setal tip, we revealed the rubberlike protein resilin in rather high concentrations [29, 30] whereas at the base of the seta the sclerotised cuticle is dominating. Previous CLSM analysis of the setal tips has clearly demonstrated the presence of the rubberlike protein resilin in rather high concentrations [28, 31]. Both central and proximal parts of the setae were dominated by green, yellow and red autofluorescences due to the presence of other presumably sclerotised proteins and very likely chitin (Fig. 9.7). Between the resilin-dominated distal part and more sclerotised basal part of the seta, a rather pronounced longitudinal gradient of material composition was revealed. AFM-nanoindentation experiments have revealed rather low elasticity modulus at the setal tip (1.2 ± 0.3 MPa), but the high one at the setal base (2.43 ± 1.9 GPa) [28]. This information about the gradients of material properties in real beetle setae was used in the numerical model presented below.

Between tip and the base, there is a gradient of material composition revealed by CLSM. This gradient is hypothesized to be an evolutionary optimization enhancing adaptation of adhesive pads to rough surfaces, while simultaneously preventing setal clusterisation. Such an optimisation presumably increases the performance of the adhesive system in general. However, this hypothesis is difficult to prove experimentally using native biological specimens. In this paragraph we ask following questions: (1) Does the presence of the material gradient along the setae contribute to the proper contact formation, if compared to setae with different bulk material properties without such gradient? (2) Does the gradient reduce clusterisation of setae?

To model gradient material properties of insect setae, minimalistic, but realistic model, has been constructed. For an efficiency it only partially involves the features of two previous models and includes following elements.

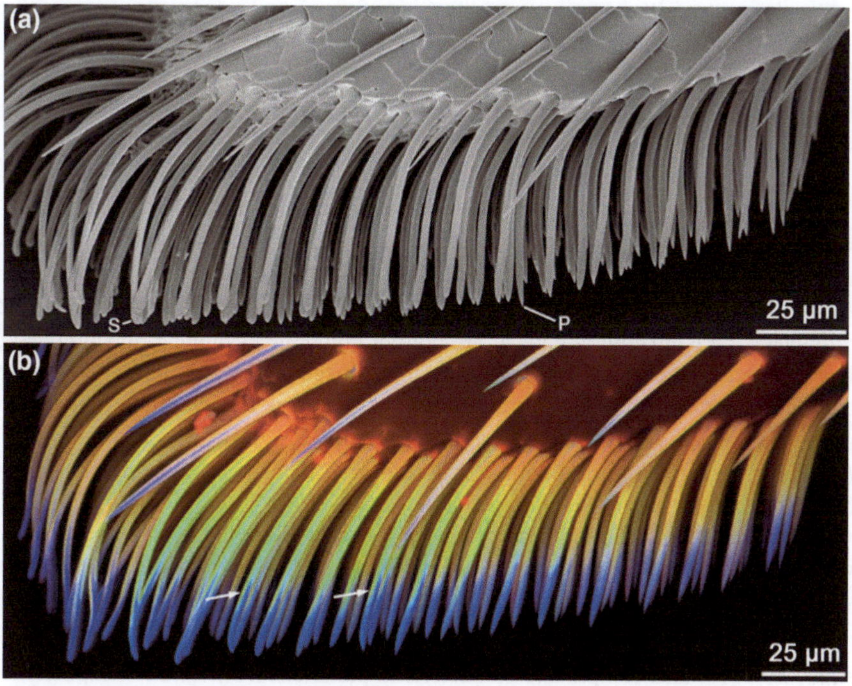

Fig. 9.7 Morphology and material composition of adhesive tarsal setae. Ventral part of the second adhesive pad of a foreleg of a female *Coccinella septempunctata*, lateral view. **a** Scanning electron micrograph (the specimen was dried using 1-propanol). **b** CLSM maximum intensity projection showing an overlay of the four different autofluorescences mentioned in the text. The *arrows* indicate the dorso-ventral material gradient in exemplary setae. *S* Exemplary spatula-like seta; *P* Exemplary seta with a pointed tip. *Scale bars* 25 mm. From [28] (Nature Publishing Group)

An array of initially parallel fibers attached to a hard planar base. Elasticity of the fibers $F_{elastic}$ is continuously varied along their length and can be changed from very soft one to much stiffer or even almost rigid one (but still with some degree of flexibility). Longitudinal $\vec{F}_{jk}^{\parallel}$ and transversal $\vec{F}_j^{\perp}$ stiffness of the fibers are simulated by the following interaction between the segments $\vec{F}_{jk}^{\parallel} = K^{\parallel}(\vec{r}_j - \vec{r}_k)\left[1 - (\vec{r}_j - \vec{r}_k)^2/\mathrm{d}r^2\right]$, and $\vec{F}_j^{\perp} = K^{\perp}(2\vec{r}_j - \vec{r}_{j+1} - \vec{r}_{j-1})$.

Here we limit ourselves by two-dimensional model, where values $\vec{r}_j = \{x_j, y_j\}$ are the coordinates of the beginning of the segment j; $k = j \pm 1$. Longitudinal force, $\vec{F}_{jk}^{\parallel}$, is described by a two-minima potential, which tends to keep a distance between the points $\vec{r}_j$ and $\vec{r}_{j\pm1}$ close to the equilibrium length of the segment dr. Transversal force, $\vec{F}_j^{\perp}$ keeps $\vec{r}_j$ close to the mean value between its nearest neighbors, $(\vec{r}_{j+1} + \vec{r}_{j-1})/2$ and tends to hold the angle between the neighboring segments close to 180°.

Counting on discussed above applications to the animals using forces of different nature, adhesive surface at the ends of the fibers attracts the ends by a sum of molecular and capillary force.

For the sake of numerical simplicity we simulate it by the gradient of Morse potential $U_{\mathrm{VdW}}(r) = U_0(1 - \exp(-r/r_0))^2$ with physically reasonable amplitude $U_0 = 10\,\mathrm{nN\,nm}$ and the minimum located at the distance $r_0 = 0.01\,\mu\mathrm{m}$ from the surface.

As in both previous parts of this Chapter, rigid surface of the substrate, where the fibers attach to, has semi-fractal structure with given Fourier spectrum and amplitude of roughness [24]. It can be simulated similar to the approach we previously used by the self-affine fractal surface given by real part of $Y(x) = A\iint \mathrm{d}q_x C(q)\exp(iq_x x + \zeta)$ with scaling spectral density amplitude of surface roughness A, i is imaginary unit, q_x are Fourier components along x direction and ζ is a random phase. Details of the generation procedure for the profile $Y(x)$ have been described above. In the current literature [11] it is accepted that majority of physical surfaces have scale-invariant spectrum $C(q) = 1/q^{\beta}$ with exponent $\beta \approx 0.9$. The amplitude of the numerical "surface" is taken to be comparable with the radius of Van der Waals interaction $A = r_0$.

Soft parts of every fiber, which normally are physically thin and flexible, interact with corresponding regions of other fibers of the array. Interaction force has the same origin as their attraction to the hard wall. Here for a simplicity and for the numerical efficiency it is modeled by the Morse potential $U_{\mathrm{interact}}(r) = U_o(1 - \exp(-r_{jk}/r_0))^2$ and comparable characteristic parameters U_0, r_0. For simplicity of the model, we reduce mutual interaction of the fibers by the interaction of the nearest neighbors: $r_{jk} = |r_j - r_{j\pm1}|$.

As quite often for the problems on such a scale, one can neglect effects of inertia and treat the system as over-damped. In this approximation equation of motion does not contain second time derivative and can be formally written in the form

$\partial r/\partial t = F$, where force accumulates all above interactions $F = F_{elastic} + F_{VdW} + F_{interact}$.

Conceptual structure of the model is illustrated in Fig. 9.8. Stiff surface of the substrate is shown by upper solid curves. In order to understand the potential functional role of the material gradients found in beetles [28], we study three different kinds of fiber arrays: (a) long stiff fibers with short elastic ends; (b) long elastic fibers connected to the basal plate by the short hard roots; (c) relatively stiff fibers with short soft elastic filaments connected to the base. All these variants are shown in subplots (a), (b) and (c) respectively.

In all the cases the stiffness of the fibers is continuously varied along vertical coordinate. To simulate it we apply smooth step function $\Theta(y) = 1/[1 + \exp(-(y - y_0)/\Delta)]$ with regulated position of bend y_0 and width Δ. This function tends to unit, when $y \ll y_0$, and gradually goes to zero in the opposite limit. This allows to model all above mentioned cases in common approach.

To illustrate different stiffness of fiber segments shown in Fig. 9.8, we formally divided the stiffness into three regions: (1) close to the maximal stiffness, (2) less than half of the maximal stiffness (a region around y_0 with the width Δ), and (3) less than 0.1 of the maximal stiffness. These parts are conditionally shown in the plots

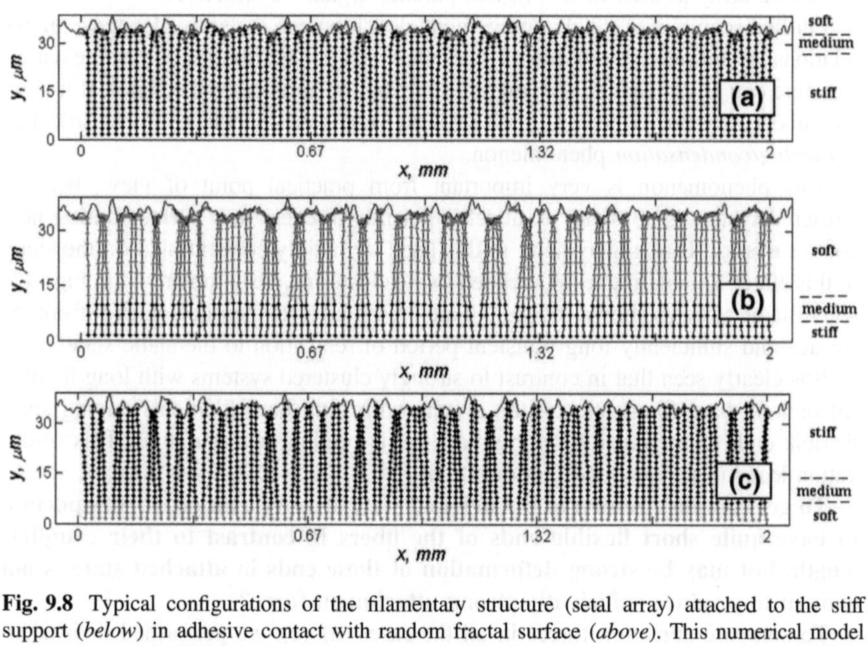

Fig. 9.8 Typical configurations of the filamentary structure (setal array) attached to the stiff support (*below*) in adhesive contact with random fractal surface (*above*). This numerical model was used to mimic biological setal arrays shown in Fig. 9.7. Three types of fibers, (1) stiff fibers with short elastic ends, (2) long elastic fibers connected to the base by short stiff roots, and (3) stiff fibers with soft elastic segments near the base are shown in subplots **a**, **b** and **c**, respectively. Different stiffness of segments is conditionally shown by *circles* with different intensity of *gray colour*. Stiff, medium and soft segments are marked by *black*, *dark gray* and *light gray circles*, respectively

by different intensity of gray. Stiff, medium and soft segments are marked by black, gray and light gray circles respectively.

The numerical procedure is organized as follows. Originally unperturbed arrays of parallel fibers is attached to the horizontal hard base. Computer routine brings them into contact with numerically generated rigid fractal surface. The fibers distort due to interaction between them and surface as well as due to their mutual interaction with the neighbors. Due to nonuniform configuration of the surface many of the fibers attract to the same individual asperities of the surface. This attraction enhances their mutual interaction in contrast to original unperturbed state.

One can record time-depending distortions of the fibers as well as variation of the interaction forces, to control the process of contact formation and stop it, when the system reaches certain stationary configuration. After this, one can "remove" rigid substrate surface and allow the system to relax spontaneously to some new stationary state. This procedure answers on the question how real system restores its original configuration being detached from the surface.

Many of fibers, which were preliminary attracted to the same asperities of the surface, still strongly interact and remain physically close one to another. The collect into local *clusters*. Mutual attraction between the fibers competes with the elastic forces inside the fibers which try to return them to the straight position and the whole array as well to its original parallel-organized structure.

Further scenario of setal arrangement development certainly depends on the relationship between these forces and their spatial distribution. In some cases, structure can return back to the original state, but in some cases it can not. If it is so, the fibers remain collected into strongly confined bunches. It is so-called *clustering/condensation* phenomenon.

This phenomenon is very important from practical point of view, because clustered system is not ready to attach efficiently to every new surface during next contact events. That is why here, in this part, we mainly concentrated on the study of this effect. Qualitative results related to the clustering are summarized in Fig. 9.9, where same systems shown in Fig. 9.8 are presented after their detachment from the surface and sufficiently long transient period of relaxation to the static state.

It is clearly seen that in contrast to strongly clustered systems with long flexible (b) or long hard filaments (c), the system with long hard filaments having short flexible ends (a) practically returns back to its original configuration. This observation leads to a very important question:

To get complete return to the original state after relaxation, it is important to have quite short flexible ends of the fibers in contrast to their complete length, but may be strong deformation of these ends in attached state is not enough to produce sufficiently strong attachment force?

To compare attraction forces in all the cases (a)–(c), we performed their accumulation over complete system during entire time interval of the attachment (Fig. 9.10). Maximal forces in the cases (a) and (b) are comparable. Moreover, the potential barrier (the difference between maximum of force at the beginning and its minimum, which system gets after good adaptation to the rigid surface) is even higher in the case (a).

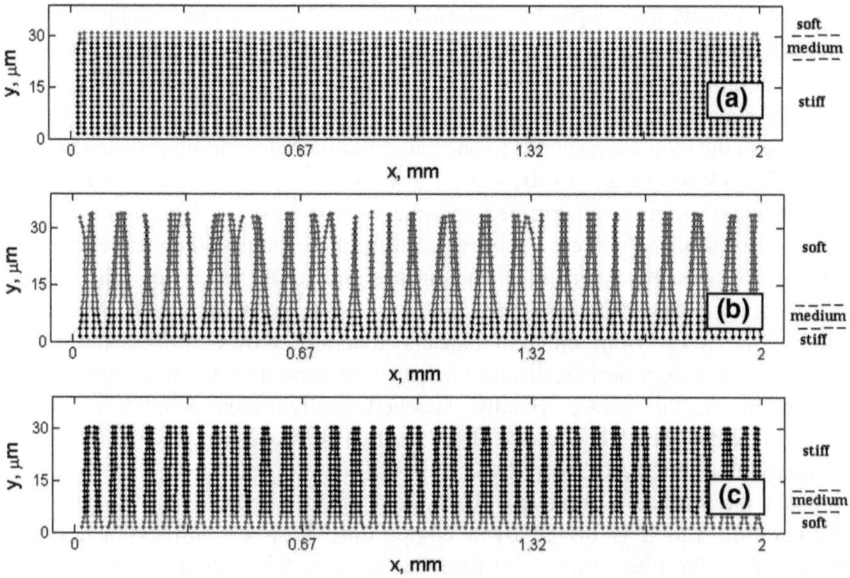

Fig. 9.9 The same system as presented in Fig. 9.8 shown after detachment from the fractal surface and sufficiently long relaxation to the static state. The difference between strongly clustered systems having either long elastic (**b**) or long stiff filaments (**c**) and the system with short soft ends (**a**), which practically returns back to its original configuration, is clearly seen. See also complementary Movies 1, 2, 3, which correspond to the cases **a**, **b**, **c**, respectively

Fig. 9.10 Time depending vertical forces developed during attachment of initially unperturbed systems to the hard surface. *Solid*, *dashed* and *dash-dotted lines* correspond to the cases **a**, **b** and **c** of previous figures, respectively

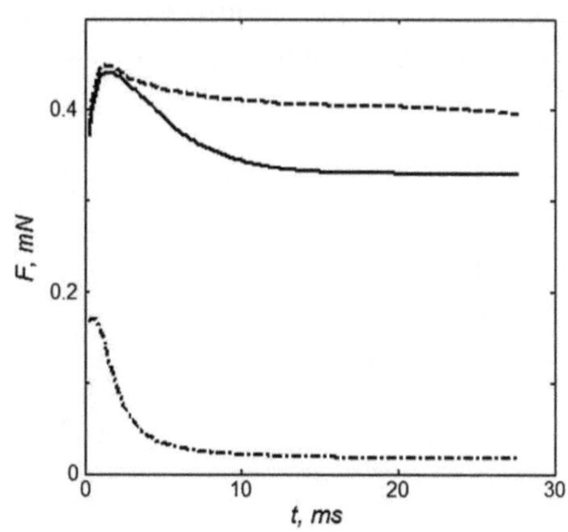

Qualitatively this effect appears, because flexible filaments are too long in the case (b). Last case (c) with long hard filaments rotating around their flexible roots, cannot perfectly adapt to the surface. As result, maximum of the attachment force here remains much lower than in two previous cases (a) and (b).

To accumulate time-dependent information about deformations of the fibers we calculate array $\{dx_j\}, j = 1, 2, \ldots N_x$ of the distances between contact ends of the nearest neighbors $dx_j = x_{j+1} - x_j$. Time evolution of every such array during complete attachment-detachment cycles for all (a)–(c) cases is shown in Fig. 9.11. Each line in the plots corresponds to one particular time-depending distance between a pair of the closest neighbors $dx_j = x_{j+1} - x_j$. All these distances are normalized to the distance of original unperturbed periodic system, so $dx_j = 1$ at $t = 0$.

The history of the process is clearly seen from the plots. When some fibers are attracted to the same asperities of the surface and form the clusters, the distance between their ends goes to zero $dx_j = x_{j+1} - x_j \to 0$. At the same time, the distance between the fibers from different clusters generally grows. This distance must correlate with a characteristic distance between the asperities, but it remains random for random fractal surface. Finally, attached configuration delivers a complex compromise between: (1) stiffness of the fibers; (2) fractal structure of the surface; (3) strengths of all the interactions.

When the surface is removed, the system of fiber array relaxes to new final configuration which is driven by a compromise between stiffness and mutual interaction of the fibers only. If stiffness dominates, the system can return to the original unperturbed state. Time-depending history of this process is clearly recorded in the subplot Fig. 9.11a. It is interesting to note that stiff fibers of case (c), having strong elastic energy cannot completely return back to initial state. They remain glued by their top ends.

To analyze the results statistically one can calculate histograms of the probability $P = P(dx)$ to find a particular value of the distance $dx_j = x_{j+1} - x_j$ between nearest fibers. It is done for a sequence of discrete time steps and summarized in Fig. 9.12. The cases (a)–(c) in this figure are the same as above.

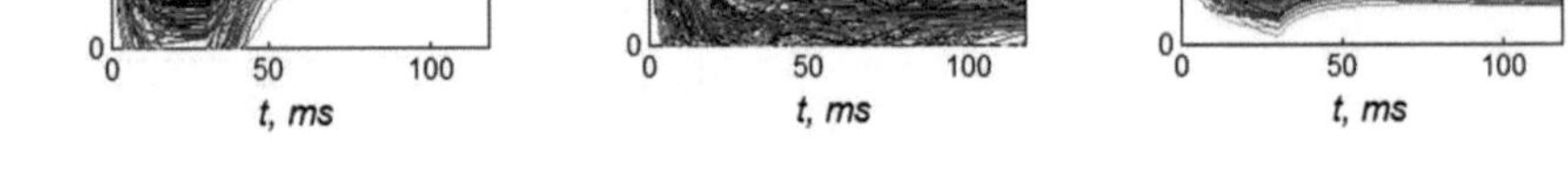

Fig. 9.11 Time evolution of arrays $\{dx_j\}$ of distances $j = 1, 2, \ldots N_x$ between ends of nearest neighbors $dx_j = x_{j+1} - x_j$ during single attachment-detachment cycle for the same cases **a–c** as before. All the distances are normalized to the distance of original unperturbed periodic system: $dx_j = dx_0$ at $t = 0$. *Each line* in the plots corresponds to a time-depending distance between one pair of the neighbors $dx_j = x_{j+1} - x_j$. In attached state, all the filaments tend to a configuration, which represents certain compromise between stiffness, adhesion to the surface and mutual interaction of the fibers. After detachment, the system relaxes to asymptotic configuration corresponding to a compromise between the stiffness and mutual interaction of the fibers only

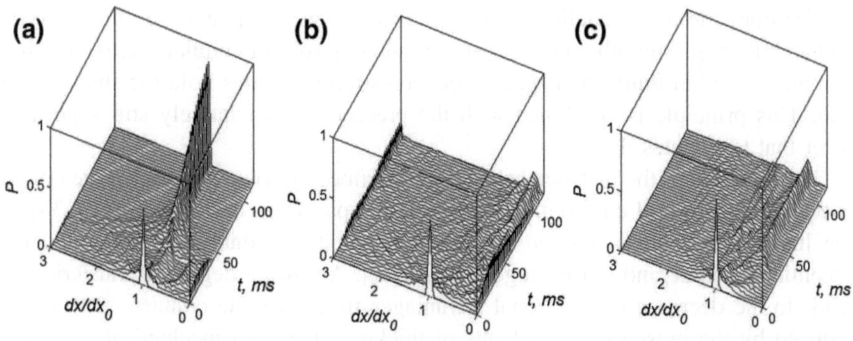

Fig. 9.12 Statistical analysis of the plots presented in Fig. 9.5. The sequences of the histograms show time evolution of the probability $P = P(\mathrm{d}x)$ to find a particular value of the distance $\mathrm{d}x_j = x_{j+1} - x_j$ between nearest fibers. The cases **a–c** are the same as above. Starting from unperturbed configuration (initial peak of probability around $\mathrm{d}x_j = \mathrm{d}x_0$) all the systems evolve to the smooth distributions $P(\mathrm{d}x)$. In clustered attached state (see cases **a** and **b**) the probabilities have well pronounced maximums at $\mathrm{d}x \approx 0$. After detachment from the surface all the systems tend to the distributions $P(\mathrm{d}x)$, which perfectly agree with the observed final states shown in the Figs. 9.3 and 9.5

These data make information presented in Fig. 9.11 clearer. Initial peak of the probability around $\mathrm{d}x = 1$ corresponds to almost unperturbed configuration at small time, just the first contact with the surface. During the time all three types of systems deform their filaments into the configurations with smooth distribution $P(\mathrm{d}x)$. It means that different distances $\mathrm{d}x_j = x_{j+1} - x_j$ appear with comparable probabilities.

With the time, in systems with soft ends [(a) and (b)], many fibers are attracted to the same asperities of the surface. As result the probabilities get well pronounced maximums near $\mathrm{d}x \approx 0$. After detachment all the systems tend to asymptotic probability distributions which perfectly agree with the observed final configurations shown in Figs. 9.9 and 9.11.

Let us add few words about biological significance of the results obtained. Pure bulk materials are absent in biology: biological materials are always composites. Also material gradients are well known in biological systems, where particular change in composition of different bulk materials along a biological structure may lead to novel and often unexpected properties. This has been previously shown for insect cuticle [32, 33], snake skin [34], human teeth [35, 36], and other biological composites.

The gradients have been also recently reported from smooth attachment devices of insects [37]. Interestingly, the gradients in smooth pads of locusts and bushcrickets are different gradients reported in hairy pads of ladybird beetles [28]. Smooth adhesive pads consist of a softer core covered by a stiffer layer, whereas hairy pads have opposite arrangement: stiffer bases combined with softer distal part. Both types of gradients combine conformability to the surface roughness of the substrate and resistance to the environment.

The opposite directionality of gradients can be well explained by difference in pad architecture. Smooth pads consist of branching rods or cellular foams, which in combination with fluid-filled spaces between solid structures hold the shape of the pad. This principle is combined with the presence of a relatively stiff superficial layer that terminates the fibers.

The layer keeps the distance between tips of fibers at some constant value (and in species living in arid environments protects the pad from desiccation) [28, 38]. In the hairy pads, adhesive setae are not terminated by continuous layer and can potentially buckle and cluster together [26, 27]. As strong degree of clusterisation leads to the decrease of functional advantages from multiple contacts [23], this is reduced by the presence of gradients of thickness [18] and mechanical properties [28].

Whereas disadvantages of purely stiff and purely soft fiber arrays are intuitively clear, it is difficult to judge about the advantages of various gradients from the fiber base to the fiber tip (soft-to-stiff/downstream and stiff-to-soft/upstream).

The numerical experiment of the present study has clearly demonstrated that gradient-bearing fibers with short soft tips and stiff gradients (*short upstream gradient*) has greater advantage in maximizing adhesion and minimizing clusterisation in multiple attachment-detachment cycles, if compared to the fibers with longer soft tips on the stiff bases (*long upstream gradient*) and fibers with stiff tips on the soft bases (*downstream gradient*). Such *short upstream gradients* were recently described in beetles [28], however, we can predict that similar gradients must have been convergently evolved in various lineages of arthropods.

9.4 Spatial Model of the Gecko Foot Hair: Functional Significance of Highly-Specialized Non-uniform Geometry

As we already told before one of the important problems appearing in experimental realizations of artificial adhesives inspired by Gecko foot hair is so-called clusterization. If artificially produced structure is flexible enough to allow efficient contact with natural rough surfaces, after few attachment-detachment cycles the fibers of the structure tend to adhere one to another and form clusters. Normally such clusters are much larger than original fibers and, since they are less flexible, form much worse adhesive contacts especially with the rough surfaces. The forces responsible for the clusterization are the same intermolecular forces, which attract fibers to fractal surface of the substrate. However, arrays of real gecko setae are much less susceptible to this problem. One of the possible reasons for this is that ends of the seta have more sophisticated non-uniformly distributed 3D structure than that of existing artificial systems.

The ventral side of gecko toes bears so-called lamellae with arrays of 3–5 μm thick setae, which are further subdivided at their tips into 100–1000 single nanofibers ending with flattened tips (spatulae) of both width and length of about 200 nm [1, 2, 39, 40] and thickness about 15 nm [11] (Fig. 9.1).

Such subdivision of large adhesive contact into many single separate contacts leads to the enhancement of adhesive force of this fibrillar system due to the variety of reasons [4, 23, 41]. This effect is also enhanced by the specific spatula-like shape of singe contacts [24, 42, 43].

One of the important problems of the synthesis of artificial fibrillar adhesives, inspired by the gecko foot hair, is so-called clusterization [26]. If artificially produced structure is flexible enough to allow efficient contact with natural rough surfaces, after few attachment-detachment cycles the fibers of the structure tend to glue one to another and form clusters [9, 44]. Normally such clusters are much larger than original fibers and, and due to their lower flexibility, they form much worse contacts with the surfaces, especially with rough surfaces [4].

Main problem here is that forces responsible for the clusterization are the same intermolecular forces, which attract fibers to the uneven surface of the substrate. The optimisation between such variables as fibre radius, fibre aspect ratio, material properties, and contact shape recently provide an optimal solution, which has been described in a recent paper [43].

However, arrays of real gecko setae are much less susceptible to this problem, in spite of their very high aspect ratio. One of the reasons is that every single seta is a hierarchical structure (Fig. 9.1b, c), and each level of such a multilevel architecture is not sensitive to clusterisation [45, 46].

Another possible reason for low clusterisation ability of real setae is that the end of the seta have more sophisticated non-uniformly distributed 3D structure (Fig. 9.1) than that of existing artificial systems [9, 44] and than that considered in existing models [45].

Below we simulate 3D spatial geometry of non-uniformly distributed branches of nanofibers of the setal tip numerically, studied its attachment-detachment dynamics, and discuss its advantages versus uniformly distributed geometry. Discrete numerical model is organized as follows.

We describe the dynamics of a set of elastic fibers attached to one wider rigid root fiber, which is initially placed at some angle $\varphi_{t=0} = \varphi_0$ and on some distance from the contact surface. The conceptual structure of the model is shown in Fig. 9.14. For definiteness, a number of the elastic fibers is equal to ten $N_x = 10$ and each fiber is constructed of a fifty elastic segments $N_y = 50$, each having a length of $\mathrm{d}R = 0.04$ μm, to get total length of each fiber according to the measurements on SEM images (Fig. 9.13). The fibers are provided with longitudinal $\left(K^{\parallel}\right)$ and transverse $\left(K^{\perp}\right)$ stiffness, $K^{\parallel} = K^{\perp}$.

A deformation of fibers produces elastic forces proportional to their stiffness. Longitudinal force, $\vec{F}^{\parallel}_{jk}$, is described here by a double-well potential, which constructed to keep a distance between the nodes $\vec{R}_j$ and $\vec{R}_{j\pm1}$ close to the equilibrium length of the segment dR:

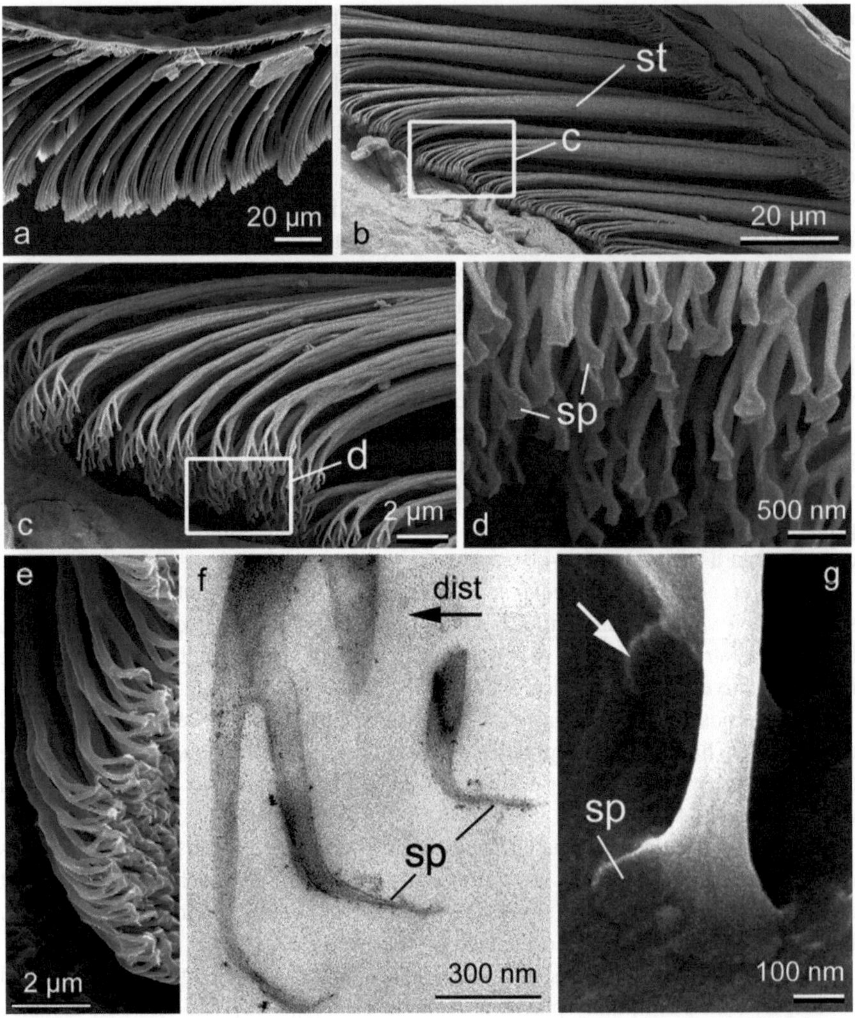

Fig. 9.13 Hierarchical organization of the gecko attachment system: **a** Longitudinal section of the gecko toe with the lamella (*horizontal thin keratinous film*) covered with setae in non-contact state; **b** Setae (st) in contact with the substrate; **c**, **d** Setae branching into single nanofibers terminated with spatulae (sp); **e–g** Spatulae orientation in a non-contact state (**e–f**) and contact state (**g**). *White rectangles* indicate areas, which are zoomed in follow up images. *Black arrow* (dist) indicates distal direction of the toe in all images in this plate. *White arrow* shows spatula flipping from non-contact orientation to the contact orientation

$$\vec{F}^{\parallel}_{jk} = K^{\parallel}(\vec{R}_j - \vec{R}_k)\left[1 - \left(\frac{\vec{R}_j - \vec{R}_k}{dR}\right)^2\right], \tag{9.17}$$

where $\vec{R}_j$ is a position vector of the middle of the segment (the node) j; $k = j \pm 1$. This form of the equation for the longitudinal force was chosen, since it is linear at small displacement and increases non-linearly at large displacements. Combination of these two impacts causes a minimum of effective potential at equilibrium length dR.

Second force, $\vec{F}_j^{\perp}$ is directly proportional to the lateral deflection and tends to keep $\vec{R}_j$ close to the mean value between its nearest neighbors $\vec{F}_j^{\perp} = K^{\perp}(2\vec{R}_j - \vec{R}_{j+1} - \vec{R}_{j-1})$. Transverse stiffness tends to hold the angle between the neighboring segments close to 180° and longitudinal one is responsible for its reaction on the extensions along the fibers. It must be noted, that the "transverse" force $\vec{F}_j^{\perp}$ is chosen in the present form because it is easy to realize numerically, in order to prevent bending of the fibers, but it is not purely bending force, since it may include a longitudinal component.

Initial structure of the fibers is constructed in a manner to mimic as much as possible the form of such a structure of real animal shown in Fig. 9.13. Each of the fibers is elastically attached to the rigid root fiber (bold straight line in the conceptual image). The fibers have different lengths and orientations. As result, their ends are shifted one from another in three directions: parallel to the line of the root and in two orthogonal directions. To mimic better natural structure, we generated individual initial positions of every single segment numerically.

Resulting configuration is shown in the conceptual image (Fig. 9.14). It reproduces also a correct orientation of the terminal parts (spatulae) of the fibers, which are turned backward to the root in equilibrium Gecko hair. According to the main hypothesis of the study, different positions of the fiber ends of the realistic fascicle-like spatial structure of the seta and its orientation to the surface are important for the functional efficiency of the structure.

So, the goal was to study how attachment scenario depends on the 3D configuration and as a result, how practically important parameters (fraction of attached spatulae, and tendency to the clusterization at an equilibrium) correlate with the 3D spatial structure of the seta.

In real system, an additional degree of freedom comes from the rotation of the relatively rigid root fiber. To reproduce it in the model, rotational stiffness, B, is introduced, which dynamically tends to keep the root angle φ close to φ_0. In the first approximation, the rotational force acting on it is linearly proportional to the difference $\varphi_0 - \varphi : f^{\varphi} = B(\varphi_0 - \varphi)$.

At $B > 0$, the force tends to attract the angle φ to its equilibrium value φ_0. As a result, the whole system rotates towards the substrate, when the fibers are attracted to it. If a force between the surface and the fibers is absent the system gradually returns to a shape close to initial one depicted in Fig. 9.14.

As before, real surfaces of the substrates is simulated by the self-affine fractal given by real part of the function $Z(x,y) = A \iint \mathrm{d}q_x \mathrm{d}q_y C(q) \exp(iq_x x + iq_y y + \zeta(x,y))$ with scaling spectral density $C(q) = 1/q^{\beta}$ and exponent $\beta \approx 0.9$.

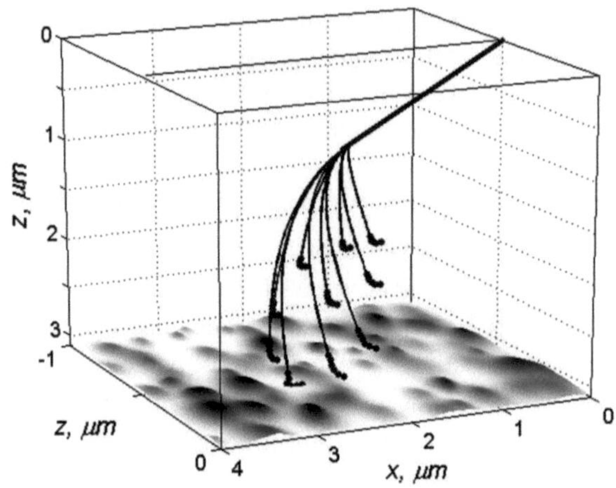

Fig. 9.14 Conceptual 3D structure of the model system. All the segments are numerically defined to mimic real terminal structure of Gecko foot hair. *Bold straight line* corresponds to the root rigid segment (seta). *Thin lines* reproduce flexible filamentous nanofibers terminated by spatulae. *Bold points* at the ends of *curved lines* mark extremely flexible spatula regions. Fractal roughness of the contact substrate is generated according to the standard procedure described in the text and shown by the gray-scale in the XY plane. Z coordinate of the whole plot is flipped up-down, in order not to shadow the filaments by the surface and make easier perception of the spatial 3D structure

The attraction to the surface is caused by an intermolecular (Van der Waals) force. For the sake of simplicity it is simulated below by the gradient of potential $U_{VdW}(r) = U_o(1 - \exp(-r/r_0))^2$, where $r_j = |\vec{r}_j - Z(x,y)|$ with physically reasonable amplitude $U_0 = 10\ nN \cdot nm$ and the minimum located at the distance $r_0 = 0.01\ \mu m$ from the surface. Let us remind also, that the amplitude of the numerically generated surface is taken to be comparable with the radius of Van der Waals interaction $A = r_0$.

Flexible and thin parts of every fiber interact with corresponding regions of other fibers of the array. Interaction force has the same origin as their attraction to the stiff substrate $U_{\text{interact}}(r) = U_o(1 - \exp(-r_{jk}/r_0))^2$ and comparable characteristic parameters U_0, r_0. This force is responsible for the clusterization of the fiber structure and our goal here is to study whether specific spatial distribution of the fibers can prevent it.

Numerical experiment is organized as follows. We place fascicle-like seta in the position, where distance from the last segment (closest to the surface) is comparable with r_0 and allows it to move according to standard equations of motion $\partial r_j/\partial t = F^j$, where total force accumulates all above interactions $F^j = F^j_{elastic} + F^j_{\text{VdW}} + F^j_{\text{interact}}$.

When the process starts, a closest segment is normally attracted to the surface first. Due to mutual elastic and molecular interactions with all other segments, it tends to cause consequent attraction of all of them. From the other side, all the segments are attached to the root and transfer the interaction to it. Being rigidly fixed by one of its ends at zero point, this segment can't completely move with all other ones, but it can rotate to minimize a distance between its free end and the surface.

To simplify visualization and plotting of the results, let us first discuss this process for reduced 2D system, which moves in {x, z} plane. In this case, instant positions of the segments at some discrete time moments, as well as trajectories of all the joint points connecting the segments, can be simultaneously depicted in quite transparent planar plot. Typical scenario of the motion is presented in Fig. 9.15, where instant positions and the trajectories are shown by black and gray lines, respectively.

Rotation of the root segment is the key difference between this scenario and ordinary attraction of regular grid of vertical fibers. As it is seen directly from Fig. 9.15, this rotation leads to a directed proximal shift of the attachment point along the surface and results in an effective healing effect. In other words, additionally to the vertical attraction to the surface, all the fibers simultaneously slightly move along it. Previously, it was experimentally and theoretically demonstrated that such shear movements applied to the spatulae enhance adhesion of thin tapes [1, 24].

Such a motion must have an important effect in the real biological system, because it automatically stimulates more efficient increase of a fraction of the attached spatulae in contact process, and most importantly it does not require active control by animal. Also, it is important to note that due to this process **all terminal spatulae turn oppositely** to the direction of motion. When they attach to the surface by their very tips, shear motion shifts them against already established contact points. As results, the spatulae flip around them. This rotation produces

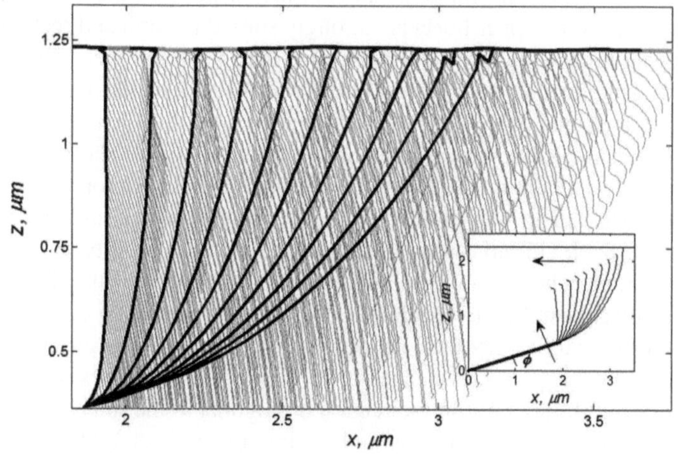

Fig. 9.15 Conceptual picture of the model in 2D plane {x, z}. The fibers and their trajectories are shown by *bold black* and *thin gray lines*, respectively. Initial configuration is presented in the *insert*. The *arrows* show additional directions of the motion, which appear due to the attraction of the fibers to the surface: rotation of the rigid root (seta) and shear along x-coordinate

typical configurations, that are clearly seen in SEM images of real Gecko hairs attached to the surface (Fig. 9.13g), where spatulae are usually turned in one direction, which is opposite to their original orientation (Fig. 9.14e, f).

As mentioned above, mutual shift of the spatulae in vertical direction should prevent their clusterisation after detachment from the surface. The main problem, causing clusterisation, is that spatulae tend to attract to the same regions of the surface (mainly to the asperities protruding from the surface into direction of the fibers). When the fascicle of the nanofibers starts to detach from the surface, the nearest fibers are located already close one to another and will remain attracted even after complete detachment. As result, in further attachment cycles, the whole system becomes already modified by preliminary process, and will not operate properly.

The 3D spatial configuration of the seta with its branches of nanofibers and terminal spatulae has been optimized by the natural selection. Our numerical experiment confirms intuitive expectation that if flexible ends of the nanofibers are shifted one from another in the vertical direction, stored elastic energy tending to return them to their original position, easily separates them. On the other side, at given trial angle φ_0, very strong shift of some of the fibers in vertical direction must prevent their attachment to the surface. It leads to a formal problem of optimization.

To solve this problem, one can compare numerical solution of the equations of motion with and without mutual interaction between the fibers and found time-depending fraction of the attached nanofibers and rotation angle φ are presented together with their final values at different shift Δ In the case of the non-interacting fibers the results are quite trivial. The fraction of the attached fibers gradually decreases with the shift Δ. Time-depending behavior of the angle φ is also quite predictable here.

When the very first fibers attach to the surface, the angle φ starts to grow. Elasticity works against this rotation and in principle stops the angle φ at some final value at $t \to \infty$. If the upper fibers are strongly shifted in vertical direction, they do not participate in the attraction, and it prevents the rotation as well. The final angle φ at $t \to \infty$ is defined by a balance of the forces caused by the very few attached lowest fibers and falls down at some shift Δ.

Quite different picture appears, when the fibers interact. The system demonstrates some clear optimum Δ. The main reason for it is that for very small shift $\Delta \to 0$ the fibers attract one to another in the space over the surface and tend to build clusters. Such clusterisation reduces their attachment ability to the substrate. In the other limit, when $\Delta \to \infty$, obviously, the interaction of the fibers becomes weaker and attachment fraction decreases. However, attraction of the upper fibers to the attached ones still favors the rotation, and the final angle φ grows.

One of the important disadvantages of 2D model is that it does not reproduce a distance between the spatulae in y-direction orthogonal to the rotation plane {x, z}. However, this coordinate becomes extremely important, when all the fibers contact the surface. As mentioned above, rotation of the structure leads to the effect of shear, which aligns all the spatulae in one direction. As result, in many places they formally overlap on the surface and their mutual attraction and respectively, a tendency to the clusterization becomes overestimated by the 2D model.

In reality, they are moved apart, as it is shown in the conceptual 3D picture (Fig. 9.14) in {x, y} projection and remain separated, when come in the contact with the substrate. Moreover, numerical experiments show that the specific configuration of forces, which appears in the particular system in contact process, turns the spatula {x, y} plane to the opposite radial orientation around common point, where all of them remain attached to the root segment. Two different views on typical final attached configuration are shown in Fig. 9.16a, b.

Being separated even on the surface the spatulae can easily return to their initial positions shown in Fig. 9.14. This effect prevents their clusterization and allows practically infinite number of repeats of the attachment-detachment cycles. Overall, this model shows that in addition to the hierarchical organisation of the seta [45, 46], its non-symmetric structure and non-uniform distribution of nanofibers lead to the preventing of clusterisation. Such a sophisticated non-uniformly distributed 3D structure additionally provides rotation-induced proximal shear of spatulae, which is crucial for contact formation of spatulae and formation of adhesive contact in thin tapes [24]. Since these effects are entirely based on particular geometry of the seta, it certainly would have practical application, due to simple, purely mechanical, effect for the enhancement of the performance of artificially made fibrillar adhesives.

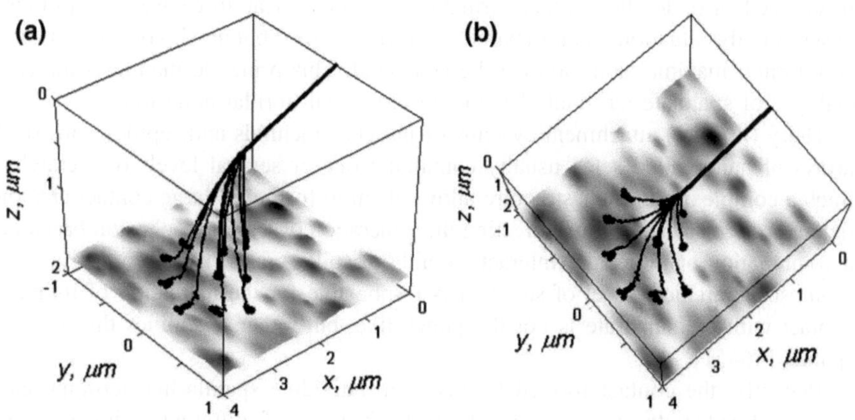

Fig. 9.16 Two different views of the same attached configuration of the modeled seta, shown in isometric and almost vertical projections in subplots **a** and **b**, respectively. All *symbols* are the same as in Fig. 9.14

9.5 Shear Induced Adhesion: Contact Mechanics of Biological Spatula-Like Attachment Devices

Most biological hairy adhesive systems of insects, arachnids and reptiles, involved in locomotion, rely not on flat punches on their tips, but rather on spatulate structures. As it was discussed above, Several hypotheses have been previously proposed to explain the functional importance of such a contact geometry: enhancement of adaptability to the rough substrate [11]; contact formation by shear force rather than by normal load [1]; increase of total peeling line due to using an array of multiple spatula [23]; contact breakage by peeling off [42]. The Kendall peeling model [55] was recently employed to explain the role of multiple spatulate tips in the enhancement of the total peeling line [23].

It is well known that application of a normal force can enhance adhesion [57]. However, for hairy attachment systems, the adhesion force always remains smaller than the initially applied normal force [58]. This would be not enough to enable walking on the ceiling. Another possibility of enhancing the adhesion force is application of a shear force. In this part of the study, we use numerical approach to study dynamics of spatulate tips during contact formation on rough substrates.

In particular, a model will be constructed, which clearly demonstrates that the contact area increases under applied shear force, especially when spatulae are misaligned prior to the contact formation. Applied shear force has an optimum describing the situation when maximal contact is formed but no slip occurs. At such equilibrium maximal adhesion can be generated. This principle manifests the crucial role of spatulate terminal elements in biological fibrillar adhesion.

Hairy (fibrillar) attachment systems of insects, arachnids and reptiles consist of arrays of hairs called setae usually containing one or several levels of hierarchy. Such a complexity in their structure allows them to form of a large contact area on flat and rough surfaces, thus, effecting the generation of a strong adhesion based on a combination of molecular interaction and/or capillary attractive forces [59]. The topmost hierarchical level of seta that is responsible for the formation of intimate contact with the substrate is not flat punch-like, but rather resembles thin film or spatula [46–51].

Recently, the contact formed between an individual spatula-like terminal element and substrate was visualized on fresh pads of different animals using Cryo-SEM technique [23]. Spatulae bear a gradient of thickness from the base to the tip of the spatula (fly: [52]; gecko: [53], beetle: [54]) (Fig. 9.1b, c). In contact, spatulae are aligned and orientated to the distal direction of the pad (Figs. 9.1a, b and 9.17).

It has been previously shown that applied shear force to a particular direction of the spider's spatulated setae results in an increase of real contact area [52] (Fig. 9.1e, f). Also, flies employ shear movements during contact formation by their attachment devices [53]. Other authors revealed strong shear dependence of measured pull off force in the gecko attachment system and even called this effect "*frictional adhesion*" [1, 2]. Since the effective elastic modulus of thin plates is very

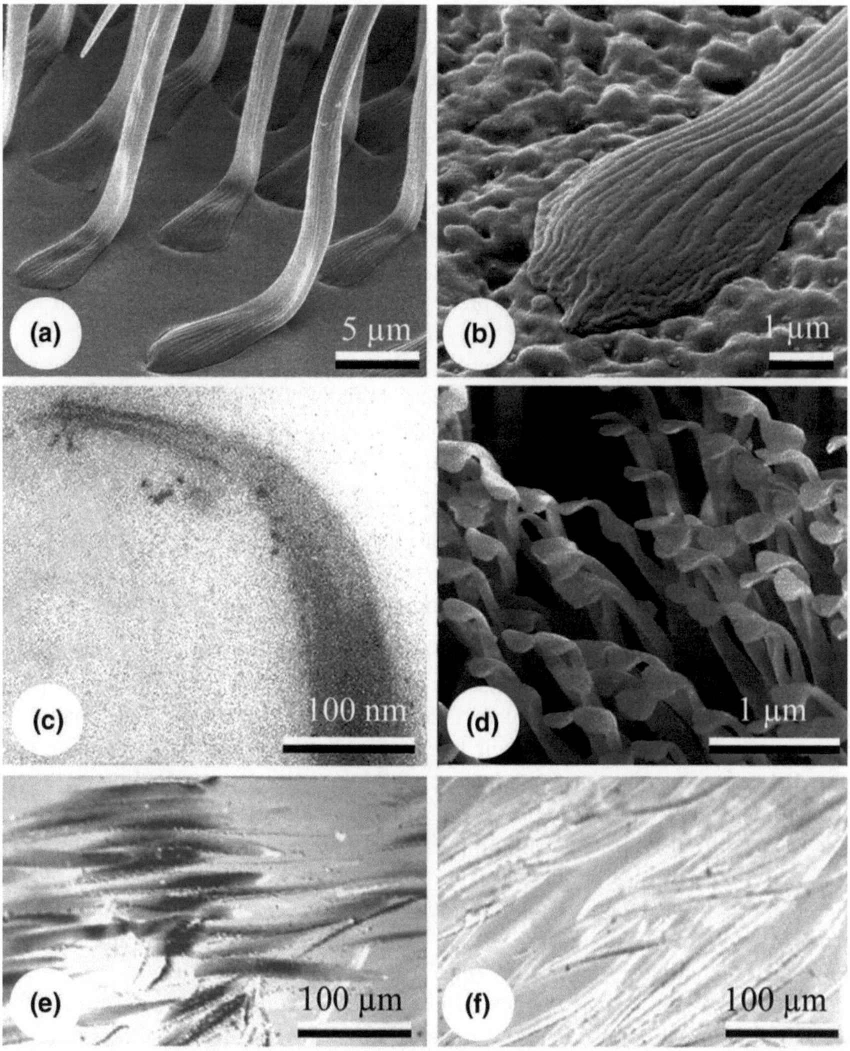

Fig. 9.17 Light microscopy (**e**, **f**), Cryo-SEM (**d**), SEM (**a**, **b**), and TEM (**c**) images of spatula-shaped thin-film terminal elements in hairy attachment pads of various animals. **a** Spatulae of the beetle *Gastrophysa viridula* in contact with the flat substrate. **b** Single spatula of the same species in contact with substrate having average asperity dimension in the range of 300 nm. **c** Longitudinal section of the single spatula of *Gekko gecko*. **d** Spatulae of *G. gecko*. **e**, **f** Setae of the spider *Cupiennius salei* in the reflection light microscope during distal (**e**) and proximal (**f**) sliding on a glass substrate (*black areas* correspond to the sites of contact between spatulae and glass surface)

small, even for relatively stiff materials such as keratin or arthropod cuticle, this geometry is of fundamental importance for adhesion on rough substrates [11] due to the low deformation energy stored in the material during contact formation. I was also previously shown that adhesion in such systems depends on the nature of the substrate roughness [50] and it is stronger in attachment devices containing predominantly spatulate tips [21].

The important role of spatulate contact elements in the generation of adhesion has recently been experimentally demonstrated on artificial, bioinspired surfaces with similar geometry [60–64]. Recent theoretical considerations supported the importance of applied shear force (pre-tension) for an increase of the peel-off force [65].

However, it remains unclear how the contact on a rough substrate can be generated by shear, especially if the spatulae are initially not aligned in the plane of the substrate. Here we use a numerical approach to study the dynamics of spatulate tips during contact formation on rough substrates and ask the following questions:

1. What is the role of the thickness gradient during contact formation?
2. Does applied shear force contribute to the enhancement of contact area on the rough substrate?
3. Is there any optimal shear distance/force for single spatula?

Examples of the spatulate contact geometry of different animals are shown in Fig. 9.1. The SEM images demonstrate similar contact geometry of setal tips in different animal lineages. Spatulae, independent of their dimensions, make flat contact with the substrate with their free ends oriented in the distal direction [23] (Fig. 9.1a, b). Spatulae have a gradient of thickness from distal (thinner) to the basal (thicker) part (Fig. 9.1b, c) [54]. On rough substrates, spatulae are partially able to adapt to the surface profile (Fig. 9.1b). In the non-contact state, spatulae are preferably, but not ideally, aligned (Fig. 9.1d). When shear force is applied in one direction (direction depends on the species and on the site of the leg), spatulae make contact with the substrate (Fig. 9.1e). When shear is applied in the opposite direction, spatulae lose the contact (Fig. 9.1f).

To simulate shear caused adhesion in a typical spatula-like structure following configuration of the system under consideration was applied. In a spirit of the previous parts terminal part of the spatula is treated as a flexible elastic plate with the width (and corresponding gradient of flexural stiffness) gradually varying along the x-coordinate.

For the initial condition, the plate is brought into an initial touch with a rigid rough "ceiling" surface along a terminal line which is parallel to the y-coordinate. For the regular study, it is supposed to be initially inclined to the horizontal plane with a trial tilt angle α which varies in the interval $0<\alpha<\pi/2$ from a horizontal to a vertical orientation.

Simplified structure of the model is symbolically presented in Fig. 9.18. The figure qualitatively shows a 2-dimensional projection of the system onto the x-z-plane. It is expected that the very thin end part of the plate is flexible enough to quite easily attract to the rigid rough surface by Van der Waals force, simulated, as before, by gradient of Morse potential $U_{VdW}(r) = U_o(1 - \exp(-ar))^2$ with a physically reasonable amplitude $U_0 \simeq 10\,\mathrm{nN\,nm}$ and the minimum located at the distance $a \simeq 1\,\mathrm{nm}$ from the surface. Rigid ceiling surface is as ususally self-affine fractal surface with scaling density $C(q) = 1/q^{\beta}\, Z(x,y) = A \iint dq_x dq_y C(q) \exp(iq_x x + iq_y y + \zeta)$, and exponent $\beta \approx 0.9$.

Here we concentrate on a single spatula. Accordingly, we restrict the amplitude A of the surface profile to $A = 1\,\mathrm{nm}$. With this amplitude the procedure generates random realizations of the rigid surface a having typical roughness up to 4 nm, which is typical for the considered scale.

Van der Waals attraction to the surface competes with the resistance of the spatula to bending. According to the theory of elasticity [56] elastic energy of the flexible plate is given by the following integral:

$$W_{elastic} = \frac{E}{24(1-\nu^2)} \iint dxdy h^3(x,y) \left\{ \left(\frac{\partial^2 z}{\partial x^2} + \frac{\partial^2 z}{\partial y^2} \right)^2 + 2(1-\nu) \left[\left(\frac{\partial^2 z}{\partial x \partial y} \right)^2 - \frac{\partial^2 z}{\partial x^2} \frac{\partial^2 z}{\partial y^2} \right] \right\} \tag{9.18}$$

where E is the Young's modulus of the plate material and ν is the Poisson ratio which is typically equal to $\nu = 1/3$.

In considered systems, adhesive force is comparable to the weight of the whole animal. Thus, the gravitational force acting on the spatula is negligible. Further, due to strong internal damping in the spatula material, the system can be treated as over-damped. The over-damped dynamics of the system along vertical z coordinate is described by the equation of motion

$$\gamma \frac{\partial z(x,y)}{\partial t} = -\frac{\partial W_{elastic}[z]}{\partial z} - \frac{\partial U_{VdW}[z]}{\partial z}, \tag{9.19}$$

where γ is damping constant which determines a characterictic time scale of the process ($\gamma = 1$). Van der Waals bonding also produces horizontal force $F^x_{VdW} = -\partial U_{VdW}[z(x)]/\partial x$. This force competes with an external shear force F^x. When F^x exceeds the total resistance of all instantly bonded segments

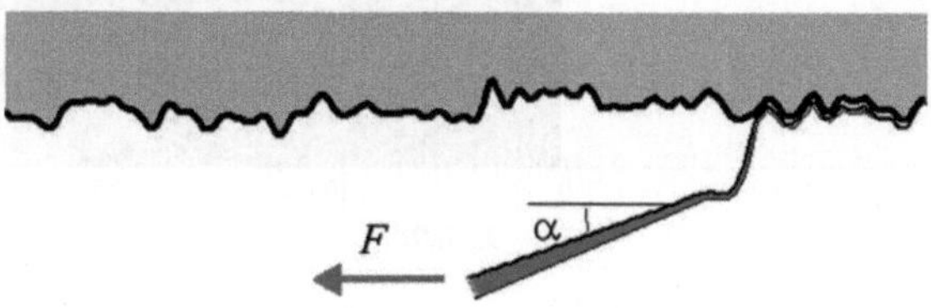

Fig. 9.18 Conceptual structure of the numerical model. We consider a terminal part of the "spatula" which is modeled by an elastic plate with variable thickness. It is brought into contact with the rough rigid surface at an initial inclination angle α and pulled in a horizontal direction by an external force F

$\int dxdyF^x_{VdW} > |F^x|$, the whole spatula moves along the x-direction according to the following relation:

$$\gamma \partial x/\partial t = F^x - \int dxdyF^x_{VdW} \tag{9.20}$$

The (9.19) and (9.20) form complete system, which can be solved numerically. A typical numerically found intermediate configuration of the system described by these equations is shown in Fig. 9.19.

The dynamic behavior leading to picture Fig. 9.3 is as follows. The "spatula" plate is initially attached to the surface by Van der Waals force by one of its end segments. It then relaxes in the course of time to an equilibrium state in which it adheres to the surface by additional segments. The rate of attachment depends on the angle α between the plate and surface, normally being faster for smaller angles α.

If external force is nonzero $F \neq 0$, it pulls the plate to the left and competes with Van der Waals attachment F_{VdW} of the "glued" segments. If the total attachment force is stronger than the horizontal component of the external force $\int dxdyF^x_{VdW} > |F^x|$, the spatula does not slide along the x-axis. However, the part which remains unattached can rotate and approach the surface $z(x, y) \rightarrow \langle z \rangle$ (here

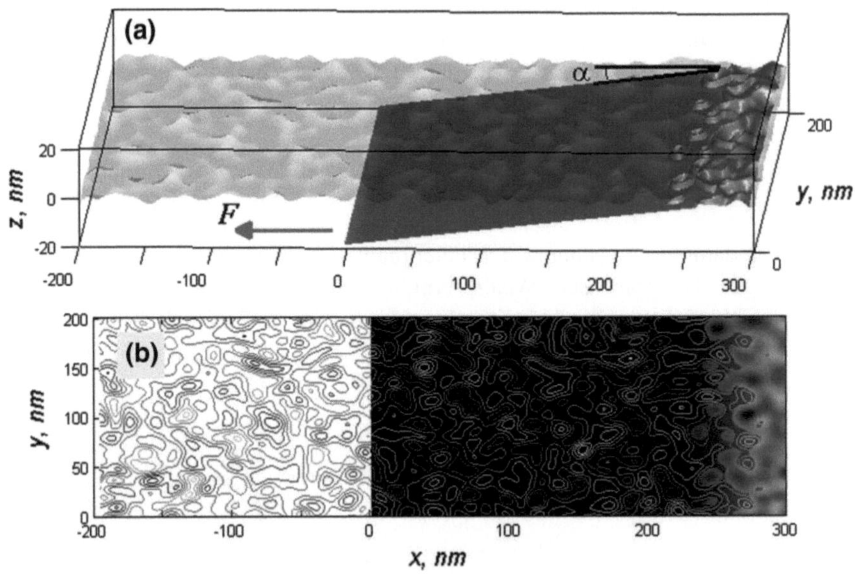

Fig. 9.19 Typical 3-dimensional configuration of the elastic plate found at an intermediate stage of motion (**a**). Rigid ceiling is shown by semi-transparent upper surface in the picture. Instant configuration of the flexible surface is presented by the gray-*scale map* (*darker color* corresponds to deeper values of z-coordinate). The picture is completed by a contour plot of the rough surface (**b**) where instant contact areas shown by the *gray-scale map* are seen directly

symbol $\langle \ldots \rangle$ denotes mean value) due to the action of the vertical component of the force $F^z > 0$; $F^z \sim z$.

This rotation reduces mutual distances between hard and flexible surfaces and can greatly enhance the total adhesion. Generally, one can expect that a stronger shear force will cause faster attachment. However, if the shear force is too strong it can exceed the Van der Waals locking and even lead to detachment of previously attached segments. In this case, the plate will start to slip along the surface, its rotation stops and additional segments do not adhere to the ceiling. These qualitative considerations gives rise to the optimization problem: up to which extent can one vary shear force to stimulate attachment without rupturing the contact?

Two sets of numerical simulations have been performed: with a fixed initial inclination angle α and varying force F and with a fixed force and varying angle. The results are summarized in Figs. 9.20 and 9.21.

The numerical experiment is organized as follows. In each experiment first a rigid surface $Z(x, y)$, with the same fractal properties, is generated as a two dimensional array. Each array has a size of $[500, 200]$ cells (corresponding to a region of 500×200 nm, while the terminal part of the spatula moving against the

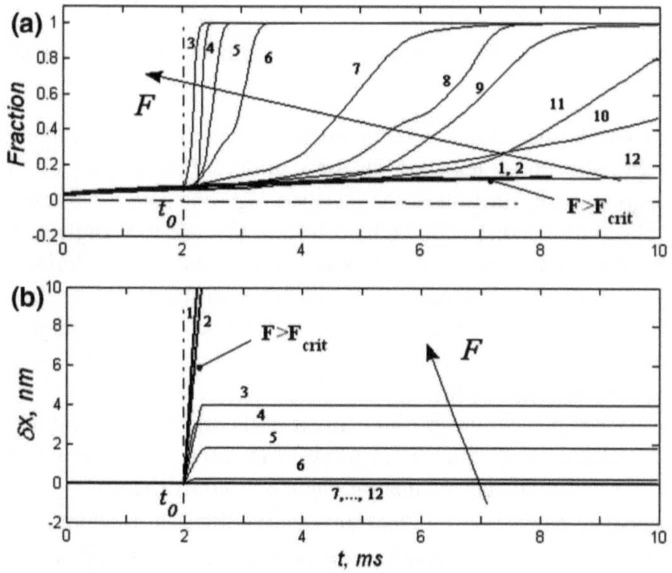

Fig. 9.20 Time dependencies of the attached fraction (contacting part of the plate normalized to its total area) calculated at different values of the pulling force (**a**) and corresponding drift δx of the attached end (**b**). The *arrows* in both subplots mark the tendencies in fraction and drift changes at increase of the force F. Bold curves 1 and 2 correspond to the forces $F = 22$ nN and $F = 20\,\text{nN} \geq F_{crit}$ higher than critical value F_{crit}, which is sufficient to rupture initial attachment and cause permanent slip of the plate. Other curves 3–12 correspond to the forces uniformly decreasing inside the interval $0 < F < F_{crit}$. All the dependencies are typical for an intermediate trial angle α (and obtained for a definiteness at fixed $\alpha = 0.2\pi$)

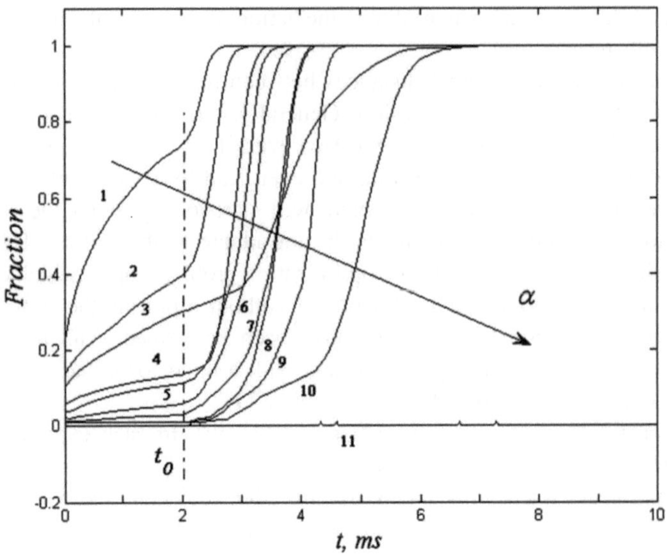

Fig. 9.21 Time dependencies of the attached fraction calculated at fixed value of the pulling force $F = 10\,\text{nm}$ and trial inclination angle which is gradually varied between almost horizontal and vertical orientations: $0.05 \cdot \frac{\pi}{2} \leq \alpha \leq 0.95 \cdot \frac{\pi}{2}$. The curves 1 and 11 correspond to the limiting cases of this interval, respectively. The *arrow* marks a tendency in fraction decrease at angle growth

rigid surface has the size $200 \times 200\,\text{nm}$. Such a size of the array is found to be large enough to provide good statistics for substantial self-averaging of the integral values (like total force, fraction of the attached segments, etc.). Now we bring one end of the plate into contact with the hard surface at some trial angle α and wait for a short period of time (about $t_0 = 2\,\mu\text{s}$ in dimensional units), during which the spatula starts to adapt spontaneously to the rigid surface (at zero external force $F_{t<t_0} = 0$). At $t = t_0$ we "turn on" the pulling. The process is continued up to the maximal time $t_{\max} = 10\,\text{ms}$, which corresponds to the observed time for contact formation of the gecko toe [42].

Subplot (a) in Fig. 9.20. shows how a part of the plate contacting with the rigid surface (normalized to its total area) grows with time at different values of the pulling force F at fixed trial inclination angle $\alpha = const$ (here we use a representative intermediate angle $\alpha = \frac{\pi}{5}$). One can see, that different random realizations of the substrate surface $Z(x, y)$ manifest themselves only in small deviations (and sometimes mutual intersections) of the subsequent solutions in Fig. 9.4 but do not affect the general monotonous trend. This observation is important to be sure that the obtained results are representative.

As expected, the higher force F leads to a faster decrease of the inclination angle. If the force is smaller than some critical value for detachment $F < F_{crit} \approx 20\,\text{nN}$, which is in the range of forces previously experimentally measured for single gecko spatula [57], then the plate gradually tends to a horizontal orientation for $t \rightarrow \infty$.

When the force approaches the critical value F_{crit} it becomes capable of breaking some of the already attached bonds and to slightly shift the plate in a horizontal direction. Corresponding small horizontal displacements $\delta x = x - x_0$ of the system from its initial position x_0 are clearly seen in the subplot (b) (curves 3–6), where time dependencies of the shift $\delta x(t)$ are presented for different forces.

The arrows in both subplots show how the time dependencies change with increasing external force F. Finally, if the force exceeds a threshold F_{crit} (curves 1 and 2) it completely breaks the initial attachment and causes permanent sliding of the plate. In this regime the plate does not rotate and does not further approach the horizontal orientation, thus not leading to any increase of adhesion. It is natural that in a biological system the animal cannot control the state (attachment and orientation) of each individual spatula, but presumably can monitor a total resistance force of the entire spatulla array, keeping it close to but not exceeding the critical value.

As mentioned above, the critical force F_{crit} depends on an initially attached area and thus on a trial inclination of a particular spatula. It is important to study variations of the time-depending scenarios at changes of the initial angle α. The results of this study are summarized in Fig. 9.5, where time dependencies of the attached fraction calculated at fixed value of the pulling force $F = 10\,\text{nN}$ are shown for different trial angles α, which is varied with a constant step from almost horizontal to almost vertical orientation: $0.05 \cdot \frac{\pi}{2} \leq \alpha \leq 0.95 \cdot \frac{\pi}{2}$. Two limiting cases of this interval are presented by the curves 1 and 11, respectively. The arrow shows how a time dependant scenario varies, increasing the angle in the interval $0.05 \cdot \frac{\pi}{2} < \alpha < 0.95 \cdot \frac{\pi}{2}$. It is clearly seen that the shear enforces the attachment process, even at very acute angles (curves 1–3), when the fraction of the attached segments quickly spontaneously grows even at $t < t_0$.

This study demonstrates three additional functional aspects of spatulate geometry on adhesion:

The first one is that contact area grows on smooth and rough surfaces under applied shear force, especially when spatulae are misaligned prior to contact formation, which is the case in the gecko system.

The second important finding is that applied shear force has an optimum, which describes the situation when the maximal contact is formed but no slip occurs. At such equilibrium maximal adhesion can be generated.

The third interesting observation is that, due to the thickness gradient, the initial contact is formed by the spatula tip. This initial contact, together with applied shear, results in a further increase of the contact area which is responsible for stronger adhesion.

These principles manifest the critical role of spatulate terminal elements in biological fibrillar adhesion. However, even when the minimum free-energy state corresponds to complete contact, the elastic plate may be trapped in a metastable state because of friction. In this case, because the kinetic friction is smaller than the static friction, sliding or vibrating the plate may increase the contact. This effect is known experimentally: by sliding of spatulae for a short distance the contact area and friction/adhesion increase.

To summarize this part, let us mention once more that contact formation, not by applying normal load but rather by shear force, is one of the possible reasons why most biological, hairy, locomotory attachment systems rely on spatulate structures. Shear force applied to spatulae, initially oriented at various angles, results in the proper alignment of the spatulae and an increase of the contact area to smooth and rough substrates. This principle generalizes the critical role of terminal elements in fibrillar adhesion. Finally, we can suggest that bio-inspired adhesives based on spatulate geometry has to be actuated according to the scheme preload-shear-peel, in contrast to mushroom shaped geometry of terminal tips. In a combination with material gradients [66] and non-trivial spatial 3D distribution of the fibers and spatulas on them [67] it could be mostly efficient construction of the perspective biologically inspired adhesive systems.

References

1. K. Autumn et al., Nature **405**, 681 (2000)
2. K. Autumn et al., Proc. Natl. Acad. Sci. U S A **99**, 12252 (2002)
3. B. Cappella, G. Dietler, Surf. Sci. Rep. **34**, 1 (1999)
4. B.N.J. Persson, J. Chem. Phys. **118**, 7614 (2003)
5. B.N.J. Persson, J. Phys.: Condens. Matter **17**, R1 (2005)
6. M. Scherge, S. Gorb, *Biological Micro- and Nano-tribology* (Springer, Berlin, 2001)
7. M. Sitti, R.S. Fearing, J. Adhes. Sci. Technol. **17**, 1055 (2003)
8. A.K. Geim et al., Nature **390**, 259 (2003)
9. M.T. Northen, K.L. Turner, Nanotechnology **16**, 1159 (2005)
10. B. Yurdumakan, N.R. Raravikar, P.M. Ajayan, A. Dhinojwala, Chem. Commun. 3799 (2005)
11. B.N.J. Persson, S.N. Gorb, J. Chem. Phys **119**, 11437 (2003)
12. D. Campolo, S.D. Jones, R.S. Fearing, in *Proceedings of International Conference IEEE Nano 2003 Aug 12–14*, San Francisco, 2003, p. 856
13. K. Autumn et al., J. Exp. Biol. **209**, 3569 (2006)
14. A.E. Filippov, V. Popov, Phys. Lett. A **358**, 309 (2006)
15. C. Majidi et al., Phys. Rev. Lett. **97**, 076103 (2006)
16. A.E. Filippov, V. Popov, J. Phys. Cond. Matter A **19**, 096012 (2007)
17. S.N. Gorb, *Attachment Devices of Insect Cuticle* (Kluwer Academic Publishers, Dordrecht, 2001)
18. S.N. Gorb, R.G. Beutel, Naturwissenschaften **88**, 530 (2001)
19. S.N. Gorb et al., Integr. Comp. Biol. **42**, 1127 (2002)
20. C. Creton, S.N. Gorb, MRS Bull. **32**, 466 (2007)
21. D. Voigt, J.M. Schuppert, S. Dattinger, S.N. Gorb, J. Insect Physiol. **54**, 765 (2008)
22. C.Y. Hui, N.J. Glassmaker, T. Tang, A. Jagota, J.R. Soc, Interface **1**, 35 (2004)
23. M. Varenberg, N.M. Pugno, S.N. Gorb, Soft Matter **6**, 3269 (2010)
24. A.E. Filippov, V.L. Popov, S.N. Gorb, J. Theor. Biol. **276**, 126 (2011)
25. F.M. Borodich, E.V. Gorb, S.N. Gorb, Appl. Phys. A **100**, 63 (2010)
26. A. Jagota, S.J. Bennison, Integr. Comp. Biol. **42**, 1140 (2002)
27. R. Spolenak, S.N. Gorb, E. Arzt, Acta Biomat. **1**, 5 (2005)
28. H. Peisker, J. Michels, S.N. Gorb, Nat. Commun. **4**, 1661 (2013)
29. T. Weis-Fogh, J. Exp. Biol. **37**, 889 (1960)

30. T. Weis-Fogh, J. Mol. Biol. **3**, 648 (1961)
31. J. Michels, S.N. Gorb, J. Microsc. **245**, 1 (2012)
32. J.F.V. Vincent, Appl. Sci. Manuf. **33**, 1311 (2002)
33. N. Barbakadze, S. Enders, S. Gorb, E. Arzt, J. Exp. Biol. **209**, 722 (2006)
34. M.-C. Klein, J. Deuschle, S. Gorb, J. Comp. Physiol. A **196**, 659 (2010)
35. R.Z. Wang, S. Weiner, J. Biomech. **31**, 135 (1997)
36. H. Fong, M. Sarikaya, S.N. White, M.L. Snead, Mater. Sci. Eng. C **7**, 119 (2000)
37. P. Goodwyn et al., J. Comp. Physiol. A **192**, 1233 (2011)
38. S.N. Gorb, in *Advances in Insect Physiology: Insect Mechanics and Control*, ed by J. Casas, S.J. Simpson, Elsevier Ltd. (Academic Press, London, 2008), vol. 34, p. 81
39. U. Hiller, Z. Morphol, Tiere **62**, 307 (1968)
40. G. Huber, S.N. Gorb, R. Spolenak, E. Arzt, Biol. Lett. **1**, 2 (2005)
41. E. Arzt, S. Gorb, R. Spolenak, Proc. Natl. Acad. Sci. U S A **100**, 10603 (2003)
42. H. Gao, H. Yao, Proc. Natl. Acad. Sci. U S A **101**, 7851 (2004)
43. R. Spolenak, S. Gorb, E. Arzt, Acta Biomater. **1**, 5 (2005)
44. B. Yurdumakan, N.R. Raravikar, P.M. Ajayan, A. Dhinojwala, Chem. Commun. **3**, 3799 (2005)
45. T.W. Kim, B. Bhushan, Ultramicroscopy **107**, 902 (2007)
46. B. Bhushan, A.G. Peressadko, T.W. Kim, J. Adhes. Sci. Tech. **20**, 1475 (2006)
47. N.E. Stork, Zool. J. Linn. Soc. **68**, 173 (1980)
48. N.E. Stork, J. Nat. Hist. **17**, 583 (1983)
49. S.N. Gorb, Proc. R. Soc. Lond. B **265**, 747 (1998)
50. S.N. Gorb, *Attachment Devices of Insect Cuticle* (Kluwer Academic Publishers, Dordrecht, 2001)
51. Y. Tian et al., PNAS **103**, 19320 (2006)
52. S. Niederegger, S.N. Gorb, J. Insect Physiol. **49**, 611 (2003)
53. S. Niederegger, S.N. Gorb, J. Comp. Physiol. A **192**, 1223 (2006)
54. T. Eimüller, P. Guttmann, S.N. Gorb, J. Exp. Biol. **211**, 1958 (2008)
55. K. Kendall, J. Phys. D Appl. Phys. **8**, 1449 (1975)
56. L.D. Landau, E.M. Lifshits, *Theory of Elasticity*, 2nd edn. (Pergamon Press, Oxford, 1981)
57. V.L. Popov, *Contact Mechanics and Friction. Physical Principles and Applications* (Springer, Berlin, 2010)
58. M. Schargott, V. Popov, S. Gorb, J. Theor. Biol. **243**, 48 (2006)
59. G. Huber et al., PNAS **102**, 16293 (2005)
60. S. Kim, B. Aksak, M. Sitti, Appl. Phys. Lett. **91**, 221913 (2007)
61. S. Kim, M. Sitti, Appl. Phys. Lett. **89**, 261911 (2006)
62. S. Kim et al., IEEE Nano **1**, 780 (2008)
63. T.W. Kim, B. Bhushan, J. Adhes. Sci. Technol. **21**, 1 (2007)
64. M.P. Murphy, B. Aksak, M. Sitti, J. Adhes. Sci. Technol. **21**, 1296 (2007)
65. B. Chen, P.D. Wu, H. Gao, J.R. Soc, Interface **6**, 529 (2009)
66. A.E. Filippov, S.N. Gorb, Beilstein J. Nanotechnol. **5**, 837 (2014)
67. A.E. Filippov, S.N. Gorb, Interface Focus **5**, 20140065 (2015)

Chapter 10
Hierarchical Models of Engineering Rough Surfaces and Bio-inspired Adhesives

Feodor M. Borodich and Ovidiu Savencu

Abstract Friction, wear, adhesion and energy dissipation during sliding are strongly influenced by deformations of asperities, which, in turn, depend on the surface profile. At the nanoscale, the effects of surface roughness and the underlying physical phenomena, such as adhesion between contacting objects, have a considerable influence on the interaction between surfaces. Here various models of rough surfaces, including multi-level models, hierarchically structured models, and appropriate multi-scale models of contact interactions between rough surfaces are reviewed and discussed. A new model for numerical simulations of dry friction between rough engineering surfaces is introduced. The main features of the new model based on the use of a multi-level and multi-scale, hierarchically structured slider are described. Although the surface topography of the biological attachment devices is rather different from the topography of engineering surfaces, some existing models of bio-inspired adhesives are classified using terminology introduced for models of engineering rough surfaces.

10.1 Introduction

Adhesion is a universal molecular phenomenon that has been studied for a long time. Usually adhesion between contacting bodies has a negligible effect on surface interactions at the macroscale. However, modern nanotechnology considers surfaces whose roughness is below the micro-metre scale. With the miniaturization of components, such as micro-electromechanical systems (MEMS), having

F.M. Borodich (✉) · O. Savencu
Cardiff School of Engineering, Cardiff University, Cardiff CF24 3AA, Wales, UK
e-mail: borodichfm@cardiff.ac.uk

O. Savencu
e-mail: ovidiusps@yahoo.com

O. Savencu
Austrian Center of Competence for Tribology, 2700 Wiener Neustadt, Austria

L. Heepe et al. (eds.), *Bio-inspired Structured Adhesives*,
Biologically-Inspired Systems 9, DOI 10.1007/978-3-319-59114-8_10

ever-smoother surfaces, and the growing use of nanoparticle arrays, effects that may be unimportant at the macroscale have now become crucially important at the nanoscale. Adhesion plays a key role in the development of various modern technologies, especially in such applications as the contact of clean surfaces, information storage devices, MEMS, and nanoindentation. Although the influence of roughness on tribological phenomena has been studied for many years, significant challenges remain if a fundamental understanding of atomic-scale surface effects on these macro scale phenomena is to be obtained. These effects may appear to be rather controversial, e.g. roughness can increase the adhesion of a micro/nano size particle to a surface, or it can decrease it, depending on the roughness scale, the location on the surface where contact is made, and the size and geometry of the particle. The mechanical response of such a system has to be considered along with adhesion caused by the van der Waals forces, the chemical bonding forces between atoms, and the applied loading conditions.

The study of the various biological adhesive systems that were developed during the evolution of nature to allow animals to vary the strength of adhesive contact may be very useful for the development of many modern technologies, spanning from the automotive to the biomedical industry, from micro/nano-technology to robotics [1]. For engineering of bio-mimetic surfaces with superior adhesive performances, one needs to understand mechanisms of contact between attachment devices of animals and rough surfaces, and to take into account specific features of the biological object structures. Indeed, the ability of insects and certain reptiles to adhere to vertical walls and walk on ceilings is due to specific hierarchical structure of their attachment elements at micrometre and nanometre scales [2]. Setae of many insects and reptilians are covered with minute hairs (tenent setae or bristles). The hairs increase the contact area between an animal foot and rough surfaces.

Although the surface topography of the biological structures is rather different from the topography of engineering surfaces, it is interesting to consider the existing models of engineering rough surfaces and to use these models for the classification of the models used for studying biological attachment devices. Here we discuss various models of rough surfaces, including multi-level models, hierarchically structured models and appropriate heterogeneous multi-scale models of contact interactions between the surfaces.

10.2 Preliminaries

There are often rather different interpretations of terms: scale, level, hierarchy and so on. Hence here we describe the meaning of the terms in the framework of our models related to engineering surfaces. The preliminary version of the models was presented by Savencu and Borodich [3].

10.2.1 Some Characteristics of Surface Topography Models

The term 'scale' is related to the capabilities of the system to reflect different physical-chemical mechanisms of interactions between surfaces. For instance, a model where the nano-scale asperities are mainly responsible for molecular and chemical interactions and the micro-scale asperities are responsible for modelling the mechanical interlocking of asperities in friction would be multi-scale, because the system is capable to model more than one mechanism of interaction. The considered scale can be changed by changing the considered characteristic length of the processes involved in friction. The following distinctions will be used:

1. Atomic scale reflects the phenomena specific to the characteristic length $l \leq 1$ nm, namely chemical interactions between surfaces.
2. Nanoscale—will reflect phenomena specific to the characteristic length $1\ \text{nm} \leq l \leq 1\ \mu\text{m}$, mainly the van der Waals interaction between surfaces.
3. Microscale—will reflect phenomena specific to the characteristic length $1\ \mu\text{m} \leq l \leq 1\ \text{mm}$, namely the mechanical interlocking of asperities.
4. Macroscale—will reflect phenomena specific to length scale over 1 mm, including the behaviour of the bulk of the body, which will couple the micro-asperities together.

The model is iterative if its vertical construction consists of several generations of asperities. The question of whether a surface topography model may be characterized as multilevel is related to its structure and to whether the asperities of the k-th generation have the same height or not. If all asperities of the same generation have the same height, then the model would be single-level, even if it was multi-scale, i.e. the configuration of the model would allow the single-level asperities to be engaged in mechanisms of interactions between surfaces specific to at least two length-scales. If the asperities of the k-th generation have different heights, then the model is multi-level.

A model will be called hierarchical if it is iterative, i.e. it has at least two asperity generations, and the 'parent' asperities have at least two descendants. In other words, to be hierarchical a model has to have at least one subdivision in the asperity generation. It is thus not enough for the asperities to be an iterative structure, i.e. to be situated on top of each other, to have a hierarchy.

10.2.2 Modelling of Dry Friction Between Rough Surfaces

The above described terminology is very useful for modelling dry friction between engineering surfaces. We would like to remind that in 1699 Amontons (see e.g. [4]) presented his experimental observations on the polishing of optical lenses in the form of the following statements (the Amontons laws): (i) the polishing force is independent of the dimensions of the lens; (ii) the polishing force is proportional to the applied force. In 1774 Kotelnikov, one of the eight former students of Euler, studied friction under external load P and he wrote: "If the friction content F to the mentioned force P one puts as unknown that is equal to μ:1 then the friction will be $F = \mu P$." Hence, Kotelnikov introduced the notion of the coefficient of friction μ, and presented the law as a formula. The Amontons-Kotelnikov law for the frictional force F_f can be written as

$$F_f = \mu P \tag{10.1}$$

In 1785 Coulomb proposed a two-term formula for the force of friction [4–6]

$$F_f = A_1 + \mu P \tag{10.2}$$

where A_1 is an adhesiveness characteristic for two contacting bodies. It follows from the Coulomb law (10.2) that the coefficient of friction (COF) does not remain constant and it depends on the applied load.

We define friction as the ratio between dissipated energy $U_{dissipated}$ and sliding distance x

$$F_f = \frac{U_{dissipated}}{x} = \frac{U_{chemical} + U_{vdW} + U_{mechanical} + U_{acoustic} + U_{capillary} + \cdots}{x} \tag{10.3}$$

The above equation has reflected the fact that the energy dissipation may consist of several components. The components may be caused by different physical mechanisms: breaking of chemical (metallic) bonds $U_{chemical}$, van der Waals interactions U_{vdW}, mechanical interlocking of asperities $U_{mechanical}$, propagation of acoustic waves and vibrations of crystal structures $U_{acoustic}$, capillary interactions $U_{capillary}$ and so on. Each of these mechanisms corresponds mainly to a specific scale of the model, e.g. breaking of chemical bonds corresponds to the atomic scale; van der Waals interactions are specific to the nano-scale, while mechanical interlocking of asperities and plastic deformations of metallic asperities correspond to the micro-scale. The friction coefficient μ can then be expressed as:

$$\mu = \frac{F_f}{(N+A)} = \frac{U_{dissipated}}{(N+A)x} \tag{10.4}$$

$$\mu = \frac{U_{chemical}}{(N+A)x} + \frac{U_{vdW}}{(N+A)x} + \frac{U_{mechanical}}{(N+A)x} + \frac{U_{acoustic}}{(N+A)x} + \frac{U_{capillary}}{(N+A)x} + \cdots \tag{10.5}$$

$$\mu = \mu_{chemical} + \mu_{vdW} + \mu_{mechanical} + \mu_{acoustic} + \mu_{capillary} + \cdots \tag{10.6}$$

where N is the normal force, and A is the force of adhesion. One can see that we have not used the classic one-term Amontons-Kotelnikov law but rather the two-term Coulomb law in the form suggested by Derjaguin [7]. However, we considerably extend these classic ideas by connecting the physical mechanisms at the different scales of the model.

In the interaction between metallic engineering surfaces, the behaviour of the micro-asperities reflects the mechanical interlocking component of energy dissipation, and the behaviour of the nano-asperities reflects the energy dissipated by breaking of chemical (at the atomic scale) and van der Waals bonds (at the nano-scale).

Let us first discuss the dissipation of energy through the breaking of chemical bonds. For very small contacts of chemically active surfaces, the participating atoms engage in chemical interactions [7]; when the body slides, these bonds are broken and a significant amount of energy is dissipated. This phenomenon was called 'cold-welding' by Bowden and Leben [8], who considered the strength of the bonds established by cold welding is equal to the shear strength of the material. Hence, they interpreted the phenomenon at the macro-scale, whereas the current approach uses it at the atomic scale. Indeed, in our model we assume that only the tip of the contacting nano-asperities will be engaged in chemical bonding, whereas a greater number of atoms of the nano-asperity will be engaged in van der Waals interactions.

Evidently, the van der Waals forces have been considered in many models (see, e.g., [9–11]). For example, Heise and Popov [9] incorporate the JKR theory of adhesive contact into the Zhuravlev-Greenwood-Williamson approach (see below) to account for the adhesive component of friction. This component of the friction force is caused by the asymmetry of the adhesive contact: contact begins when the non-deformed asperities come into mechanical contact, but it ends when the critical snap-off force is reached. The novelty of our approach is that adhesion not only increases the area of true contact, but also represents a resistance that has to be overcome when sliding even when surfaces are not in mechanical contact, leading to energy dissipation. The contribution of the van der Waals forces to the shearing of the bonds between particles is a very recent concern in physical chemistry [12].

10.2.3 Simulations of Friction by a Multi-scale Non-hierarchical Model

To illustrate different concepts of scale, level and hierarchy, a single rod model is first considered. The model consists of elements at different scales having the same width but no subdivision from scale to scale, as shown in Fig. 10.1. The model is multiscale, however it is not multilevel, as the asperities of the same generation are on the same level (being only one). In addition, this model is not hierarchical because a model needs more than one micro-asperity or more than one nano-asperity to be called hierarchical.

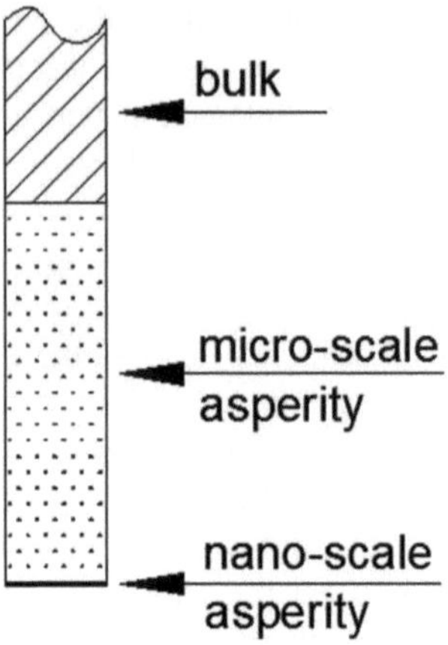

Fig. 10.1 Multiscale, single-level, non-hierarchical model example

This can be a model of one asperity rubbing against a counter-surface. Suppose the model is formulated for a fixed gap between the top of the rod and the mean-line of the counter-surface. The known characteristics are:

- H_1, H_2, H_3—the heights of the nano-asperity, micro-asperity and the considered bulk section of the slider
- E_1, E_2, E_3—the values of the Young's modulus of the nano-asperity, micro-asperity and the considered bulk of the slider
- A—the cross-sectional area of the three sections of the slider
- G—the gap between the top of the slider and the mean-line of the counter-surface
- h—the height of the asperity of the counter-surface.

A sketch of the deformed shape of the asperity in contact with the counter-surface can be seen in Fig. 10.2. From the deformed geometry, it can be written:

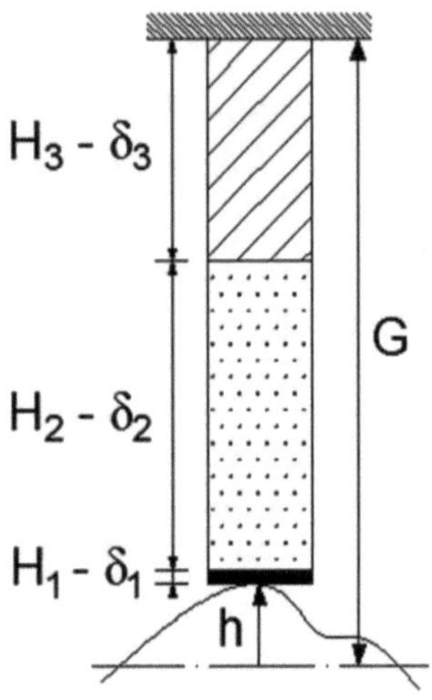

Fig. 10.2 Deformed state of a single-asperity model in contact with the counter-surface

$$\mathrm{h} + (\mathrm{H}_1 - \delta_1) + (H_2 - \delta_2) + (H_3 - \delta_3) = G \tag{10.7}$$

The force equilibrium equation can be written as:

$$\delta_1 \frac{E_1 A}{H_1} = \delta_2 \frac{E_2 A}{H_2} = \delta_3 \frac{E_3 A}{H_3} \tag{10.8}$$

From (10.8), δ_2 and δ_3 can be expressed in terms of δ_1 and substituted in (10.7) to obtain:

$$\delta_1 = \frac{h + H_1 + H_2 + H_3 - G}{1 + \frac{E_1 H_2}{E_2 H_1} + \frac{E_1 H_3}{E_3 H_1}} \tag{10.9}$$

If δ_1 is known, δ_2 and δ_3 can be found by:

$$\delta_2 = \delta_1 \frac{E_1 H_2}{E_2 H_1} \quad \text{and} \quad \delta_3 = \delta_1 \frac{E_1 H_3}{E_3 H_1} \tag{10.10}$$

This model may be developed by incorporating the Polonsky-Keer effect, saying that nano-scale asperities do not have plastic deformations [13, 14]. The further development is due to the influence of the force of adhesion and the energy dissipated by breaking the chemical and van der Waals bonds.

Even though very simple, this model can be valuable in describing the interaction between one asperity and a counter-surface. However, it cannot describe the interaction between two rubbing surfaces as it cannot reflect the interplay between the asperities in contact.

10.3 The Early Non-hierarchical Models of Rough Surfaces

10.3.1 The Prandtl-Tomlinson Model

Even though widely attributed to [15], the model we are going to discuss here cannot be found in his paper, which is about the contribution of molecular forces to friction. As [16] also notes, the model has been introduced by Prandtl [17] and was first used to model plastic deformations in crystals. For this reason we will refer to it

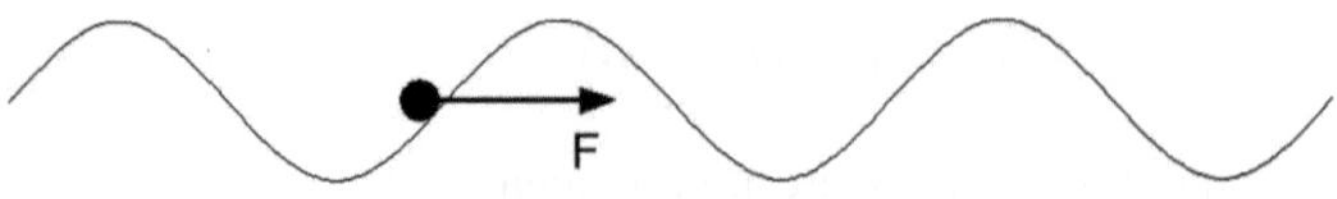

Fig. 10.3 Scheme of the Prandtl-Tomlinson model

as the 'Prandtl-Tomlinson model.' The simplest formulation describes the movement of a point mass in a periodic potential, pulled by a constant force (Fig. 10.3).

The equation of motion of the point mass is:

$$m\ddot{x} = F - \eta\dot{x} - N\sin kx \tag{10.11}$$

where x is the coordinate of the body, m is the mass of the body, F is the external constant force, η is the damping factor, N is the amplitude of the periodic force and k is the wave number. The Prandtl-Tomlinson model reflects the distinction between static and kinetic friction. The minimum force that will produce movement of the point mass is the force of static friction (N) in (10.11). After the body starts to move it will keep sliding even for a smaller force (the force of kinetic friction), because of the energy it already possesses due to its inertia and due to the damping component.

The model thus offers a qualitative description of the friction phenomena. The point mass would represent a surface asperity and the periodic potential would represent the resistance that the asperity has to overcome rubbing against a counter-surface. Because such a uniform roughness profile can hardly be met in a real life scenario, [18] extended the model for a more general shape.

Prandtl [17] also introduced another version of the model where the mass is not pulled by a constant force but through a spring moving horizontally. In this case the equation of motion becomes [16]:

$$m\ddot{x} = c(x_0 - x) - \eta\dot{x} - \frac{\partial U}{\partial x} \tag{10.12}$$

where c is the stiffness of the pulling spring, and x_0 is the position of the free end of the spring.

The spring-generalization of the Prandtl-Tomlinson model has been often used in nanotribology to model the friction between an AFM tip and a surface. The spring takes into account the stiffness of the cantilever of the AFM. This model has been further extended in order to account for both the stiffness of the cantilever and the stiffness of the tip by introducing another spring and mass in the system, as can be seen in Fig. 10.4.

After a straight-forward analytical analysis comparing the two models (see Fig. 10.4) under quasistatic conditions, [19] have reached the conclusion that the solutions given by the two models are identical.

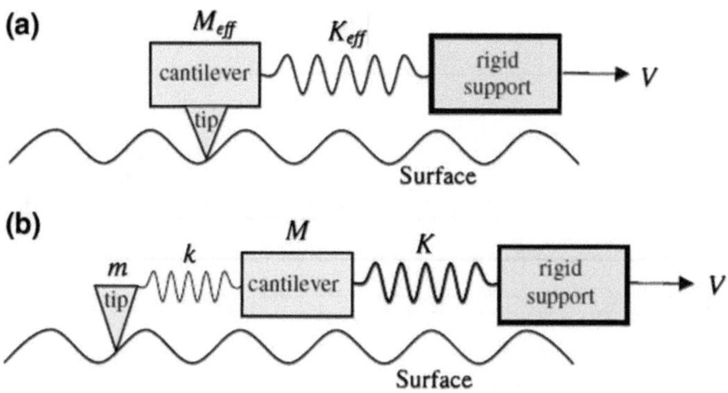

Fig. 10.4 One spring and two springs versions of Prandtl-Tomlinson model. From Lu et al. [19]

10.3.2 The Zhuravlev and Greenwood-Williamson Models

Zhuravlev (1940) model. After Tomlinson's paper was published, Zhuravlev was assigned by his supervisor Davidenkov to explain the Amontons-Kotelnikov friction law [20]. As a result of his studies, a model of dry friction for nominally flat surfaces was introduced, where asperities were represented as independent elastic spherical caps, of the same radius, with random height distribution, that deform according to Hertz contact theory [21]. Figure 10.5 offers a schematic representation of the model.

Zhuravlev's findings were that the true contact area and the friction force are approximately proportional to the external compressing force, and the influence of the external load on the coefficient of friction is very small. Let us briefly go through the argument of Zhuravlev's paper. The meaning of the terms involved will be explained and illustrated in Fig. 10.6.

Using Hertz's solution for elastic contact, the compressing force P and the true contact area S are expressed as follows:

Fig. 10.5 Representation of rough surfaces according to the Zhuravlev model

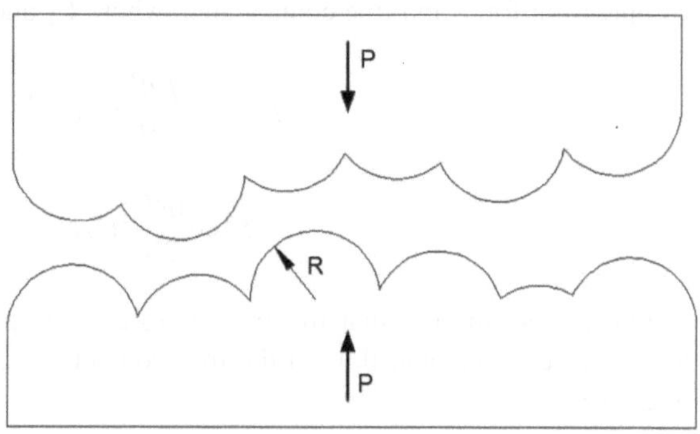

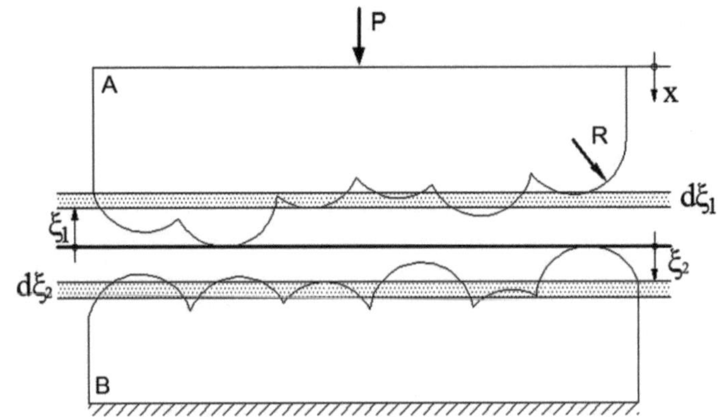

Fig. 10.6 Contact between asperities of the Zhuravlev model

$$P = \frac{\sqrt{2R}}{3\pi k} \int_0^{x/2} \int_0^{x/2} [x - (\xi_1 + \xi_2)]^{3/2} \frac{n(\xi_1)n(\xi_2)d\xi_1 d\xi_2}{N} \tag{10.13}$$

$$S = \frac{\pi R}{2} \int_0^{x/2} \int_0^{x/2} [x - (\xi_1 + \xi_2)] \frac{n(\xi_1)n(\xi_2)d\xi_1 d\xi_2}{N} \tag{10.14}$$

It is obvious that the two integrals in the above expressions are to calculate the total sum penetration for all the contacting asperities within the $d\xi_1$ and $d\xi_2$ layers. The term $\frac{n(\xi_1)n(\xi_2)d\xi_1 d\xi_2}{N}$ expresses the expected number of contacting asperities within these layers because $\frac{n(\xi_2)d\xi_2}{N}$ gives the probability for an asperity on surface *A* to meet an asperity of surface *B* in the layer $d\xi_2$. $N = \int_0^x n(\xi)d\xi$ is the total number of asperities from the level of the tallest asperity down to the depth *x*, and *n* (ξ) is the distribution function of the spherical cap summits.

In order to be able to integrate analytically the expressions (10.13) and (10.14), Zhuravlev assumed the distribution function is linear: $n(\xi) = \beta\xi$. After solving the integrals in (10.13) and (10.14), the following expressions are obtained for the total compressing force and true contact area, where C_1 and C_2 are integration constants:

$$P = \frac{\sqrt{2R}\beta^2}{3\pi kN} C_1 x^{11/2} \tag{10.15}$$

$$S = \frac{\pi R \beta^2}{2N} C_2 x^5 \tag{10.16}$$

On the assumption that the friction force *F* is proportional to the number of molecular contacts and thus to the true contact area, Zhuravlev expresses friction force as:

$$F = \alpha S = \frac{\alpha \pi R \beta^2}{2N} C_2 x^5 \tag{10.17}$$

where α is a coefficient depending on the cohesive forces. Eliminating x from (10.15) and (10.17) the relation between the friction force and the normal compressing force is obtained:

$$F = C \frac{\alpha R^{1/11} \beta^2}{N^{1/11}} k^{10/11} P^{10/11} \tag{10.18}$$

The coefficient of friction can be further expressed as:

$$\mu = C \frac{\alpha R^{1/11} \beta^2}{N^{1/11}} k^{10/11} P^{-(1/11)}$$

The criticism brought to this model was directed to a number of features [20]. Kragelsky [22] noted that the natural height distribution of asperities is rather Gaussian than linear, as considered by Zhuravlev. However, Zhuravlev chose a linear asperity height distribution as an initial approximation, as this allowed him to solve analytically the integrals in order to express the compressive force and the true area of contact. Any heights distribution function can be used in expressions (10.13) and (10.14).

Giving a high appraisal of Zhuravlev paper, [5] also made the following two criticisms: (a) Zhuravlev assumed that the specific frictional force is constant, i.e. $\mu = F_f/P$, while only one term of two in the Amontons–Coulomb law does not depend on the pressure; and (b) the linear distribution of spherical cap heights is not observed in reality.

Greenwood and Williamson (1966). Greenwood [23] notes that the Greenwood and Williamson model (1966) *improves on Zhuravlev by assuming a Gaussian height distribution (still of identical spherical caps).* As Greenwood noted, they introduce not only a Gaussian asperity height distribution, but also plastic deformation of asperities. Greenwood and Williamson [24] consider the contact between a rough surface and a plane (Fig. 10.7).

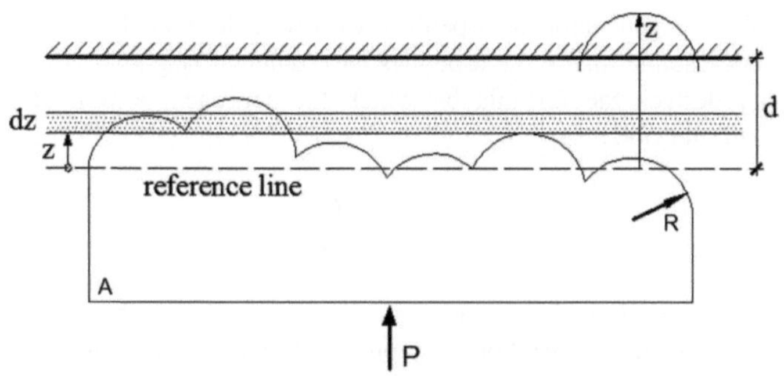

Fig. 10.7 Contact between asperities of the Greenwood-Williamson model

The expected total compressive force and the expected total area of contact are expressed as follows:

$$P = \frac{4}{3} N E^* R^{1/2} \int_d^\infty (z-d)^{\frac{3}{2}} \phi(z) dz \quad (10.19)$$

$$A = \pi N R \int_d^\infty (z-d) \phi(z) dz \quad (10.20)$$

In these expressions N is the total number of asperities, $\phi(z)$ gives the probability that an asperity is within a layer dz and E^* is the contact modulus. The term $N \int_d^\infty \phi(z) dz$ expresses the number of asperities that are expected to make contact when the surfaces are at a distance d apart.

The Greenwood–Williamson model has found good agreement with a variety of experimental systems. However, the model has important drawbacks. One of the most important drawbacks is that the asperities are uncoupled; in reality the asperities are connected through the bulk and influence one-another. The resulting friction force is constant for the pair of rubbing surfaces and does not reflect the local effects and interplay between the asperities of the interacting surfaces.

Thus, any of the Zhuravlev-Greenwood-Williamson type models is a multi-asperity, multi-level model. The model is not multi-scale, as it does not consider multiple physical mechanisms of energy dissipation, it is not iterative, and hence not hierarchical, as it has only one asperity generation.

10.3.3 The Kragelsky Rod-Assembly Model

Trying to improve the Zhuravlev model, [22] developed a multi-asperity model using an assembly of elastic rods of constant cross section [4], as illustrated in Fig. 10.8. According to the above classification, this model is not multi-scale because only one physical mechanism is considered, i.e. the deformation of the rods. It is not hierarchical, as it has only one generation of asperities. However, the model can be considered multi-level because the tips of the asperities have different heights.

Kragelsky derives the formula by which the approach between the surfaces relates to load and obtains:

$$N = A_c \gamma_1 \gamma_2 \frac{K a^3}{6} \quad (10.21)$$

N is the normal load, A_c is the contact area and a is the approach between surfaces. γ_1 and γ_2 are factors that depend on the representation of the distribution function of the asperities, and K stands for the stiffness of the rods. One can see that each

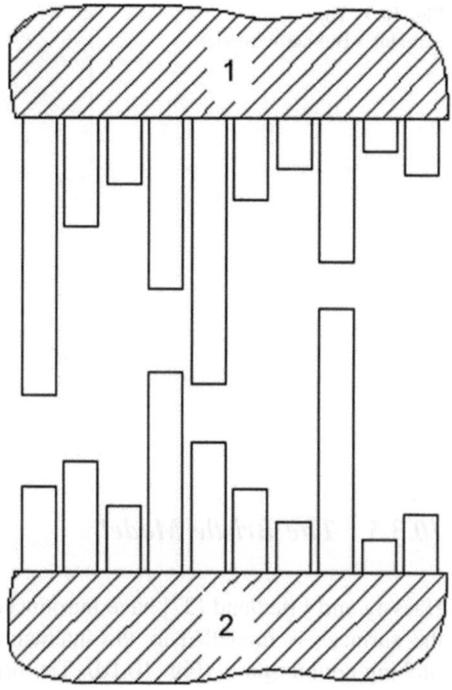

Fig. 10.8 Kragelsky rod assembly model

asperity deforms independently, without any influence on the neighbouring asperities.

10.3.4 The Sergienko-Bukharov Model

In order to develop a model that would reflect the interrelation of normal and tangential micro-oscillations between two rubbing surfaces, Sergienko and Bukharov [25] have studied a model that is a development of an earlier work by Kudinov. As the sketch in Fig. 10.9 illustrates, the model consists in a rigid body resting on an assembly of springs which simulate the asperities. The model reflects oscillations both normally and tangentially to the movement direction because any horizontal impulse given to the rigid body will be accompanied by oscillations in the normal direction. As the body descends, due to the normal oscillations, it will come in contact with more springs. Greater oscillations will give smaller friction force.

One good feature of this model is that it takes into account the dynamic lift of the slider, which is an important phenomenon, as first noted by Tolstoi [26]. Like Kragelsky's, this is a multi-level model, but not multi-scale, nor hierarchical. The asperities are uncoupled as well, because the springs deform independently.

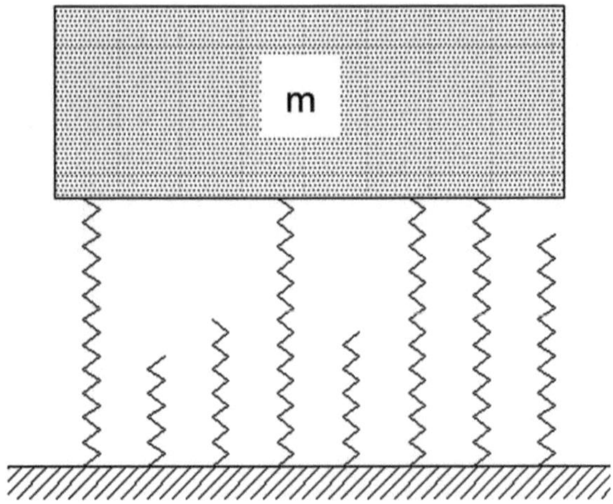

Fig. 10.9 The Sergienko-Bukharov model

10.3.5 *The Bristle Model*

Haessig and Friedland [27] have introduced a model where the interaction between the surfaces is modelled as the interaction between bristles extending from the rubbing counter-parts (Fig. 10.10). The bristles do not represent asperities, but the bonds created between the surfaces. The bristles on one of the surfaces are rigid, whereas the bristles on the counter-surface are deformable. When the slider moves the energy in the contacting bristles increases. When an individual bristle reaches a critical deflection, it snaps off contact, and a new bond is created between the rigid bristle and another deformable bristle. The friction force is the resistance that the slider experiences due to all the bristles in contact and can be expressed as:

$$F_f = \sum_{i=1}^{N} k(x_i - b_i) \tag{10.22}$$

where N is the total number of bristles, k is the stiffness of the bristles, x_i is the position of the bristles, and b_i is the position where the bond with the rigid bristle was formed. The main ideas of the Bristle model may be traced back to classic papers by Euler and Coulomb (see discussions of the models presented by

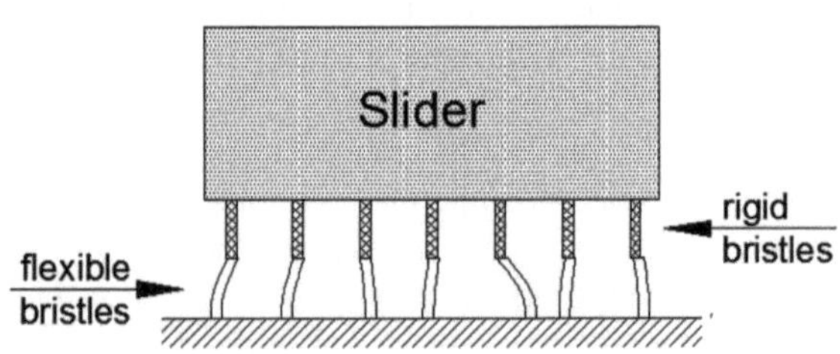

Fig. 10.10 The bristle model

Kragelsky and Schedrov [5], Bowden and Leben [28]). It is quite evident that the behaviour of one deformable bristle is very similar to the behaviour of the elastic pile in the [29] model discussed below in Sect. 10.4.2.

The Bristle model is a multi-asperity model, but it is not multi-scale, multi-level or hierarchical.

10.3.6 The Al-Bender Model

Al-Bender et al. [30] introduced a multi-asperity model of friction that formally reflects mechanisms of creeping, adhesion and elastic-plastic deformations. The model has been developed to describe phenomenologically other mechanisms like pre-sliding behaviour, lift-up hysteresis [31] and wear [32]. A schematic representation of the model is shown in Fig. 10.11.

The model is comprised of a slider and a counter-profile. The slider is perfectly rigid and it contains the flexible surface made of a representation of asperities. One asperity is modelled as a mass connected to the slider through a spring with a certain normal and tangential stiffness, damping and length. The system of uncoupled normal and tangential springs used to model the asperity (shown in the snippet in Fig. 10.11) is equivalent to the elastic pile connected through a torsion spring of [29], discussed in Sect. 10.4.2. The asperities deform independently without any interaction with each other and they are given a length distribution which reflects the original surfaces. The counter-surface has a much higher stiffness than the asperities and its geometry reflects the overlap between asperities during contact. In order to reflect elastic-plastic behaviour, the springs are modelled as individual Maxwell-Slip elements. The equations of motion for the main mass M are as follows:

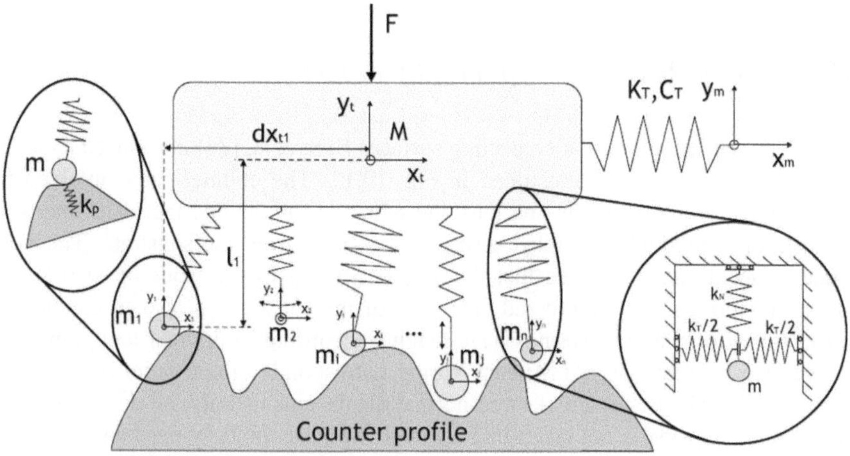

Fig. 10.11 Al-Bender et al. model (from 33)

$$M\ddot{x}_t = K_T(x_m - x_t) + C_T(\dot{x}_m - \dot{x}_t) + \sum_{i=1}^{N} k_{T,i}(x_i - x_t - dx_{T,i}) + \cdots \sum_{i=1}^{N} c_{T,i}(\dot{x}_i - \dot{x}_t) \tag{10.23}$$

$$M\ddot{y}_t = \sum_{i=1}^{N} k_{N,i}(y_i - y_t + l_i) + \sum_{i=1}^{N} c_{N,i}(\dot{y}_i - \dot{y}_t) - F \tag{10.24}$$

The meaning of the terms involved in the equations is as specified in Fig. 10.11. The equations for the asperity masses are given below.

$$m_i\ddot{x}_i = k_{T,i}(x_t - x_i + dx_{T,i}) + c_{T,i}(\dot{x}_t - \dot{x}_i) + \gamma k_{p,i}(x - x_i) + F_{adh,i,x} \tag{10.25}$$

$$m_i\ddot{y}_i = k_{N,i}(y_t - y_i - l_i) + c_{N,i}(\dot{y}_t - \dot{y}_i) + \gamma k_{p,i}(y - y_i) + F_{adh,i,y} \tag{10.26}$$

The meaning of most terms is illustrated in Fig. 10.11. The value of γ is 1 if the asperity is in contact and 0 otherwise. The terms $F_{adh,i,x}$ and $F_{adh,I,y}$ are the tangential and normal components of the force of adhesion acting upon asperity i.

According to our classification, this is a multi-asperity, multi-level model, without features of multi-scale or hierarchy. It is, indeed, a generic model that describes friction-related phenomena only qualitatively. Therefore some of the features of the model lack physical meaning. For example, the force of adhesion in the model is not exactly the force of molecular adhesion, but rather it is a local, asperity level, tangential resistance force, that the asperity has to overcome to be set in motion. However, this force is not physically connected to any chemical or inter-molecular phenomena and therefore, the model cannot be considered multi-scale. The damping coefficients also are artificially imposed parameters that lack physical meaning.

10.3.7 The Real Nano-topography Model

To reflect real topography of contacting surfaces Bora et al. [34] introduced a model whose main features are illustrated in Fig. 10.12. The counter-parts involved in contact are represented by two regions: the substrate and the interface. The interface consists of individual uncoupled springs that extend from the substrate. The substrate is an elastic half-space that deforms according to the Boussinesq solution when the springs are compressed. The role of the substrate is also to couple the springs on top of it. The springs have heights according to the real topography of the surface measured by AFM; each spring corresponds to one point of the measured profile. The nodes are allowed normal displacements only, all deformation is elastic and adhesion is not taken into account. Because the only mechanisms taken into account are the deformation of the asperities represented by springs and their

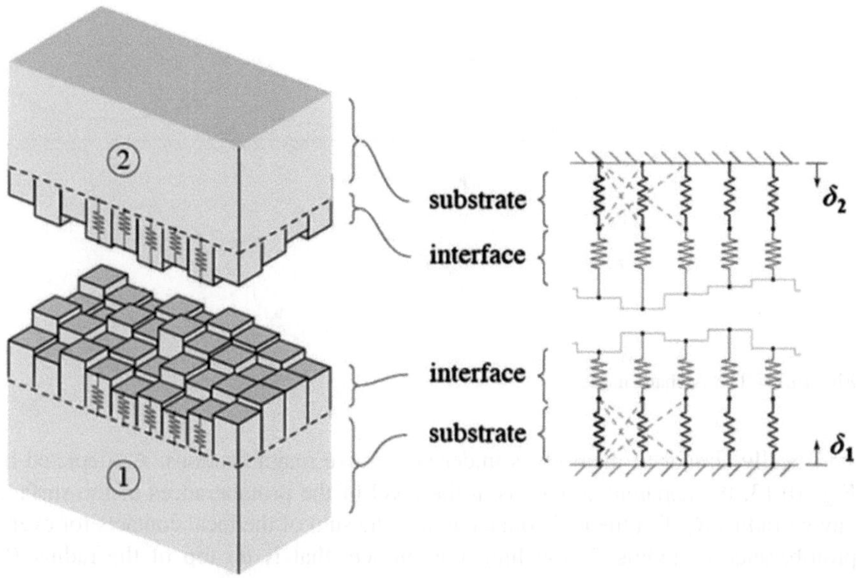

Fig. 10.12 The real nano-topography model (from [34])

coupling through the substrate, this is not a multi-scale model. The model is not hierarchical either because the substrate is a half-space, which cannot reflect the behaviour of individual asperities at the micro-scale. However, the model can be considered multi-level because the nano-asperity generation has different heights. It is a coupled multi-asperity, multi-level model of contact between rough surfaces.

10.4 Hierarchical Models of Engineering Rough Surfaces

We have seen a number of multi-asperity models that are multi-level but feature uncoupled asperities. A class of models that offer asperity coupling is the hierarchical models, as it will be shown in the current section.

10.4.1 The Archard Model

Using Hertz contact theory, Archard [35] introduced the first hierarchical model of roughness, popularly referred to as the "bump on bump" or "asperity on asperity" model. Experimental studies later confirmed Archard's observation, by showing that, at some intervals, the power spectral density functions of real surfaces have power-law character [36]. This was interpreted to mean that rough surfaces are

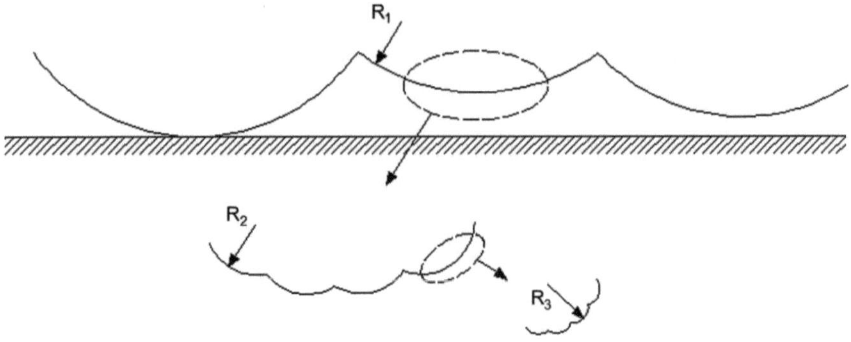

Fig. 10.13 The Archard model

statistically similar to themselves under successive magnifications. As depicted in Fig. 10.13, the apparent contact is at the level of the protuberances of the surface having radius R_1. But the real contact area is the sum of the local contacts for every protuberance of radius R_2 touching the surface, that is on top of the radius R_1 asperity. But if we further magnify the profile, each of these R_2 protuberances has its own protuberances; the model thus reflects the multiscale nature of the surface that is found from macroscale down to atomic scale.

The model is hierarchical because the asperities of generation k are located on the asperities of generation $k-1$. As a consequence, the deformation of the asperity at generation k will influence its neighbours, as well as the asperities at previous generations. Even though it is hierarchical and multi-level, Archard's model is not multi-scale as it takes into account only one physical contact mechanism, namely the elastic deformation of asperities.

10.4.2 The Cantor-Liu and Cantor-Borodich Profiles

A number of scientists have suggested that fractals are a good representation of topography [37, 38]. As [39, 40] noted it is still often believed that the fractal dimension is a scale-invariant characteristic of surface topography and the fractal approach can be very helpful in solving problems of contact between rough solids. The early papers on fractal contact were based on a naive belief that solutions to contact problems for punches having the same fractal dimension would have the same asymptotic behaviour (see, e.g., [41]). In fact, it was shown as early as in 1993 that this is not the case, and the fractal dimension alone cannot characterize the contact properties of surfaces [29, 42]. Furthermore, even if the fractal dimension alone could characterize surface roughness, one would have to solve a very difficult problem of contact between fractal rough surfaces [39, 40, 43].

To model fractal roughness [44] introduced an iterative one-level model known as the Cantor-Liu profile. The main drawback of this model is that the asperities of all generations have the same height and therefore the height of the model is not bounded.

The hierarchical Cantor-Borodich profile [41, 45] has used some features of the Cantor-Liu profile, but with decreasing heights for every asperity generation. The profile has two scaling parameters f_x and f_z such that the width L_i and height h_i of each generation of asperities are given by $L_i = L_0 f_x^i$ and $h_i = h_0 f_z^i$ respectively. The authors say that such a surface could result after unidirectional polishing or wearing, leading to this system of parallel scratches of different depths, as can be seen in Fig. 10.14.

Borodich and Onishchenko [29] have solved indentation problems between a Winkler-Fuss elastic spring foundation and the Cantor-Borodich profile and its reflection. Their conclusion was that fractal dimension alone cannot be used to describe the contact between rough surfaces.

In the same paper, another model is introduced, which uses the Cantor-Borodich profile to model friction between rough surfaces (Fig. 10.15). The asperities of the sliding surface are modelled as telescopic piles with a certain rotational stiffness of the connection with the top body. The model tries to take into account the interlocking of asperities and stick-slip mechanisms of friction. The following stages can be observed in the motion of the pile:

- The pile 'catches' the asperity of the Cantor-Borodich profile.
- The pile rotates with respect to its connection to the top body, gathering elastic energy in the joint, leading to an increase in the resistance force.
- When the pile moves past the asperity, the elastic energy in the connection is released and the resistance force decreases, until it catches another asperity.

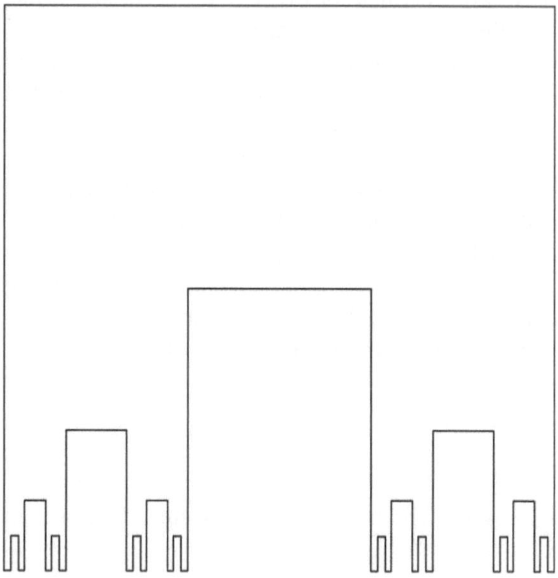

Fig. 10.14 The Cantor-Borodich profile

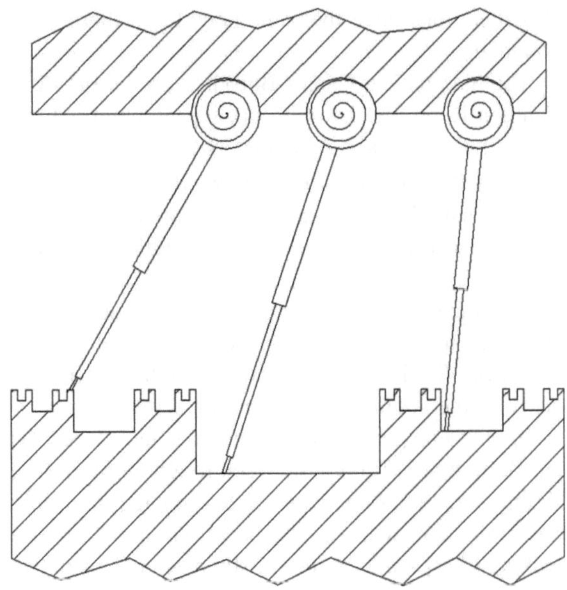

Fig. 10.15 Elastic pile model by Borodich and Onishchenko [29]

10.4.3 The Models Based on the Cantor-Borodich Profiles

There are various generalizations of the Cantor-Borodich profile. In particular, the profile can be multilevel [29], random [46], deformable [29, 47].

Warren and Krajcinovic models. The Cantor-Borodich profile always creates only two asperities for every new generation of hierarchy. However, it is evident that the generation can be such that an asperity can produce any number of descendants in the next generation. Warren and Krajcinovic [48] suggested using three descendants. The model has been used to study contact between elastic and plastic bodies. Warren and Krajcinovic [48] have also developed a stochastic version of the Cantor-Borodich profile.

The Plesha and Ni model. To study the normal stiffness of the interface, Plesha and Ni [47] developed an analytical model (Fig. 10.16) based on the Cantor-Borodich profile and its modification by Warren and Krajcinovic [48]. In this model the asperities are considered elastic springs. When subjected to a compressive load P, the model undergoes a deflection Δ which is given by summation of the deformation at each i layer:

$$\Delta = P\sum_{i=0}^{\infty}\frac{z_i}{A_iE_i} \tag{10.27}$$

A_i, E_i and z_i are the cross-sectional area, the elastic modulus and the height of layer i.

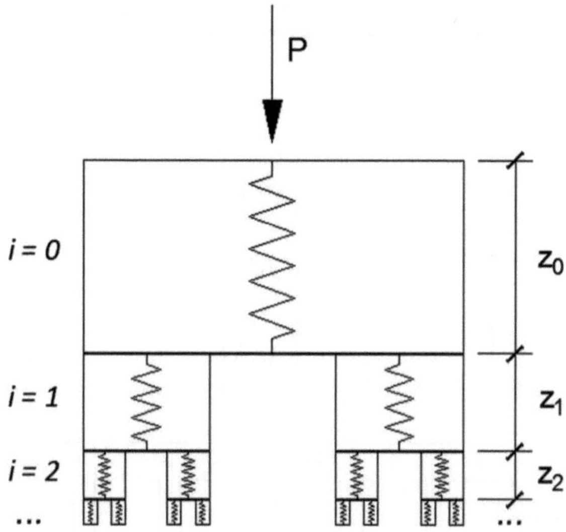

Fig. 10.16 The Plesha and Ni [47] hierarchical spring model of the Cantor-Borodich profile

The visco-elastic Cantor-Borodich profile. Abuzeid and Eberhard [49] have used the Cantor-Borodich profile in order to develop a linear viscoelastic creep model of contact. They used Radok's technique of replacing the elastic constants in the elastic solution by the corresponding operators in the stress-strain equation for the standard-linear-solid model describing the viscoelastic behaviour of the material.

The Schmittbuhl hierarchical model. Schmittbuhl et al. [50] developed a hierarchical model to account for the interaction at the contact between geologic plates. As the sketch in Fig. 10.17 shows, the body is discretized in rigid elements—finer close to the interface, to capture the small details of the slip, and coarser farther away because the stress and strain fields are expected to be smoother. These rigid elements interact through horizontal springs with the elements at the same level, and are coupled with the previous level by vertical springs. The element size increases in geometric series with the distance to the interface (the element size is multiplied by a fixed ratio N) and all the physical parameters describing the elastic interaction are rescaled accordingly.

The friction problem is solved in a recursive manner, reducing each step to a Burridge-Knopoff problem. About 50 years ago, Burridge and Knopoff [51] have proposed a model to describe the mechanics of earthquake faults. As illustrated in Fig. 10.18, the interface consists in a number of blocks of mass m_i, resting on a surface and connected by springs. The blocks are also connected to the moving body by another series of springs, which is the driving geological plate.

The equation of motion of the ith block would then be:

$$m\ddot{x}_i = k_c(x_{i+1} - 2x_i + x_{i-1}) - k_p(x_i - Vt) - F_f(\dot{x}_i) \qquad (10.28)$$

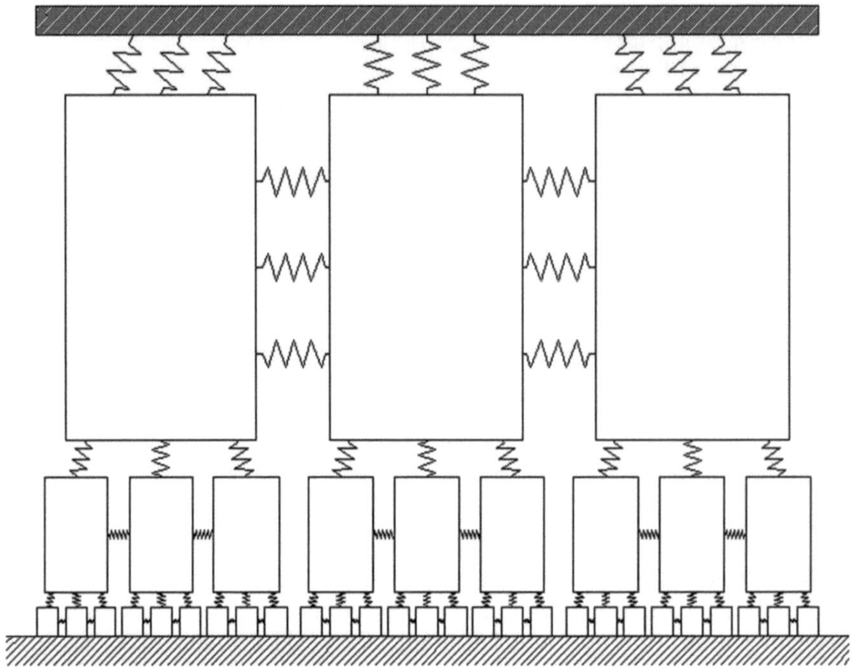

Fig. 10.17 The Schmittbuhl hierarchical model

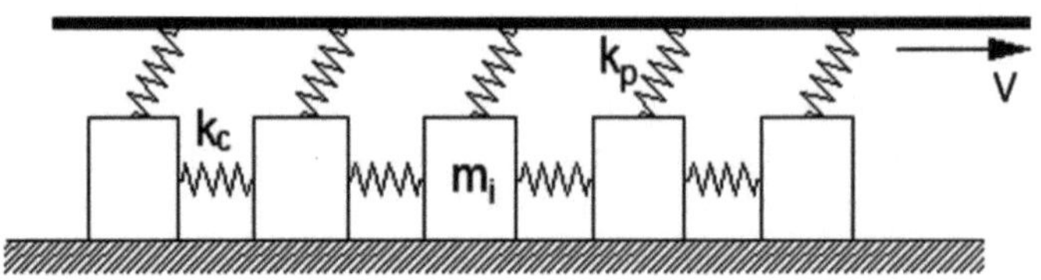

Fig. 10.18 Burridge-Knopoff model

where k_c is the stiffness of the horizontal springs, k_p is the stiffness of the driving springs, and V is the velocity of the moving plate. F_f is a velocity weakening friction force, which means the system will show stick-slip behaviour; the individual block will stick to the surface until the pulling force matches the static friction threshold. After the block will be set in motion, the friction force will decrease.

One can also see the similitude between the Schmittbuhl model and the Cantor-Borodich model with 3 descendants for every generation. Figure 10.19 shows a rearrangement of the Cantor-Borodich profile using blocks and springs to show the similitude in the hierarchical structure between the two models. This hierarchical structure takes into account the lateral stiffness of the system, modelling the coupling between the bulk elastic response of the slider and the friction at the interface level. As it has been mentioned, the hierarchical Cantor-Borodich

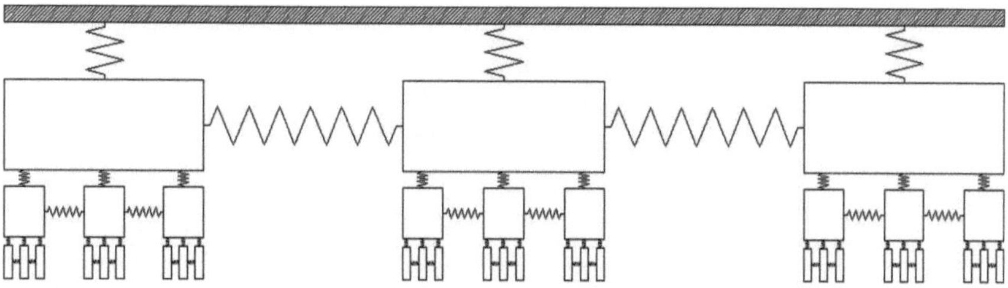

Fig. 10.19 Cantor-Borodich profile using blocks and springs

profile is a single-level model as every block generation is situated at the same level and has constant height, hence the same is applied to the Schmittbuhl model. The model is not multi-scale because it does not feature other physical phenomena than the deformation of the sliding body.

The procedure developed by Schmittbuhl et al. [50] may be described as follows. They initially impose a velocity weakening friction law for the interaction between the rigid interface and the elastic slider. Then the Burridge-Knopoff model is used to solve the problem at this generation and the obtained results are used to formulate the friction law for the next iteration at the superior block generation. Therefore, the problem is reduced to iteratively solving Burridge-Knopoff problems for the chain of blocks at every generation. A drawback of the model is that the friction law is artificially imposed and does not follow from the physical mechanisms involved in friction.

10.4.4 The Borodich-Onishchenko Multilevel and Multiscale Hierarchical Profiles

Borodich and Onishchenko [29, 52] proposed to modify the Cantor-Borodich profile and construct multilevel hierarchical profiles where asperities or gaps of the next generation lie on the tops of all asperities of the previous generations, as it can be seen in Fig. 10.20. The multiscale nature of friction is implied when they say they 'expect that some subset of the surface profile would be responsible for the adhesive component of the contact interaction while another one, having possibly a different fractal dimension, determines the deformational component.'

10.5 Simulations of Friction by Multi-scale Models

The hierarchical models can be used for modelling of various tribological processes, e.g. friction, adhesion, wear and so on. Here we introduce a new model that has many features of the above described models. However, our model gives new meaning for many well-known concepts and is used for the simulation of dry friction.

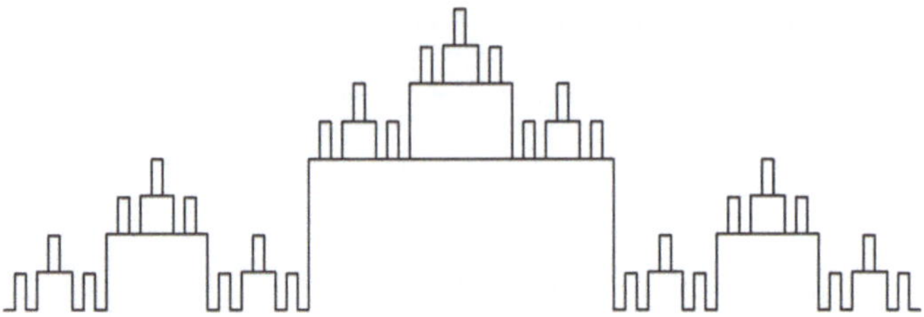

Fig. 10.20 Multilevel extension of the Cantor-Borodich profile

10.5.1 Multi-asperity Models of Surface Roughness

Borrowing elements from the above reviewed models, the author has developed several multi-scale hierarchical models of friction (see a preliminary version of the model presented by Savencu and Borodich [3]). In the first quasistatic, multi-scale, hierarchical model of friction the bottom surface is considered perfectly rigid and contains information about geometry. Its profile is the measurement of the surface roughness by a profilometre. The top surface is represented by a body, from here on referred to by 'slider'. The slider contains information about both the geometry of the sliding surface and the mechanical properties of the counterparts. As one can see in Fig. 10.21, the geometry of the slider is very similar to the Cantor-Borodich profile.

The model also has some features of the Kragelsky rod assembly model, as the asperities are represented as elastic rods. However, as Fig. 10.22 shows, in the case of two copper surfaces, the aspect ratio of the asperities is quite different from Kragelsky's model. The measured surfaces that the slider approximates are quite smooth, therefore the statistical interpretation of the roughness profile has resulted

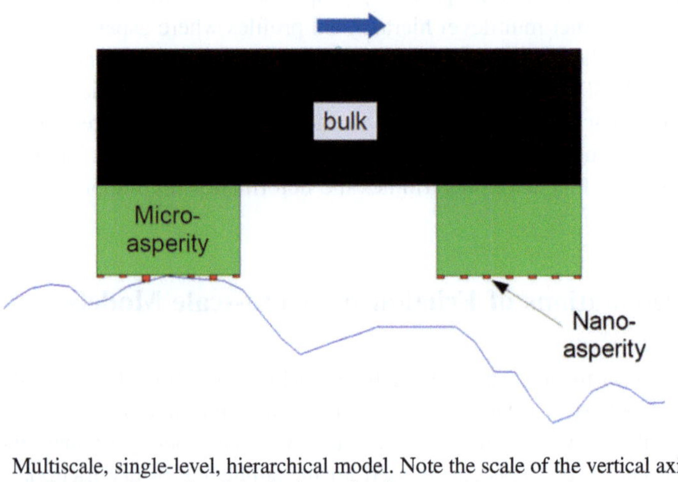

Fig. 10.21 Multiscale, single-level, hierarchical model. Note the scale of the vertical axis is about 1000 times smaller than the scale of the horizontal axis

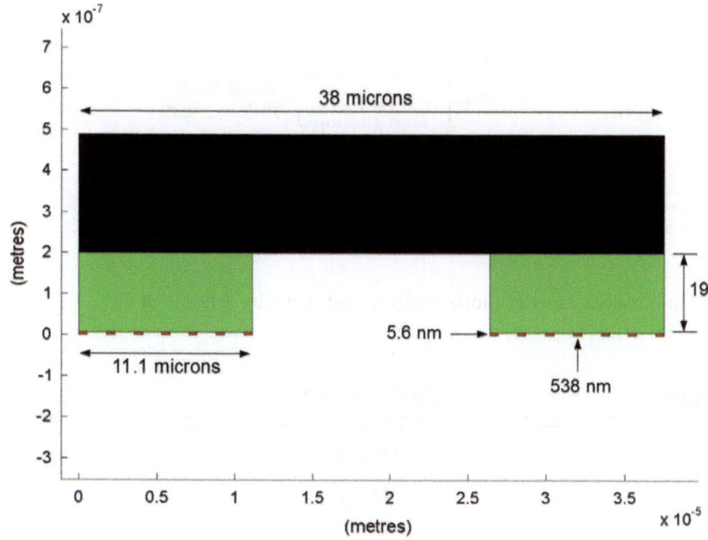

Fig. 10.22 Geometry of the slider. Note the different scales of the axes

in the asperities being more like slabs, rather than vertical rods. The height and width of the micro-asperities, as well as the distance between them were determined by analysing the slider's surface at micro-level, using a profilometre. The geometric characteristics of the nano-asperities, as well as their density, were determined by analysing the slider's surface at nano-level, measured by AFM.

In our model, the nano-asperity can be engaged in 2 types of contact. If the nano-asperity is within the adhesive layer, then it will establish contact due to van der Waals forces. In this case w_{vdW} is the diameter of adhesive contact and w_{chem} is the diameter of chemical contact. Due to modelling both the chemical and the vdW interactions, the nano-asperity has two characteristic widths and two characteristic heights (Figs. 10.23 and 10.24).

It should be mentioned that the geometrical characteristics for both the micro-asperities and the nano-asperities were obtained on the corresponding roughness profiles. For the nano-asperities, the geometry was obtained by analysing the nano-roughness, whereas the geometry for the micro-asperities was obtained by analysing the micro-roughness.

The geometrical dimensions we need to obtain from the surface measurement are the height of the nano-asperity ($h_{vdwI} + h_{chem}$), the width of the chemical interaction domain (w_{chem}), the width of the van der Waals interaction domain (w_{vdW}), and the distance between nano-asperities (d_n).

The height of the van der Waals interaction slab was computed using root mean square of all asperity heights with respect to the mean line

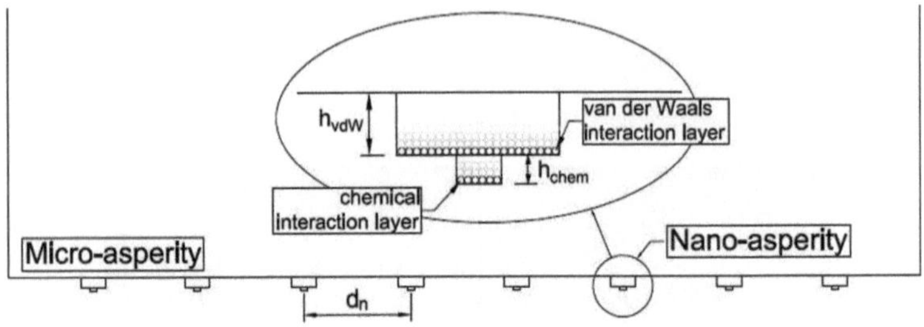

Fig. 10.23 Tip of micro-asperity (horizontally scaled down by a factor of 25)

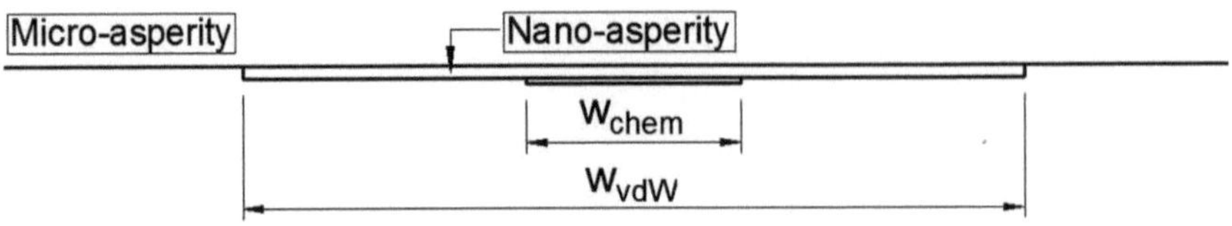

Fig. 10.24 1:1 Representation of one nano-asperity

$$h_{vdW} = \mathrm{R}_q = \sqrt{\frac{1}{n}\sum\nolimits_{i=1}^{n} h_i^2} \tag{10.29}$$

where h_i are the heights of the discrete points of the measured nano-roughness, and n is the number of points.

The width of the van der Waals interaction domain was computed by equally distributing the asperity volume bounded by the mean-line and the height of the van der Waals interaction slab, to the number of asperities in this window. So, as illustrated in Fig. 10.25, the hatched area is the area equally distributed to the number of asperities intersected by the mean-line. The resulting approximation is calculated according to (10.30) and illustrated in Fig. 10.26.

$$\mathrm{w}_{\mathrm{vdW}} = \frac{A_{vdW}}{h_{rms} n_{mean}} \tag{10.30}$$

where A_{vdW} is the area hatched in Fig. 10.26, h_{rms} is the quadratic mean of all asperity heights and n_{mean} is the number of asperities intersected by the mean-line.

The distance between nano-asperities is computed according to (10.31).

$$\lambda = \frac{L_{nano}}{n_{mean}} \tag{10.31}$$

where L_{nano} is the length of the sampling profile.

The height of the chemical interaction slab was computed as the arithmetic average of all asperity heights above the height of the van de Waals interaction slab.

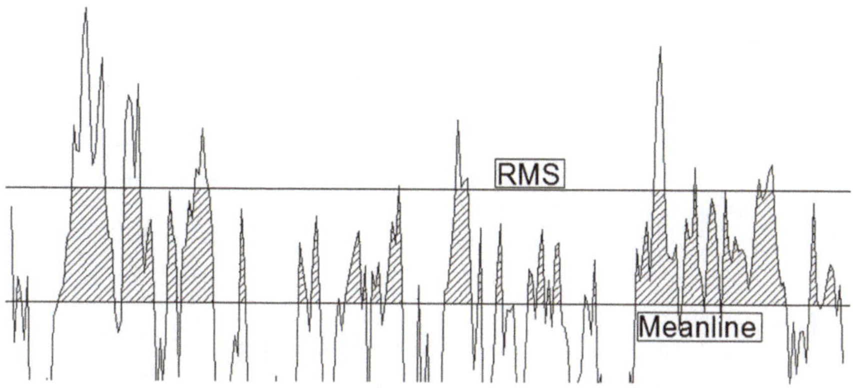

Fig. 10.25 Computation of the width of the van der Waals interaction domain. Note the different axis scales

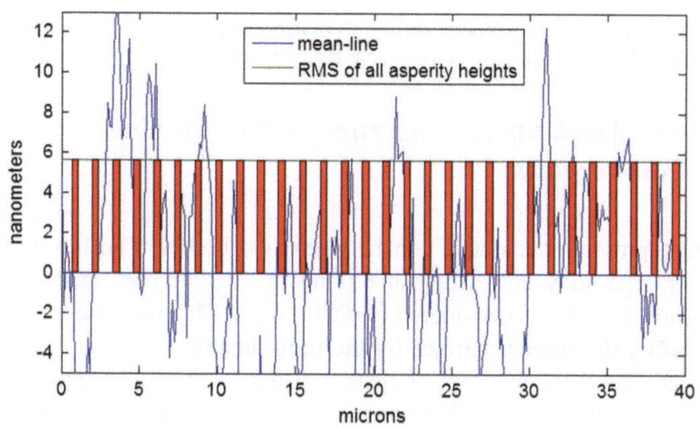

Fig. 10.26 Computation of the width of the van der Waals interaction domain. Note the different axis scales

The width of the chemical interaction domain is computed in a similar way to the width of the van der Waals interaction domain.

We will here illustrate the configuration of a slider derived from analysing a surface as described above, but with improved geometry of the micro-asperities. While the first version of the slider, illustrated in Figs. 10.21 and 10.22, had micro-asperities of the same height, width and distribution, according to the statistical features of the measured roughness, the new model reflects the real roughness profile in a manner similar to the model by Bora et al. [34]. Our input data are two roughness profiles: one profile obtained by measuring the surface using a profilometre and one roughness profile obtained by measuring the surface by AFM. The resulting geometry is sketched in Fig. 10.27.

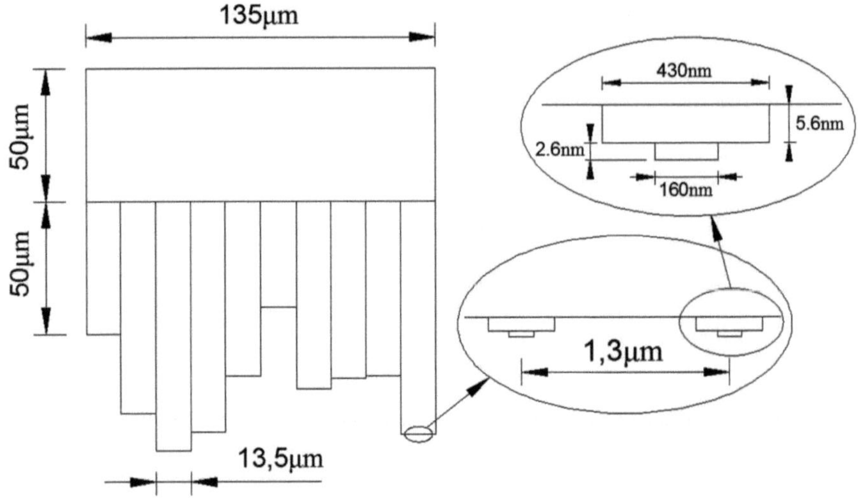

Fig. 10.27 Sketch of a slider geometry derived from measured profiles. Note the snippet sketches have been vertically scaled up by a considerable factor to make geometry distinguishable

10.5.2 The Mechanical Properties of the Rubbing Counterparts

The model represents the interaction between two bodies. The top body will be called slider. The body onto which the slider moves is represented solely by the roughness profile and it is considered perfectly rigid. This roughness profile comes from measuring the support surface by the profilometre.

The slider reflects the mechanical characteristics for the rubbing pair. The model can work with any material characteristics, as it can work with any measured roughness profiles. The section representing the bulk of the slider has the mechanical properties of the bulk of the material. The micro and nano-asperities are characterized by an effective contact modulus for the interacting bodies.

In this model, the adhesive zone was modelled as an attractive layer on top of the support surface which brings into contact any slider nano-asperity that enters it. Concerning the thickness of the adhesive layer, there is experimental evidence of the force of adhesion influencing friction and the jump-off forces up to 150 nm from the surface [53]. It is also known that when the pair potentials corresponding to the London, Keesom and Debye forces are summed up, the result is significant even for separations of 100 nm or more [54]. However, the magnitude of the force of adhesion rapidly decays over this distance. One way to model this using the idea of Maugis would be to consider a 'multi-step' adhesion function, i.e. to consider more than one layer with varying intensity of the force of attraction. However, to minimise the computational cost, in the current work only the bottom layer will be considered, where adhesion is the greatest and can cause nano-asperities to snap into contact. Therefore the thickness of the adhesive layer is assumed here to be 5 nm.

10.6 Some Bio-inspired Adhesive Devices

The abilities of insects and lizards to adhere to vertical walls and walk on ceilings provide a source of inspiration for creation of adhesive devices. One needs to realise that the adhesion between a lizard foot and a wall is governed mainly by van der Waals forces that may be enhanced by capillary effects due to water molecules of vapour, while the insects always use capillary adhesion that is created between hairs and the surface due to secreting a sticky fluid (see, e.g. [55]). Hence, direct application of either classic Hertz contact theory or classic models of adhesive contact between smooth spheres [like Johnson-Kendall-Roberts (JKR) and Derjaguin-Muller-Toporov (DMT) theories] to contact between structured surfaces is not possible. The classic models of adhesive contact need to know the work of adhesion (w) and the effective elastic modulus of the contacting pair in order to predict the adhesion forces. However, w is not a well-known quantity for many modern materials, and to obtain experimental values of w for solids is rather difficult. The common methods for extracting adhesive characteristics of materials are based on the use of direct methods, when even highly skilled researchers need to exercise heightened caution during measurements and their interpretation [56]. In the case of a solitary contact, even the slightest damage of contact due to the presence of contamination or surface irregularities will immediately lead to contact breakage, similarly to the crack propagation in bulk material [57].

Yao and Gao have developed hierarchical gecko hair models to study the properties of contact of the hierarchical biological structures (see, for example, [58]). Their results affirm that the hierarchical structure of the attachment mechanisms is responsible for the excellent adhesive properties, while the asymmetrically aligned seta accounts for the anisotropy of adhesion.

10.6.1 Single-Level, Non-hierarchical Adhesive Microstructures

According to the above introduced definitions, a structure is hierarchical if it has at least 2 generations of asperities, and the k-th generation of asperities consists of more than just one descendant per parent asperity. The authors of micro-fabricated fibrillar adhesives use the concept of hierarchy in a different way, as it will be pointed out in the following section.

Another confusion might appear around the question of whether a structure is multi-scale. Intuitively, one may think the question is about whether the structure has different length-scales. However, our classification is focused more on whether the structure takes into account physical phenomena characteristic to those different length-scales. For example, we might say a structure is multi-scale if it engages in adhesive interactions (nano-scale specific phenomenon) and conforms to the rough counter-surface by allowing mechanical deformation (micro-scale specific phenomenon).

Finally, our third criterion concerns the geometrical levels of a structure, which is not to be confused with the generations of a hierarchical structure. If we say a structure is multi-level, we mean that the asperities at the same generation have different heights, and not that it has more asperity generations, i.e. it is iterative.

With this classification in mind, we will briefly discuss in this section some non-hierarchical adhesive microstructures, and then some hierarchical adhesive microstructures in Sect. 10.6.2.

10.6.1.1 Micro-fabricated Non-hierarchical Adhesives Mimicking Gecko Pads

The attachment mechanism of gecko pads have been well studied, and it is now well-known that they achieve their high adhesive properties due to their hierarchical structure. At the smallest length-scale, the adhesive properties are given to millions of spatulae engaging in attractive molecular interactions. These spatulae are generated on the top of many setae, which in turn are generated on the top of many lamellae that make the gecko pad. Many authors provide magnified images of gecko's pads where this hierarchy can be observed (see, for example, [59]). One can observe the gecko's attaching mechanism is made of a multi-scale hierarchical single-level fibrillar structure. It is multi-scale because the small spatulae engage in molecular forces at the nano-scale, while the higher structures ensure compliance with the rough and uneven counter-surface by deformation at the micro-scale.

Further, the articulation of the components of the hierarchy and the tilted setae allow for anisotropic adhesive properties. This means that, after establishing contact with the counter-surface, adhesion is increased if a shear force is applied in the appropriate direction, and it is decreased if the shear force is applied in the opposite direction.

One of the first adhesives to mimic the gecko pads was built by Geim et al. [60]. Their adhesive was made of a rectangular lattice of non-iterative, single-level polyimide pillars (see Fig. 10.28). The authors have achieved this geometry trying to make the pillars sufficiently flexible, resting on a flexible substrate, to be able to attach to uneven surfaces all at the same time. They have found that the adhesion of the contact is dependent on the pillars' density, and only weakly on their heights or diameter. The assumed reason was that the pillars do not establish contact on the whole surface of their tip, but only on a very small region. This means then that their adhesive worked only in the direction normal to the surface, not taking into account the shear-activated mechanism used by geckos, which was modelled in later studies.

Mimicking this mechanism, Jin et al. [61] have built an adhesive structure that is able to bear a load up to ~ 1.25 N/cm^2 in the direction normal to the surface, and a load of ~ 2.8 N/cm^2 in the direction tangent to the surface. Their 'biomimetic bidirectional switchable adhesive' adhesive is made of tilted prismatic pillars, approximately 30 μm apart, arranged in a rectangular lattice, as illustrated in Fig. 10.29.

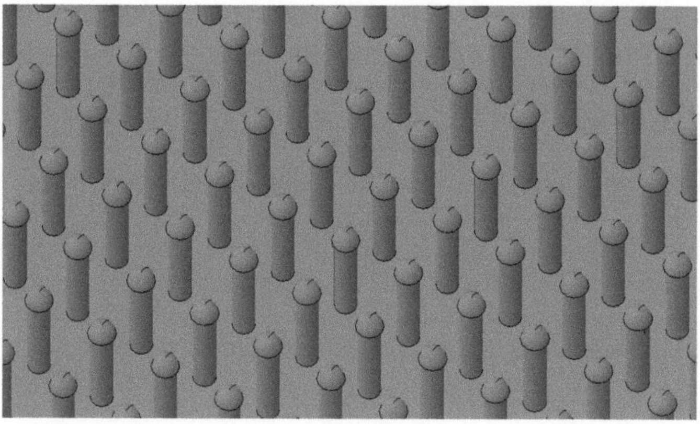

Fig. 10.28 Bio-inspired fibrillary surface. Height of pillar is ~2 μm (after [60])

Fig. 10.29 Sketch of a gecko-inspired adhesive consisting of tilted prismatic pillars with rectangular tip biased towards one of the flat edges (after [61])

The authors say their micro-structure consists of '2 levels of hierarchy', because their triangular pillars (30 μm × 25 μm × 25 μm, 80 μm high) terminate with a rectangular tip (20 μm × 20 μm × 7 μm). From our point of view, this is just a change in cross section, although we grant that, due to the change in cross-section, the rectangular tip and the triangular pillar play different roles. The triangular pillar will provide the mechanical deformation for the structure to better conform to the counter-surface, and their tilted orientation will give the anisotropic properties of adhesion, i.e. when shear will be applied in the appropriate direction the contact area will increase, whereas the contact area will decrease if shear is applied in the opposite direction. The role of the rectangular tip is to adhere to the surface due to the molecular attractive forces. The model is designed to be engaged in two physical mechanisms specific to different length-scales; it will make use of the adhesive forces at the nanoscale and it will use mechanical deformation to conform to uneven surfaces at the microscale. Therefore, the structure built using two iterations (tilted prismatic pillars and then rectangular tips) is multi-scale. The triangular cross-section was chosen so that, when the pillars are deformed due to the shear applied for attaching, they will be in contact with each-other only in one point, and therefore will not stick together.

It is known that, before the onset of sliding, the static coefficient of friction between two surfaces is higher than the kinetic coefficient of friction, and the friction force further decreases as velocity increases (see, for example, [62]). However, studies by Gravish et al. [63] show that, in the case of the gecko setae, it is exactly the opposite: friction and adhesion increase when sliding begins, and they continue to increase when the shear speed increases from 500 nm/s to 158 mm/s. To study this, Gravish et al. [63] built a gecko-like synthetic adhesive, as shown in Fig. 10.30.

It is obvious that the adhesive does not have a hierarchical structure like the gecko fibrils it tries to mimic, and all fibrils are on the same level. However, the fibrils deform mechanically and they also engage in adhesive interactions, two physical mechanisms that make the model multi-scale.

Qu et al. [64] have created an adhesive structure where the fibrils are represented by a single-level, non-hierarchical array of carbon nanotubes, with a straight body and curly entangled tip. The structure showed very good adhesive qualities, with a force about 10 times greater than that of a gecko's pad. Another advantage of this microstructure is that the friction force is much larger than the normal adhesion force, which makes it easy to detach from the counter-surface. However, the very slender nano-tubes (diameters between 10 and 15 nm, and length of about 150 μm) can hardly play any role in micro-scale specific phenomena like deformation and surface compliance mechanisms. Therefore their sole function is to enable the structure to interact with the counter-surface at the nano-scale, engaging in short range van der Waals interactions, which makes the structure single-scale.

Fig. 10.30 Sketch of the fibrillar synthetic adhesive geometry (after [63])

10.6.1.2 Mushroom-like Non-hierarchical Fibrillar Adhesive Microstructures

Even though geckos are very popular for their abilities to climb on vertical or inverted surfaces (see e.g. [65]), they are not the only ones geared with sophisticated adhesive mechanisms. Attachment abilities of insects depend on various factors including the roughness of plant surfaces [66]. Hence, the attachment devices of insects are very good samples for inspiration, e.g. the adhesive introduced by Gorb et al. [67] was inspired by the attachment mechanisms of beetles. The Gorb model is made of an array of mushroom-like fibrils with quite sophisticated shape (see Fig. 10.31) and function (see Fig. 10.32). The fibrillary assembly is not hierarchical in the sense previously defined, nor is it multi-level, as all the pillars are on a flat substrate and have the same heights.

However, due to their shape and size, the pillars have good adhesive capabilities, and are able to establish a large contact area both in normal contact and when shear force is applied (see Fig. 10.32). They are also able to undertake some deformation, both in normal and tangential directions. Therefore, the fibrillar structure can be considered multi-scale.

As illustrated in Fig. 10.31 and pointed out by Gorb et al. [67], the shape of the pillars give the adhesive a good performance in many scenarios. The contact plate lip is tolerant to surface impurities, the narrow neck joint and high aspect ratio make it adaptable to uneven surfaces. Also the fact that the pillars are distributed in a hexagonal pattern rather than in a rectangular array, gives a higher packing density of the fibrils.

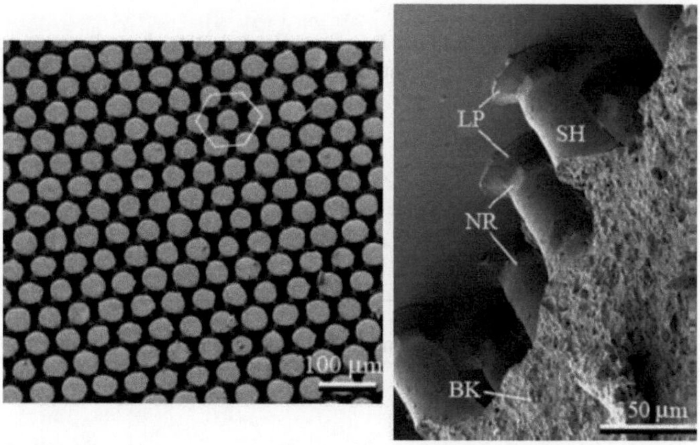

Fig. 10.31 Biomimetic mushroom-shape fibrillary adhesive microstructure made of PVS, view from above (*left-hand side*) and in contact with glass substrate (*right-hand side*). *BK* backing; *LP* contact plate lip; *NR* narrow neck; *SH* pillar shaft (from [67], used with permission)

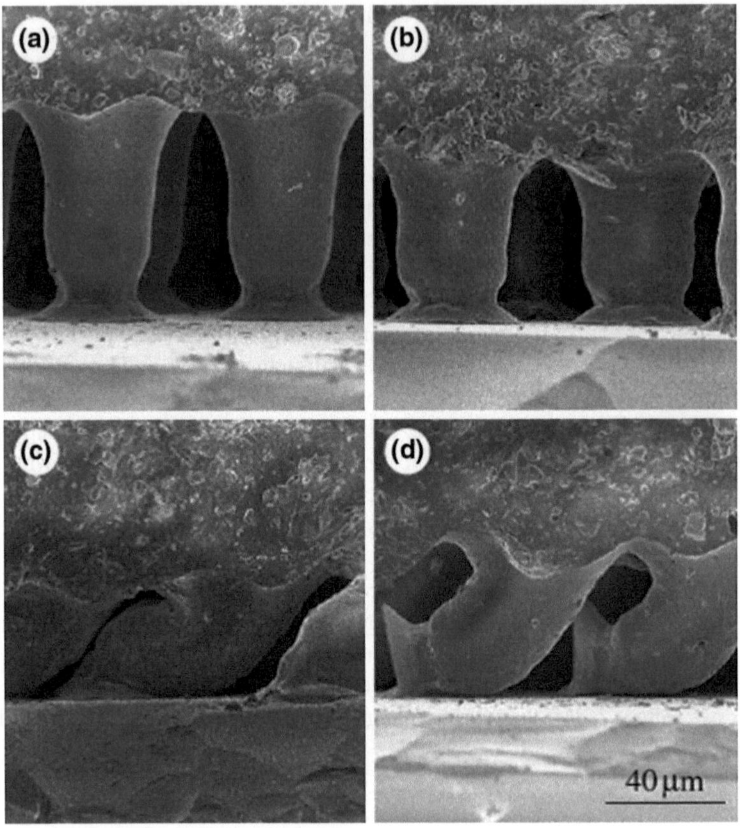

Fig. 10.32 Mushroom-shaped microfibres in **a** an unloaded state, **b** under pure compression, **c** under pure compression distorted by slight shear and **d** under large shear (from [68], used with permission)

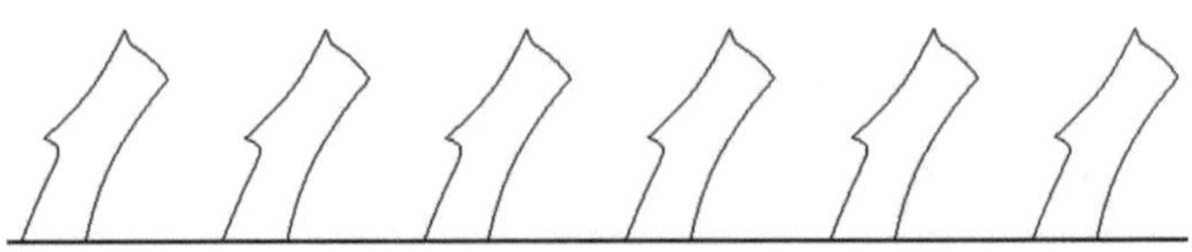

Fig. 10.33 Sketch of angled polyurethane microfibers with angled mushroom tips. Tip orientation can be controlled to form tips with varying angles (after [69])

Another variation of the mushroom-like fibrils is proposed by Murphy et al. [69]. However, as it can be seen in Fig. 10.33, the adhesive is made of a single-level array of angled mushroom tips, with no feature of hierarchy.

The idea of mushroom like fibrils was used also by Kim et al. [70] to develop an adhesive with wet self-cleaning ability. The fibrillar structure is still single-level and non-hierarchical.

10.6.2 Hierarchical Adhesive Microstructures

We have seen the mentioned models are multi-scale, but all of them lack the hierarchical nature we can find in the biological attachment mechanisms. In the current section we will analyse some multi-scale hierarchical adhesive microstructures.

10.6.2.1 Micro-fabricated Hierarchical Adhesives Mimicking Gecko Pads

Tian et al. [59] also point out that most of the gecko-inspired adhesives have limited hierarchy, and the nano-pillars rest on a continuous substrate, which is different to the hierarchical structure of the real gecko. In order to couple the mechanical deformation specific to the macro-scale with the adhesive interactions specific to the nano-scale, they have introduced a hierarchical, multi-scale model (Fig. 10.34). Note similarities with a single generation model by Schargott et al. [11].

The proposed model is hierarchical because, as the biological structure it mimics, it has the small springs (representing the setae) populating the larger spring (representing the lamellar skin). As a result, the larger spring sustains most of the normal load during preloading and provides the soft support for the setae to achieve good contact with the counter-surface. At the nano-scale, the spatulae that populate the setae are responsible for the adhesive properties of the gecko pad. Therefore the resulting mechanism is multi-scale.

Jeong et al. [71] have built two types of dry adhesive with slanted nano-hairs. The distinction is that in one of the adhesives the hairs were supported by a flat layer, whereas in the other, the hairs were supported by micro-pillars, resulting in a hierarchical structure (see Fig. 10.35).

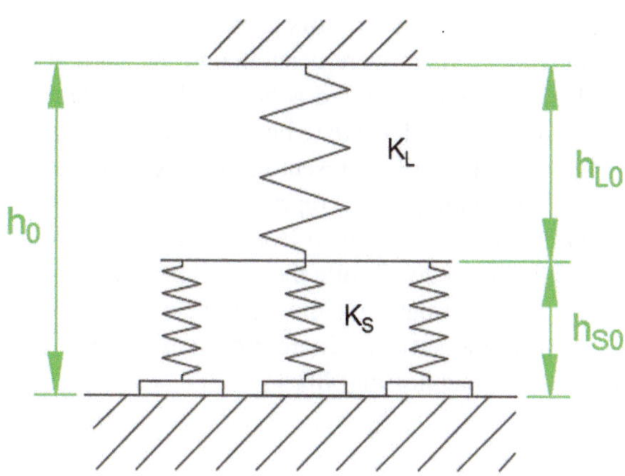

Fig. 10.34 Sketch of the spring model at initial contact. The lamellar skin has an initial height h_{LO} and a spring constant K_L, whereas the setal array has an initial height h_{SO} and an effective spring constant K_S (after [59])

Fig. 10.35 Sketch of 2-level hierarchical PUA hairs formed on 5 μm pillars with 5 μm spacing and 25 μm height (after [71])

Both types of adhesives show anisotropic adhesive properties, due to the nano-hairs being tilted. So, when shear is applied in one direction the contact area and the adhesion increases (to a maximum of ~26 N/cm^2), whereas when shear is applied in the opposite direction adhesion decreases (to ~2.2 N/cm^2) allowing for easy detachment.

The hierarchical adhesive proved superior in maintaining adhesion force even on a rough surface, whereas the simple nano-hairs lost their adhesion strength. The reason for this is that the hierarchical adhesive was more compliant to the counter-surface and therefore made better contact. This reflects multi-scale behaviour; at the micro-scale the micro-pillars allowed for more deformation so that a greater number of nano-hairs came in adhesive contact at the nano-scale. However, all pillars and hairs were generated at the same level, all having the same heights.

The last gecko-inspired adhesive that we discuss was probably the first hierarchical reversible adhesive to be built, and was developed by Northen et al. [72]. The polymeric nanorods representing the gecko's spatulae are generated on the surface of flexible nickle paddles representing gecko's setae (see sketch in Fig. 10.36). The polymeric fibrils have the role of sticking to the counter-surface due to the forces of molecular attraction, while the nickel paddles can enhance or decrease adhesion by controlled deformation in magnetic field, which can turn the fibrillar face towards or away from the surface. Therefore, the system is multi-scale and hierarchical, with single-level paddles and fibrils.

Fig. 10.36 Sketch of freestanding nickel cantilevers and paddles coated with nanorods (after [72])

10.6.2.2 Micro-fabricated Hierarchical Adhesive Mimicking Spider's Leg

Spiders are another source for bio-mimetic solutions [73]. In order to analyse the adhesive mechanism of a spider's leg, Gasparetto et al. [74] proposed a three level hierarchical spring model. The first spring level was designed to model the spatulae, the second level models the setae, and the third represents the tarsus of the spider's attachment mechanism. This model is quite similar to three generations of the spring version of the Cantor-Borodich profile (see Fig. 10.16).

It has been observed that the spider has the ability of changing the stiffness of certain portions in the limb by changing the inner pressure. This changes also the stiffness of the tarsus, the third level in the hierarchical model, leading to less overall adhesion. Indeed, the fact that a soft substrate enables the small fibrils to establish better contact with the counter-surface was noted before. Whereas spiders use varying inner leg pressure to obtain a material gradient, other fibrillar adhesives have attained special adhesive properties by varying the cross-section of the fibrils (e.g. [61, 63]). We can also mention here the adhesive tarsal setae of beetles which use the same principle [75, 76].

The hierarchical springs model then enables a good coupling of the macro-scale deformation and rigidity to the nano-scale adhesive properties. Therefore the resulting system can be called multi-scale.

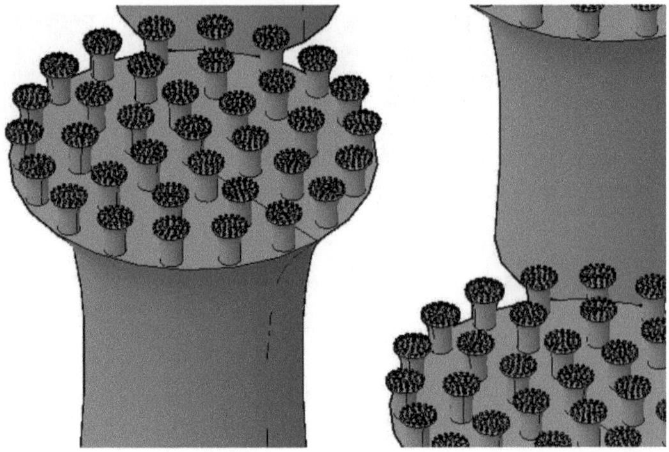

Fig. 10.37 Sketch of three fibrils generations polyurethane fibers (after [77])

10.6.2.3 Mushroom-like Hierarchical Fibrillar Adhesive Microstructures

A hierarchical adhesive made of similar mushroom-like fibrils was built by [77]. Figure 10.37 is a sketch of the hierarchical structure they managed to build, with three generations of fibrils. The authors reported the hierarchical adhesives had a superior behaviour to the simple one-generation adhesive, with increased adhesion and interface toughness, which is attributed to greater surface compliance and increased extension during detachment. Besides being hierarchical, this adhesive works both at the nano-scale, making use of the adhesive interactions, and at the macro-scale, where the superior generations of fibrils give increased conformation to the surface.

Due to small volume to surface ratios, the small fibrils of the above hierarchical structures show very good adhesive properties. However, it would be interesting to understand if the higher generations of the above bio-inspired surfaces do not wear away after a few attachment-detachment cycles.

10.7 Concluding Discussion

In this chapter, we have discussed mainly issues related to surface topography in a large number of models inspired from engineering applications and biological attachment mechanisms. The physical and chemical processes involved were considered only briefly.

One important aspect which distinguishes the engineering models from the biology-inspired ones is the aspect ratio. The engineering models representing rough surfaces are quite flat compared to the slender fibrils of the adhesive microstructures. This is due to the fact that they have different functions; the engineering surfaces are designed to be resistant to friction and wear in the form of ploughing, chipping, scratching or other forms of surface deterioration. The asperities are repeatedly subjected to very high shear loads, whereas the shear force in the adhesive mechanisms is only applied to increase the adhesive capabilities of the elements involved. In the biological mechanisms the fibrils have to have enough flexibility to conform to the counter-surface, but they have to be stiff enough not to stick to each other and be able to be used again for many cycles. However, there are exceptions where biological adhesive mechanisms do not have fibrillar structure and their aspect ratios are closer to the engineering surfaces (see e.g. the toe pad of the rock frog [78]).

We have seen that, with the exception of the nanotube micro-adhesive structure by Qu et al. [64], all the discussed adhesive structures are multi-scale. In addition to the small fibrils engaging in the short range van der Waals interactions at the nano-scale, all adhesives relied either on a soft support layer or on the higher generations of the hierarchical structures to provide the mechanical deformation needed for compliance with uneven or rough surfaces.

We have also seen that all biologically-inspired models are single-level. While the engineering surfaces need to be multi-level to achieve larger contact area with the rough counter-surface, the biological attachment mechanisms rely on the compliance of a soft support layer, or on the compliance of the higher generations in the case of the hierarchical structures, just like the biological attachment mechanisms they mimic.

Some of the authors of the biomimetic adhesives call their adhesive microstructures 'hierarchical'. However, they use the term in a different sense than as defined in the beginning of our discussion. In their terminology, changes in the cross-section of one fibril are different 'levels of hierarchy.' For instance [67] considers the structure is hierarchical because the mushroom shaped fibrils end with a plate, and [61] considers their microstructure is hierarchical because the fibril changes the cross-section from triangular to rectangular. Although we grant these changes in the cross-section of the fibrils enable the different sections to play distinct roles (for example, the section change at the tip maximises the area of adhesive contact), they do not make the structure hierarchical as we have defined the term. In our discussion, a structure is hierarchical if it has more than two generations of asperities/fibrils, and the number of descendants for each parent asperity/fibril is more than one.

Acknowledgements This work was fulfilled in the framework of the CARBTRIB International Network on Nano-phenomena and Functionality of Modern Carbon-Based Tribo-Coatings. Thanks are due to the Leverhulme Trust for financial support of the Network. The materials of the paper have been presented at the 1st International CARBTRIB Workshop (Cardiff University, 19–20 January 2016). The authors are grateful to their partners within the CARBTRIB International Network: Professor S. Gorb from University of Kiel, Germany and Dr. A. Vernes from Austrian Excellence Center for Tribology (AC2T), Wiener Neustadt, Austria for their valuable comments on the work. One of the authors (O.S.) was also supported within the framework of the Austrian COMET-Program (Project K2, XTribology, No. 824187/849109) of the Excellence Centre of Tribology.

References

1. M. Scherge, S.N. Gorb, *Biological Micro- and Nanotribology* (Springer, Berlin, 2001)
2. S.N. Gorb, *Attachment Devices of Insect Cuticle* (Kluwer Academic Publishers, Dordrecht, 2001)
3. O. Savencu, F.M. Borodich, in *NSCM-27 27th Nordic Seminar on Computational Mechanics*, ed. by A. Eriksson, A. Kulachenko, M. Mihaiescu, G. Tibert (Stockholm, 2014), pp. 136–139
4. I.V. Kragelsky, M.N. Dobychin, V.S. Kombalov, *Friction and Wear: Calculation Methods* (Pergamon, Oxford, 1982)
5. I.V. Kragelsky, V.S. Schedrov, *Development of the Science About Friction* (USSR Academy of Sciences Press, Moscow, 1956)
6. D. Dowson, *History of Tribology* (Longman, New York, 1979)
7. B. Derjaguin, Kolloid Zeitschrift **69**, 155 (1934)
8. F.P. Bowden, L. Leben, Nature **141**, 691 (1938)
9. R. Heise, V.L. Popov, Tribol. Lett. **39**, 247 (2010)
10. I.G. Goryacheva, Y.Y. Makhovskaya, Dokl. Phys. **49**, 534 (2004)
11. M. Schargott, V.L. Popov, S. Gorb, J. Theor. Biol. **243**, 48 (2006)
12. J. Carrasco, J. Klimes, A. Michaelides, J. Chem. Phys. **138**, 9 (2013)
13. I.A. Polonsky, L.M. Keer, J. Tribol. ASME **118**, 335 (1996)
14. I.A. Polonsky, L.M. Keer, Proc. R. Soc. a Math. Phys. Eng. Sci. **452**, 2173 (1996)
15. G.A. Tomlinson, Philos. Mag. **7**, 905 (1929)
16. V.L. Popov, *Contact Mechanics and Friction* (Springer, Berlin, Heidelberg, 2010)
17. L. Prandtl, Zeitschrift Fur Angew. Math. Und Mech. **8**, 85 (1928)
18. A.E. Filippov, V.L. Popov, Phys. Rev. E **75**, 4 (2007)
19. P. Lu, Y.C. Loke, X.S. Tang, S.S. Kushvaha, S.J. O'Shea, Tribol. Lett. **43**, 73 (2011)
20. F.M. Borodich, Proc. Inst. Mech. Eng. Part J: J. Eng. Tribol. **221**, 893 (2007)
21. V.A. Zhuravlev, F.M. Borodich, Zh. Tekh. Fiz (J. Tech. Phys.) **10**, 1447 (1940)
22. I.V. Kragelsky, Bull. Acad. Sci. USSR Tech. Sci. Div. **10**, 1621 (1948)
23. J.A. Greenwood, in *Applied Surface Modelling*, ed. by C.F.M. Creasy, C. Craggs (Ellis Horwood, New York, 1990)
24. J.A. Greenwood, J.B. Williamson, Proc. R. Soc. Lond. Ser. A Math. Phys. Sci. **295**, 300 (1966)
25. P.V. Sergienko, N.S. Bukharov, *Noise and Vibration in Friction Systems*, 1st edn. (Springer International Publishing, Berlin, 2015)
26. D.M. Tolstoi, Wear **10**, 199 (1967)
27. D.A. Haessig, B. Friedland, J. Dyn. Syst. Meas. Control ASME **113**, 354 (1991)
28. D. Dowson, *History of Tribology* (Professional Engineering Publishers, London, 1998)
29. F.M. Borodich, D.A. Onishchenko, J. Frict. Wear **14**, 14 (1993)
30. F. Al-Bender, V. Lampaert, J. Swevers, Tribol. Lett. **16**, 81 (2004)
31. F. Al-Bender, K. De Moerlooze, P. Vanherck, Tribol. Lett. **46**, 23 (2012)
32. K. De Moerlooze, F. Al-Bender, H. Van Brussel, Wear **270**, 760 (2011)
33. K. De Moerlooze, F. Al-Bender, H. Van Brussel, Tribol. Lett. **40**, 1 (2010)
34. C.K. Bora, M.E. Plesha, R.W. Carpick, Tribol. Lett. **50**, 331 (2013)
35. J.F. Archard, Proc. R. Soc. London Ser. A Math. Phys. Sci. **243**, 190 (1957)
36. R.S. Sayles, T.R. Thomas, Nature **271**, 431 (1978)
37. M.V. Berry, J.H. Hannay, Nature **273**, 573 (1978)
38. S.R. Brown, C.H. Scholz, J. Geophys. Res. Earth Planets **90**, 5531 (1985)
39. F.M. Borodich, Encycl. Tribol. **2**, 1249 (2013)
40. F.M. Borodich, Encycl. Tribol. **2**, 1264 (2013)
41. F.M. Borodich, A.B. Mosolov, PMM J. Appl. Math. Mech. **56**, 681 (1992)
42. F.M. Borodich, C. R. Acad. Des. Sci. Paris Ser. 2 **316**(2), 281 (1993)
43. F.M. Borodich, H.P. Evans, Encycl. Tribol. **2**, 1246 (2013)
44. T. Kaplan, S.H. Liu, L.J. Gray, Phys. Rev. B **34**, 4870 (1986)

45. F.M. Borodich, A.B. Mosolov, Zhurnal Tekhnicheskoi Fiz. **61**, 50 (1991)
46. T.L. Warren, D. Krajcinovic, Wear **196**, 1 (1996)
47. M.E. Plesha, D. Ni, Int. J. Numer. Anal. Methods Geomech. **25**, 741 (2001)
48. T.L. Warren, D. Krajcinovic, Int. J. Solids Struct. **32**, 2907 (1995)
49. O.M. Abuzeid, P. Eberhard, J. Tribol. ASME **129**, 461 (2007)
50. J. Schmittbuhl, J.P. Vilotte, S. Roux, J. Geophys. Res. Earth **101**, 13911 (1996)
51. R. Burridge, L. Knopoff, Bull. Seismol. Soc. Am. **57**, 341 (1967)
52. F.M. Borodich, D.A. Onishchenko, Int. J. Solids Struct. **36**, 2585 (1999)
53. M. Lessel, P. Loskill, F. Hausen, N.N. Gosvami, R. Bennewitz, K. Jacobs, Phys. Rev. Lett. **111**, 5 (2013)
54. H.M. Pollock, Fundam. Frict. Macrosc. Microsc. Process. **220**, 77 (1992)
55. E. Gorb, S. Gorb, Ecol. Biomech. A Mech. Approach Ecol. Anim. Plants **147** (2006)
56. F.M. Borodich, B.A. Galanov, Proc. R. Soc. A Math. Phys. Eng. Sci. **464**, 2759 (2008)
57. E.V. Gorb, S.N. Gorb, Iutam Symp. Scaling Solid Mech. **10**, 243 (2009)
58. H. Yao, H. Gao, J. Mech. Phys. Solids **54**, 1120 (2006)
59. Y. Tian, J. Wan, N. Pesika, M. Zhou, Sci. Rep. **3**, 6 (2013)
60. A.K. Geim, S.V. Dubonos, I.V. Grigorieva, K.S. Novoselov, A.A. Zhukov, S.Y. Shapoval, Nat. Mater. **2**, 461 (2003)
61. K.J. Jin, J.C. Cremaldi, J.S. Erickson, Y. Tian, J.N. Israelachvili, N.S. Pesika, Adv. Funct. Mater. **24**, 574 (2014)
62. E. Rabinowicz, J. Appl. Phys. **22**, 1373 (1951)
63. N. Gravish, M. Wilkinson, S. Sponberg, A. Parness, N. Esparza, D. Soto, T. Yamaguchi, M. Broide, M. Cutkosky, C. Creton, K. Autumn, J.R. Soc, Interface **7**, 259 (2010)
64. L.T. Qu, L.M. Dai, M. Stone, Z.H. Xia, Z.L. Wang, Science **322**, 238 (2008). (80)
65. H.J. Gao, X. Wang, H.M. Yao, S. Gorb, E. Arzt, Mech. Mater. **37**, 275 (2005)
66. F.M. Borodich, E.V. Gorb, S.N. Gorb, Appl. Phys. A Mater. Sci. Process. **100**, 63 (2010)
67. S. Gorb, M. Varenberg, A. Peressadko, J. Tuma, J.R. Soc, Interface **4**, 271 (2007)
68. M. Varenberg, S. Gorb, J.R. Soc, Interface **5**, 785 (2008)
69. M.P. Murphy, B. Aksak, M. Sitti, Small **5**, 170 (2009)
70. S. Kim, E. Cheung, M. Sitti, Langmuir **25**, 7196 (2009)
71. H.E. Jeong, J.K. Lee, H.N. Kim, S.H. Moon, K.Y. Suh, Proc. Natl. Acad. Sci. U.S.A. **106**, 5639 (2009)
72. M.T. Northen, C. Greiner, E. Arzt, K.L. Turner, Adv. Mater. **20**, 3905 (2008)
73. S. Niederegger, S.N. Gorb, J. Comp. Physiol. A Neuroethol. Sens. Neural Behav. Physiol. **192**, 1223 (2006)
74. A. Gasparetto, R. Vidoni, V. Zanotto, E. Brusa, *Attaching Mechanisms and Strategies Inspired by the Spiders' Leg* (Department of Electrical, Mechanical and Management Engineering of the University of Udine, Italy, 2008)
75. H. Peisker, J. Michels, S.N. Gorb, Nat. Commun. **4**, 7 (2013)
76. S.N. Gorb, A.E. Filippov, Beilstein J. Nanotechnol. **5**, 837 (2014)
77. M.P. Murphy, S. Kim, M. Sitti, Appl. Mater. Interfaces **1**, 849 (2009)
78. D.M. Drotlef, E. Appel, H. Peisker, K. Dening, A. del Campo, S.N. Gorb, W.J.P. Barnes, Interface Focus **5**, 11 (2015)

[illegible]

Chapter 11
Manufacturing Approaches and Applications for Bioinspired Dry Adhesives

A Path to Commercialization

Dan Sameoto

Abstract This chapter covers the major reported types of manufacturing processes and resulting materials used to reproduce gecko-inspired fibrillar adhesives and similar systems that reversibly adhere based on van der Waals forces. A selection of the large number of structural and mold materials for producing these adhesives is covered, along with a comparison of relative strengths and weaknesses of each process/design from a manufacturing perspective. The challenges of scalable production and use of these adhesives for long term application are covered and suggest that silicone rubbers are most appropriate for durable high quality adhesives, and thermoplastic elastomers more suited for inexpensive, less durable or disposable products.

11.1 Introduction

Since the beginning of serious efforts to duplicate the functionality of the gecko adhesive system over a decade ago [1], researchers have dramatically improved the performance, reliability and yield of artificial fibrillar adhesives for a variety of applications. However, with the bulk of research focusing on the understanding of basic adhesion mechanisms and new methods of producing fibers, the field has had some difficulty in converging on best methods to produce these adhesives, with a wide variety of manufacturing methods, materials and shapes being introduced to the literature [2–4]. A secondary priority of this research has been in the area of manufacturability of the materials, because if they are to become useful or ubiquitous in a way similar to hook and loop fasteners eventually did, new manufacturing methods that are more scalable, controllable and precise are needed so as to produce adhesion that is strong, reliable and predictable from one batch to the next.

D. Sameoto (✉)
Department of Mechanical Engineering, University of Alberta, Edmonton, Canada
e-mail: sameoto@ualberta.ca

L. Heepe et al. (eds.), *Bio-inspired Structured Adhesives*,
Biologically-Inspired Systems 9, DOI 10.1007/978-3-319-59114-8_11

In this chapter, a variety of reported manufacturing methods, materials and geometries will be covered with a specific focus on the challenges and solutions to manufacturability of these biomimetic adhesives on larger scales. It is not meant to be a comprehensive review of all possible methods, but rather an overview of dry adhesives with a specific focus on novel and industrially relevant manufacturing processes. Additionally, while gecko-inspired adhesives have been manufactured in a variety of materials like carbon nanotubes, silicon nanowires and other inorganic nanoscale fibers, the recent trends in research publications and known commercial ventures focus primarily on polymer based fabrication, so this chapter will mostly be concerned with fabrication approaches and challenges using these materials.

11.1.1 History

Prior to the year 2005, virtually all reported gecko-inspired adhesives were simple fibers, with no particular focus on the contact tip geometry [1, 5–9]. The majority of produced designs were based on cylindrical structures, mostly made of polymers, but also including carbon nanotubes [10]. Adhesion tests were completed both with normal loading and shear, but the adhesion strengths of most were very small when considered at any scale larger than a few μm^2. The first manufactured and tested prototypes of dry adhesives were described by Metin Sitti and Ron Fearing in 2003 [1]. This work used either nanorobotics imprinting with an AFM tip to produce a negative template in wax, or using nanoporous membranes from commercial sources as the templates. This basic technique of producing a mold and using a castable polymer to reproduce fibers is by far the most common method of producing gecko inspired adhesives [2] and was also the primary method used in early efforts by a variety of research groups as well. A notable exception to the molding/casting approach was the polyimide dry adhesive fiber system produced by Geim et al. in which polyimide nanofibers were directly produced by e-beam lithography and dry etching [5]. This technique produced fibers with sub-micrometer diameter and maximum heights of 2 μm in a variety of aspect ratios. Macroscopic adhesion was only achieved by placing this polyimide array on top of a softer pressure sensitive adhesive which permitted some compliance to maximize contact area with opposing surfaces. Despite this apparent success, the authors did mention difficulties in the production (very expensive and time consuming to use e-beam lithography to pattern larger areas) and the durability (only several attachment cycles were possible before adhesive failure) of the designs. As such it was one of the first papers to note the severe issues that could arise when trying to scale up such materials for manufacturing. Importantly as well, subsequent efforts to duplicate these original results by other researchers failed to get the same manufacturing techniques to provide any significant macroscale adhesion at all [7], which may not be surprising in retrospect given the extraordinary sensitivity of adhesion strength on contact geometry which was discovered in the years following this early work and at the time was not fully appreciated or investigated.

After 2005 several groups introduced fiber designs that had a mushroom shaped tip which then showed significantly improved adhesion over flat punch geometries in normal loading [11–24]. This symmetric mushroom shape didn't have some features of a true gecko adhesive (easy release or direction sensitive adhesion strength for example), but produced normal and peel strengths far superior to those of simple fibers and was the basis for several early efforts at making climbing robots [25]. One of the best early works confirming the practical impact of contact geometry was reported by Del Campo et al. [12], in which a variety of techniques were used to produce fibers with flat-punch, rounded, concave, spatula and mushroom shaped geometries. Since that time, the majority of adhesives that require high normal adhesion strengths have been made with some type of mushroom shaped tip, while those adhesives designed to mimic the shear enhanced adhesion of geckos can be found in a variety of fiber shapes, including simple cylinders [26–33], asymmetric or tilted posts [14, 34–48], Janus pillars [49, 50] and spatula style tips [12, 18, 51–55].

11.1.2 Manufacturing Approaches

There have been dozens of different processes used by a variety of groups in the last 15 years to produce bioinspired dry adhesives. There is a great deal of variation with respect to material choices, fabrication equipment, size of fibers, areas that can be produced and scalability to industrial production. In general, the most popular structural materials have been elastomers (owing to less demanding feature sizes to produce significant adhesion) and micromolding processes to produce fibers of the desired shape and size. Whether fibers are produced directly or via a replication process, all manufacturing methods can be broadly sorted as either top down or bottom up techniques. At the time of writing, there are at least two commercial ventures focusing on top down manufacturing approaches that rely on molds, and casting or dipping approaches: Gottlieb Binder GmbH & Co. KG [56] and nanoGriptech Inc. [57], and at least one company focusing on the bottom up approach of electrospinnning—Akron Ascent Innovations [58].

11.1.2.1 Top Down Manufacturing Approaches

A top down manufacturing approach is the most commonly used method to define fiber structures directly or indirectly (through production of a mold). This technique broadly refers to manufacturing methods with discrete designs and process steps are well defined, typically through the use of traditional photolithography, deposition and etching steps. The benefit of these techniques are extremely well defined geometries but the initial costs of producing molds or fibers can be high due to the expense of masks, equipment and materials when working in traditional microfabrication environments. In comparison to modern microelectronics and microelectromechanical systems, the minimum sizes or tolerances for dry adhesives are quite large, but the cost and complexity of producing them increase dramatically as the feature sizes are reduced. Additionally, the majority of infrastructure designed for lithography has been setup for batch fabrication on silicon substrates or similar, which have maximum diameter of 300 mm with current generation technology. While exceptions exist for producing large area features like those in flat panel displays, such infrastructure has not yet made it into the literature for manufacturing dry adhesives. Additionally, no directly comparable size scale to the actual gecko foot hair nanostructures has been produced by artificial top-down means, although some specific processes come close [9, 10, 20, 31, 32, 35, 59–61].

With top down manufacturing methods, the primary tool for production of fibers is photolithography. This technology uses light to expose a photosensitive material (known as a photoresist) in a specific pattern that can be then be used as a direct feature, or as a mask layer for subsequent patterning of a non-photosensitive material like silicon or metal. Most photosensitive materials are polymers and can be positive or negative acting. A positive photoresist will become more soluble in a developer after exposure to UV light, meaning that exposed features will become holes in the resist. A negative acting photoresist will become more resistant to development after exposure so that exposed features remain on the substrate after processing. Combining either or both types of photoresist in a single process is an approach that many groups have applied to produce gecko-inspired adhesives in the past [12, 17–19, 22, 26, 38]. Once patterning of a photoresist is completed, further processing like reactive ion etching of underlying silicon can be employed to produce features that are much more mechanically robust than a typical photoresist and can act as a more durable mold. Depending on the choice of the underlying substrate, complex features including spatula or mushroom tips can be produced with a single etch step [11, 16, 39]. Beyond traditional photolithography, alternative approaches such as two-photon lithography [24, 60], micro-milling [23, 42], and laser ablation [47, 62, 63] have been used to produce either fibers directly or produce a mold for subsequent polymer replication. These techniques may permit more specialized [60] or optimized [24] adhesive performance in future designs as they can avoid many of the structural limitations from traditional lithography approaches. A non-exhaustive list of different top-down manufacturing approaches to produce gecko inspired adhesives is listed in Table 11.1.

11.1.2.2 Bottom up Manufacturing Approaches

Bottom up manufacturing is in general a method that uses either growth, self-assembly or some other statistical process that produces the features that are desired. The most common examples of bottom up manufacturing for gecko-inspired adhesives are those of carbon nanotubes [10], colloidal self-assembly [82, 83], electrospinning [84] and plasma induced etching [9, 27, 85] (Table 11.2).

Table 11.1 A selection of top down fabrication methods for dry adhesive manufacturing as reported in literature

Fabrication method	Mold material	Structural material	References
Casting	Wax	PDMS, Polyester	[1]
Casting	PC	PI, PDMS	[1]
Casting	Silicon	PDMS	[6, 16, 39, 41, 64]
Casting	SU-8	PDMS	[12]
Casting	Silicon	PU	[11, 65]
Casting	Positive photoresist	PDMS	[17, 18, 22, 53, 66–69]
Casting	Steel	PVS	[8, 70]
Casting	Silicone	PU, TPE, PDMS	[19, 51, 52, 71–75]
Casting	Wire mesh	PDMS	[21]
Casting	Silicon	SMP	[76]
Casting/deformation	PUA	PUA	[20]
Casting/deformation	Silicone	PU (ST-1060)	[40]
Dry Etching	NA	PI	[5, 16]
Two-photon lithography	NA	IP-G 780	[60]
Two-photon lithography/casting	IP-dip/mold max 20 silicone rubber	PU (ST-1060, ST-1087)	[24]
Compression Molding	Silicone	TPE	[52, 75, 77]
Dipping	Silicone	PU	[15, 37, 78]
Dipping	SU-8	PDMS	[12]
Dipping	Silicon	PDMS	[64, 79, 80]
Dipping	Steel	PDMS	[23]
Casting/dipping	Silicon/Teflon	PUA/PDMS	[81]

PDMS polydimethylsiloxane, *PI* polyimide, *PVS* polyvinylsiloxane, *PU* polyurethane, *TPE* thermoplastic elastomer, *SMP* shape memory polymer, *PUA* polyurethane acrylate

Anodic aluminum oxide molds used in some fabrication processes [86, 87] could also be considered a bottom up manufacturing technique as the production of holes is not defined by lithography but rather is a unique feature of the anodic aluminum process. Bottom up manufacturing offers the promise of far smaller features available at lower costs because precision lithography and machining is no longer required, and these techniques may eventually be used to produce fibers of desired properties in commercially viable scales. At the moment however, most bottom up processes are limited in their patterned areas, uniformity, durability and costs because the production processes are less developed and predictable than traditional top-down approaches. Especially in the case of fibers that are designed to have high

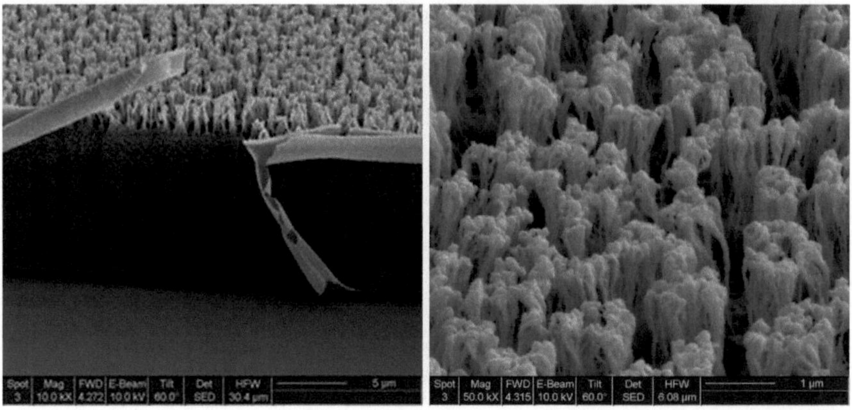

Fig. 11.1 SU-8 Nanopillars produced by a *bottom up* oxygen plasma process relying on gold micromasking [27]. *Top down* processing is used to make the larger SU-8 blocks that are etched

Table 11.2 A selection of top down fabrication methods for dry adhesive manufacturing as reported in literature

Fabrication method	Mold material	Structural material	References
CVD growth	NA	Carbon nanotubes	[10]
Electrospinning	NA	Aligned Nylon-6 nanofibers	[58, 84]
Colloidal assembly and casting	PS spheres	PDMS	[82]
Plasma etching	NA	Photoresist/silicon/silicon oxide	[9]
Plasma etching	NA	SU-8	[27, 85]
Electrohydrodynamic force	NA	PMMA	[88]

PMMA Polymethylmethacrylate

normal or peel adhesion strength, the exact fiber shape is extremely important to performance [12] so random fiber production often is less attractive. An example of nanoscale polymer fibers produced using a bottom up approach (random micro-masking effects in oxygen plasma as shown in Fig. 11.1) demonstrate how a similar scale to gecko foot hairs may be achieved, but the normal adhesion strength of similar materials is typically very low [9, 85]. The majority of bottom up manufacturing processes reported to date produce the adhesive structures directly, but a few can also be designed to produce molds for subsequent replication of more attractive structural materials [83]. Of the available options, electrospinning is likely the most scalable and industrially relevant process due to its low infrastructure requirements [84], although large volume productions may still be difficult to achieve because volumetric rates are low.

11.1.2.3 Silicon Based Mold Fabrication

The most common material for most microfabrication is silicon, owing to its ready availability and extensive use in traditional microelectronics and MEMS processes. Through a variety of means, silicon can be patterned as either a direct mold (where the polymer fibers are filled into silicon holes), or as a positive template for fibers with an intermediate negative mold, usually made of polydimethylsiloxane (PDMS). In most cases, a photolithography step combined with deep reactive ion etching (DRIE) produces the desired features with simple fibers being reported by Glassmaker et al. over a decade ago [6]. Despite the conceptual simplicity, there have been numerous variations on this silicon processing technique to produce a variety of impressive geometries and adhesive variations. For example, Kim and Sitti introduced the concept of directly producing mushroom shaped holes in a silicon-on-insulator wafer, where the etch stop of an embedded silicon oxide caused an undercut in the last etch step and produced a larger cap than fiber diameter [11, 65]. These fibers were cast in the silicon mold using a polyurethane rubber (ST-1060 from BJB Enterprises [89]) and were one of the first fiber designs to show a significant normal adhesion strength improvement over unstructured materials. The drawback to this work was that the silicon template had to be vaporized in XeF_2 gas to release the fibers, making this an expensive and time consuming process to complete. Other notable work using RIE of polysilicon was shown by Jeong et al. which used a Faraday cage inside the RIE to achieve angled etching, and nanoscale holes for the finest feature shapes [35]. An etch stop layer was again used to help flare out the fiber tip, and wet etching of the silicon oxide could also be used to increase that tip dimension further if desired. These nanoscale holes were used to mold a UV curing polyurethane acrylate (PUA) with a tunable modulus between $\sim$20 and 320 MPa which permitted the nanoscale pillars to be accurately reproduced and the material also had the ability to self-replicate from the original nanoscale fibers, to produce flexible nanofiber molds similar to the original silicon. Alternative techniques using silicon include separate etching, alignment and bonding steps, which permit the user to specify exact cap shapes and alignment if needed. Examples include papers by Davies et al. [16] and Kwak et al. [39, 64]. These processes are much more complex due to the alignment, multi-etch, bonding and grinding requirements, but can produce reusable molds and have excellent quality control owing to the silicon based processing.

As examples of silicon processing used to produce positive templates, several groups have used variations on this method to produce effective adhesive designs. Jin et al. have used a combination etch of silicon fibers to produce partially cured silicone rubber negatives that could then be angled through shearing followed by full cure. The fibers had rectangular flexible tips and a circular or triangular main fiber cross section to better mimic the shear engaged adhesive performance of the gecko [40, 43].

11.1.2.4 Polymer Based Mold Fabrication

While silicon is an extremely attractive material for making molds, it does have significant drawbacks. It is optically opaque, meaning that if complex undercut features are produced in the material, it can be extremely difficult to see the geometry without completing a polymer cast, or a destructive test of the mold. The molds themselves are brittle and rigid, meaning that any small flaws that may be introduced after repeated molding cycles could cause a catastrophic failure by fracturing the mold if there are bending stresses on it. The maximum common size of silicon substrates presently available is 300 mm in diameter, but these are typically only used in state of the art microelectronics foundries, six inch diameter or smaller are much more common in MEMS foundries and universities. Finally, in most cases, a surface treatment of a fluorocarbon coating or silanization step is required to ensure easy release of molded polymer fibers from silicon molds.

As alternative mold materials, polymers can have many advantages. First is that they can be relatively inexpensive—the infrastructure requirements for working with polymers is often much less than that of full silicon processing [19]. The mold materials may be transparent to both visible and UV light, making some polymers suitable for directly molding UV curing resins. Many polymers have very low surface energies, making them less likely to require surface treatments to ensure easy release of silicone rubbers from them. Polymers can be relatively tough in comparison to silicon, and are able to bend or flex more without facture, although this property varies widely depending on choice of material. The disadvantages of polymers include their dimensional stability—higher thermal expansion coefficients and less consistent material properties than silicon make the dimensional tolerance far less accurate. Processing is less standardized than silicon manufacturing, making it more important that individual processes be optimized at specific labs. Temperature and solvent tolerance and chemical compatibility vary widely—problems with dissolution, intermixing, cure inhibition and other unwanted interactions can occur if inappropriate material choices are made.

Photoresists are the most common polymers used in producing molds directly. They can be positive or negative acting, and be used for producing a positive feature or a negative template for directly molding the fibers that are desired. Some of the first examples of using photoresist as positive molds were that of Aksak et al. [34] and Murphy et al. [14] which used the negative acting, epoxy based photoresist SU-8 as a template material for vertical and angled fibers. These fibers templates were cast with a silicone rubber and then that silicone mold was used to cast adhesive fibers from polyurethane. In the case of Murphy et al. the polyurethane fibers then had a dipping process applied that permitted tips of various geometries to be introduced to the final adhesive [14]. At around the same time del Campo et al. used SU-8 as a direct mold material for PDMS dry adhesives [12]. In a similar technique to that of Murphy, different tips were produced by dipping the fibers produced from an SU-8 mold into uncured PDMS and then allowing the material to cure in either contact with surfaces or free to produce a variety of tip shapes. One novel modification however was to partially delaminate the SU-8 from the substrate using a quenching step after curing the SU-8, resulting in a small crack where PDMS could fill in during direct casting [12]. The result was a mushroom shaped fiber that was directly molded from a rigid photoresist without a secondary tip forming step, and was one of the first described methods to produce demoldable interlocked features directly. This marked a large improvement over dipping techniques because consistency across areas and between samples is far greater once a mold is produced.

Alternatives to SU-8 for making direct molds for PDMS fibers include thick-film positive photoresists. A bi-layer photoresist of AZ9260 and PMGI was used to produce mushroom shaped fibers [17] and asymmetric fibers [18] to investigate the exact influence of tip geometry on adhesion strength. The PMGI layer could be independently exposed and developed compared to the thicker AZ9260 photoresist to permit a wide range of cap dimensions and positions to be produced on a single wafer and help experimentally determine optimized adhesion properties. Peel strength was greatly improved with increased cap overhang [53], but reduced structural stability and demolding yields imposes a practical limit on the maximum cap overhang to thickness ratio [17]. Other efforts with these thick film positive acting photoresists eliminate the need for a second undercutting layer entirely, as a back side exposure through a transparent wafer can permit a flared base of a hole in photoresist with geometry dependent on the exposure dose and development time of the photoresist [22]. Unfortunately these molds require dissolution in solvent to release the PDMS fibers, possibly because the funnel shaped tip is not as easily deformed to fit through the smaller base hole as flat cap designs.

Other classes of polymers have also been adapted for making positive fiber templates. PMMA was introduced as a deep UV sensitive material and, when combined with a masking layer of thin SU-8, was able to produce mushroom shaped master fibers which could then yield negative silicone rubber templates with the interlocked mushroom shaped features directly cast in polyurethanes and other materials [19]. This represented the first time that mushroom shaped fibers were directly molded from a flexible rubber template, and permitted a much wider range of material modulus to be tested for dry adhesives [52, 71, 75, 90]. Another group which used PMMA as a material used an electrohydrodynamic molding process to produce high aspect ratio mushroom shaped fibers using only a single embossed layer of PMMA to begin the process [88]. Recently, two-photon lithography systems have been used to produce hierarchical [60] and optimized adhesive fibers [24]. Attractively with two-photon lithography systems, virtually any geometry can be produced without the same limitations imposed on traditional lithography. The drawback is the extremely slow writing times and small areas that can be produced in a single process. However, many design iterations can be tested in a short period of time to help experimentally optimize device geometry [24]. The technique works best as a positive template for subsequent replication of fiber designs, as the photosensitive resins normally used are too stiff for appreciable adhesion on their own [60].

In recent years, other more simple methods of producing fibers have enjoyed success, including direct mesh templates [21], and direct machining of wax [42] followed by dipping techniques to reduce surface roughness.

11.1.3 Structural Material Choice

While the production of molds for dry adhesive fabrication tends to be the dominant challenge in manufacturing, a suitable structural material for the adhesives still needs to be selected which is compatible with the molding process. To make a good gecko inspired adhesive a structural material should have the following properties:

- Easy production in micrometer/nanometer scale.
- High yield strength.
- Low surface energy (for easy demolding, self-cleaning, and superhydrophobicity).
- Relative insensitivity of material properties with environmental changes (humidity, temperature, solvents).
- Minimal plastic deformation (prevention of permanent deformation after repeated cycles)
- High tear strength (improved durability of designs, and less damage during demolding steps).
- Low sensitivity to cure inhibition (for a thermosetting polymer).

For complex fiber shapes the material class of choice for most of the last decade has been curable elastomers [2, 3]. These materials are attractive because of their relatively low costs, potential for replicating sub-micrometer features, high deformation before yield/failure and simple processing with a variety of microfabricated molds. Most of these processes can fall under the category of soft-lithography [91], using materials and methods popularized by the Whitesides group at Harvard. Curable elastomers are thermosetting materials, which mean that they can be poured into or on a mold as a liquid, and then solidify into a rubbery solid after a chemical reaction. The reaction can be triggered by a catalyst, heat, UV or other means but generally results in a non-reflowable or meltable material. Common examples of curable elastomers that have been successfully used as dry adhesives include polydimethylsiloxane (PDMS), polyvinylsiloxane (PVS), polyurethane rubbers (PU) and polyurethane acrylate rubbers (PUA). Other materials that may be used with direct molding include thermoplastics and thermoplastic elastomers. While less common in the available literature, thermoplastic materials are highly attractive from a processing point of view because they can be melted and formed much faster than curable materials, and may also be recycled with appropriate process technology, providing further advantages in environmentally friendly manufacturing of adhesives.

11.1.3.1 Silicone Rubbers

Silicone rubbers have been tremendously popular as dry adhesive structural materials owing to their low surface energies, easy curing, high temperature and solvent tolerances and high elongations before failure. All of these factors combine to make silicone rubbers an excellent material to replicate fibers from molds of polymers, metals and silicon. For dry adhesives specifically, the most popular materials have been addition cured silicones, namely PDMS and PVS. Of the types of PDMS that can be used, Sylgard 184 from Dow Corning [92] has proven to be the most popular owing to its widespread use in soft-lithography [91]. Its properties are thus probably the most studied of all silicone rubbers. PVS is much less common for microfabrication, but has been used extensively for producing dental impressions [8], and its major advantage is that it can cure within a few minutes to nearly final mechanical properties, making it a material more suited for rapid fabrication. It has been primarily used by the Gorb group [8, 13, 70] and Gottlieb Binder GmbH & Co, KG where it is known as Gecko® Nanoplast® [13, 56]. Both Sylgard 184 and the PVS used for dry adhesives are considered addition cured silicones where the cure reaction will not produce byproducts that are not part of the final silicone rubber. This type of cure can be accelerated by the application of heat, but is vulnerable to inhibition by various chemicals, including sulfur, organo-tin compounds, nitrogen containing compounds and others [92]. As a result, great care must be taken that the mold material, mixing containers and even the gloves used during handling the pre-polymers must be non-inhibiting.

Silicone rubbers can be found in variety of formulations, with durometers typically ranging from low Shore OO scale to high Shore A scale. Sylgard 184 is listed as Shore A 43 on company datasheets [92] but can easily have significant variation in these values depending on curing procedures. Additionally, it may be cured with a deficit or surplus of curing agent to provide a much wider range of material properties, but typically a maximum Young's modulus may be in the 2 MPa range [79]. Alternative silicone rubbers that have been reported for use as adhesives include Plat-Sil 73-40 [44], a moderate durometer addition cured silicone, Sylgard 170 [93] with similar durometer to Sylgard 184 but with a much faster cure time, and TC-5005, a low durometer condensation cured silicone [26]. Condensation cured silicones have more shrinkage than addition cured silicones due to the liquid byproducts of the cure reaction, and cannot be accelerated in their cure by heating. The costs for these condensation cured silicones are usually less, they are less sensitive in their final properties to the exact ratio of curing agent, and most attractively are insensitive to most inhibiting agents such as that could cause problems for addition cured silicones. This insensitivity to inhibition permits silicone/soft polyurethane composites to be cast [26].

11.1.3.2 Polyurethane Rubbers

Polyurethanes can come in a much wider variety of material properties than silicones, and can range from extremely soft rubbers to hard plastics similar to polycarbonate in modulus and hardness. Polyurethanes in general are less temperature and solvent tolerant than PDMS, will have higher surface energies, are mechanically more durable (for elongation, tear and abrasion resistance) and can often use the same soft-lithography techniques as silicone rubbers [91]. Surface treatments are required for most mold materials in order for polyurethanes to be successfully demolded, with the notable exception of silicone rubbers, where polyurethanes can usually be demolded without special surface modifications. The majority of polyurethane rubbers reported to work as dry adhesives have been relatively soft, with the same order of magnitude modulus as PDMS [11, 14, 15, 19, 34, 37, 51, 71, 78], although a few types with higher modulus have been successfully used with appropriate geometries [35, 71]. As with silicone rubbers, there are some problems with cure inhibition depending on the cure mechanism and mold materials. For several soft polyurethanes such as ST-1060 which was in common use for gecko inspired adhesives before it was discontinued by the manufacturer, the use of a tin catalyzed silicone rubber is counter indicated due to possible cure inhibition [89]. While mitigating strategies could be adopted, this leads to some problematic variability if not appropriately taken in account, and smaller features will have larger issues with this problem primarily from the shorter diffusion distances that inhibiting agents have to travel.

11.1.3.3 Thermoplastic Elastomers

Thermoplastic elastomers combine the advantages of rubbers, with the relative ease of processing of thermoplastics. This material category includes thermoplastic polyurethanes, thermoplastic polyester copolymers, styrene-ethylene-styrene copolymers, plasticized polyvinyl chloride (PVC) and others. Far less research work on these materials have been reported for microfabrication in comparison to their curable counterparts, but in theory these materials should be more scalable in production rates, potentially recyclable (depending on the specific material choice) and useful for the majority of expected temperatures in everyday applications. Unfortunately these materials can be less solvent resistant, will creep at elevated temperatures more than curable elastomers [52] and may not be as easy to form with high replication fidelity or durability [90].

11.1.3.4 Rigid Polymers

Unlike thermoplastic and curable elastomers, thermoplastics and thermosets are quite stiff, with mechanical properties much closer to that of the actual gecko adhesive fibers (β-keratin). This high stiffness requires that minimum features be much smaller than those typically seen in rubber-like biomimetic adhesives to achieve appreciable adhesion. Some thermoplastics that have been introduced as dry adhesives include, low density polyethylene (LDPE) [29], high density polyethylene (HDPE) [28], polypropylene (PP) [28], and polymethylmethacrylate (PMMA) [88]. Common thermosets have included epoxies [94], polyimides [5, 7, 16], rigid polyurethanes [19, 71] and curable acrylates [35, 95, 96]. In general, thermosetting materials will have fewer issues with creep, high temperature durability or solvent compatibility than thermoplastics. Most of the successful designs produced with rigid polymers are designed to operate in shear and only a few have demonstrated significant normal adhesion strength, although there are reports of extremely high adhesion strengths ($\sim$18 Atm) with appropriate geometries and a temperature varying modulus due to shape memory effects [76].

11.1.3.5 Composites

The use of composites in the fabrication of dry adhesives is becoming more common compared to the early years of the field [2]. These composite materials may be used in the fibers themselves, the backing layer or a combination of the two and are typically used to provide improved performance via additional mechanical, electrical or magnetic functionality. Composite fibers include those where multiple materials are bonded together (for example different modulus PDMS for skin adhesives [79]), rigid core fibers coated in PDMS [81], fabric reinforced rubber materials [44, 62, 97], gradient material fibers [98] and fluidic or phase change materials contained within a base polymer [67]. A more common approach to composites is to introduce fillers within a base polymer to provide electrical [71, 99–101], or magnetic functionality [102–104] for enhanced adhesion, sensing or adhesion control applications. Work on electro-dry adhesion used a combination of mechanical pressure and capillary filling to help wick in a viscous PDMS/carbon black mixture into a rigid mold for mushroom shaped features (Fig. 11.2a) [99]. In this case, the PDMS that wicked into the caps had a lower portion of carbon and was relatively non-conductive. Viscous polyurethane/carbon black blends can be filled into flexible molds using squeegee casting or lamination methods for higher uniformity in conductive performance (Fig. 11.2b) [105], but was often found to have bubbles trapped in the backing layer [74]. Magnetic particles can be added to pre-polymers like PDMS for making adhesives [102, 103] and wettable surfaces [106] respond to magnetic fields, and their effectiveness and performance was a function of the type of magnetic field the polymers were cured in due to pre-alignment of particles. The exact effect of performance of adhesives with fillers however is complex and dependent on location, size and interaction of fillers with the matrix [33].

Composite materials that focus on larger structures embedded in the backing layer of the adhesives have been reported by several groups. In most cases, these embedded fillers are designed to enhance or control adhesion via methods such as electrostatic attraction [44], mechanical compliance [67, 97], and surface roughness

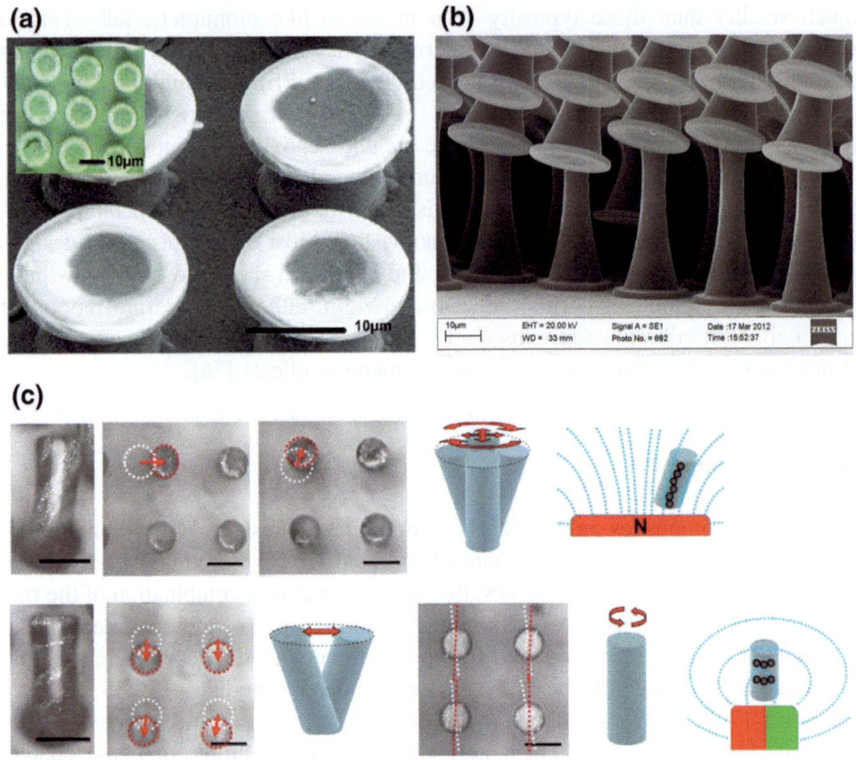

Fig. 11.2 Sample composite fiber designs. **a** PDMS/carbon black composites taken with permission from "Electro-Dry-Adhesion", J Krahn and C Menon Langmuir 2012 28 (12), 5438–5443. Copyright 2012 American Chemical Society [99]. **b** Carbon black/ST-1060 dry adhesives. **c** Carbonyl iron/PDMS composites. Figure adapted with permission from "Magnetically Actuated Micropatterns for Switchable Wettability", D Drotlef, et al. ACS Applied Materials & Interfaces 2014 6 (11), 8702–8707. Copyright 2014 American Chemical Society [106]

changes [107]. One very impressive implementation of passive reinforcement structures uses an unstructured polyurethane rubber sheet reinforced with carbon fiber to achieve macroscale shear adhesion on the order of close to 30 N/cm^2 which is well beyond most fibrillar adhesives [97]. By avoiding fibrillar structures this particular adhesive design greatly improves manufacturability and potentially durability, at the expense of directionality and high peel strength.

11.1.4 Production Scaling Issues

There are very few articles explicitly devoted to manufacturing larger areas of dry adhesives. Some earlier work demonstrated that flexible molds made of silicone rubber could be bonded together to make larger sheets of dry adhesives through casting [74]. Another article dramatically reduced complexity of molding by simply casting PDMS through either a plastic or metal mesh, which could then produce an approximation of the mushroom shaped fibers used in other processes, although the uniformity and quality control was relatively poor [21]. Some work achieved scalable production by doing away with the fibers completely and simply casting polyurethane into large fabric sheets, and demonstrated extremely high shear adhesion forces through excellent load sharing [97] although it still suffers from the long period of time needed to cure the polyurethane. Recently however there has been one group that has achieved continuous production of dry adhesives with complex mushroom shaped fibers using a roll-to-roll lithography process and a UV curing elastic PUA, to achieve production rates on the order of 40 cm^2/s [96]. The major weakness of this process is that the elastic PUA had a higher modulus (reported ~ 10 MPa) than PDMS and the resulting adhesives required relatively high preloads to fully contact another flat surface. It does however show that with appropriate mold technology, complex interlocked features can be replicated in a continuous manner. Limits on this technique are yet to be determined, but other work on the demolding of high aspect ratio fibers has demonstrated a highly non-linear relationship between the aspect ratio of fibers being demolded and the forces required to do so, with greater than 500$\times$ the force required to demold (at 180° peel) PDMS from 6:1 aspect ratio cylindrical pillars than an unstructured surface [108]. Stiffer gecko inspired adhesives with high aspect ratios may be extraordinarily difficult to demold at high rates and without damage to the master molds if the trends observed in that work are accurate.

Alternative materials may yet prove to be better suited for commercial production than the curable elastomers used in the majority of work. Despite the dramatic improvements in curing time for materials such as elastic PUA over traditional polyurethanes (seconds vs. hours), thermoplastic processing still has significant advantages, in that it can make use of infrastructure including extrusion, injection molding and hot embossing. The use of thermoplastic elastomers to produce mushroom shaped dry adhesives has been demonstrated to produce materials roughly on par in quality as that of common curable polyurethanes, at orders of magnitude less cost in materials and fabrication time [52, 75, 77]. While not quite as durable as cured elastomers [52], these materials may be acceptable for lifetimes in the hundreds of cycles and can be sufficiently hydrophobic to permit some adhesion in wet environments [75, 77].

11.1.5 Manufacturing Failure Modes

In general, once an appropriate geometry has been chosen for a dry adhesive, a material needs to be found that does two things—replicates properly from the desired manufacturing process, and survives multiple cycles of operation while maintaining desired adhesive properties. The basic failure modes in replication depend on the types of fibers being produced, the mold material properties, and the number of process steps required. Durability of the fibers is primarily a function of their material properties, the loading/unloading forces in normal and shear and the environment in which they are used.

11.1.5.1 Replication Failures

Replication fidelity is a material property that cannot be taken for granted with every polymer. Some types of polyurethanes and other curable materials will outgas, or diffuse into PDMS molds in their uncured forms and will not always produce the same surface quality as materials like silicone rubbers [72]. Other materials may be too viscous in their pre-polymer form (for thermosets/composites) or have too low a melt flow index (for thermoplastics) to properly fill in molds without applying high pressures or shear rates [105]. The surface tension of polymers in the liquid or melt form may also be a dominating factor in filling processes as the size of the fibers is reduced [31, 61]. This is a particular issue if vacuum filling/degassing is a driving mechanism to get the filling to occur, because then external pressure may only be $\sim$100 kPa to act against capillary pressures.

Replication work done with a variety of thermoplastic elastomers has shown a wide variety of quality and failure modes, ranging from incomplete filling, plastic deformation during demolding, tearing failures during demolding and unwanted interactions or degradation of silicone rubber molds [90]. If a polymer cannot replicate to a very smooth finish, it is unlikely to achieve a very high adhesion if it is structured in a mushroom shaped fiber. The reason for this is that any internal voids (caused by defects in the cap, or roughness of the mating surface) will produce stress concentrations, and the total adhesive force per fiber will be drastically reduced [109]. Broadly speaking, if a mushroom shape fiber makes full contact with another surface, its adhesion will be very high, but manufacturing defects in the cap will make this geometry mostly ineffective at supporting loads.

To demonstrate some of the possible failure modes in the manufacturing of dry adhesives (rather than in the adhesive failure) several images are shown of problems related to molds/demolding, and structural material replication challenges. Figure 11.3 shows various images of uneven backing layers, film ruptures, plastic and brittle failures of molds that can occur in casting processes. Figure 11.4 shows what can occur with an identical mold material and shapes but different structural thermoplastic elastomers. In this case, very different surface roughness and final fiber dimensions are achieved, due to interactions with plasticizers, poor tensile strengths and low yield strengths of various structural materials [90].

Unfortunately, technical datasheets are not a reliable means to ensure an elastomer will work at the level of fidelity necessary for dry adhesives and experiments are needed to validate every material when working with PDMS molds. In particular, thermoplastic elastomers with plasticizers tended to be very problematic, as the lower molecular weight components would diffuse into the PDMS and cause both unexpected surface roughness of the remaining fibers and also affect the lifetime of the mold [90]. Some of these thermoplastic elastomers may still be

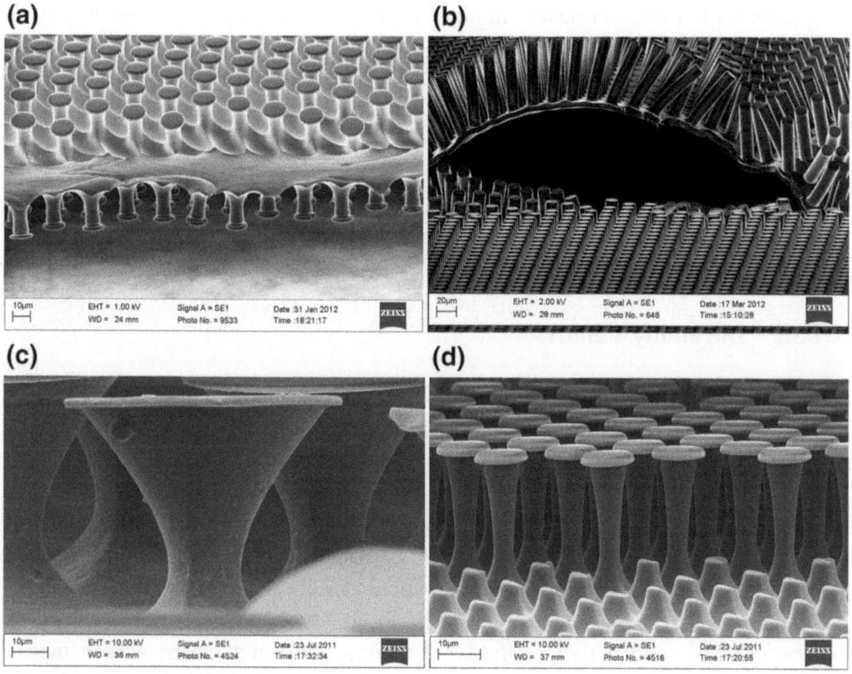

Fig. 11.3 Adhesive fibers showing different molding failures. **a** Non-uniform backing layer thickness (Loctite 3525 structural material), **b** backing layer tearing during demolding Loctite 3525 structural material), **c** plastic deformation of rigid polyurethane positive molds (TC-854 from BJB Enterprises), **d** master mold fracture during initial demolding

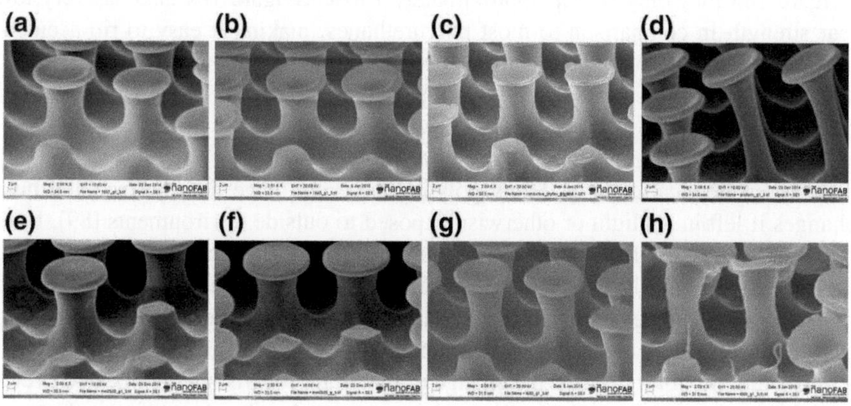

Fig. 11.4 Thermoplastic elastomer replication trials. Materials are **a** G1657 (Kraton), **b** G1645 (Kraton), **c** C3 6068 (Dryflex), **d** Ethlyene vinyl acetate, **e** MM2520 (SMP Technologies), **f** MM4520 (SMP Technologies), **g** T509 (ELASTO), **h** T909 (ELASTO). Figure is reprinted with permission from D. Sameoto and A. Wasay, "Materials selection and manufacturing of thermoplastic elastomer microfluidics", Proc. SPIE 9320, Microfluidics, BioMEMS, and Medical Microsystems XIII, 2015 [90]

acceptable with a non-permeable mold like silicon for example, because glossy flat surfaces were produced when molded against glass slides for even very poorly replicated elastomer composites like C3 6068 (Fig. 11.4c). Thermoplastic elastomers with high Shore A hardness (Ethylene Vinyl Acetate, T909 SEBS) showed a tendency to have severe plastic deformation after demolding. Better results were achieved in these cases with lower aspect ratio fibers and smaller caps, which reduces the stress on the fibers when they are removed from interlocking flexible molds.

11.1.5.2 Durability Failures

The adhesive structures may degrade from many causes: contamination, abrasion, plastic deformation, tearing, surface changes, and bulk changes. Contamination with dust or dirt particles has been the most heavily investigated field with respect to dry adhesive durability [66, 110–112], as the self-cleaning nature of gecko foot hairs has long been a driving factor in the motivation to produce these adhesives [113]. While no artificial system is perfect, lower surface energy and rigid materials are the most likely to be cleaned via dry or wet self-cleaning. Of the elastomeric materials, silicone rubbers are the best choice for dust contamination, as they can also be cleaned with active methods, including soap and water, scotch tape or molten wax [66]. Their crosslinked nature also means they can be cleaned of organic contaminants like oils by organic solvents, which may dissolve or degrade many polyurethanes or thermoplastic elastomers.

For abrasion and tear strength polyurethanes are in general better than silicones. ST-1060 polyurethane for example can stretch close to 600% before failure, while Sylgard 184 may only stretch approximately 150%. Sylgard 184 also has very low tear strength in comparison to most polyurethanes, making it easy to rip accidentally from the edge of a sheet. Other silicones may be better suited with respect to tear strength, but very few others are reported for this type of testing in the dry adhesive field.

For surface changes, silicones are relatively stable unless actively changed by plasma or chemical etching. Many polyurethanes can have both surface and bulk changes if left in sunlight or otherwise exposed to outside environments [89]. Both polyurethanes and thermoplastic elastomers will have much more temperature variance in their mechanical properties than silicone rubbers, and these changes can be particularly severe for thermoplastic elastomers, where other changes, including reflow may occur at elevated temperatures [52].

For other durability issues, the change of material properties with repeated cycles is quite important. Sometimes these changes may be due to surface effects or material transfer between adhesives and contacting surfaces, but in other cases for high adhesive strength or highly loaded adhesives, bulk changes to the polymers may be severe and can complicate the prediction or modeling of adhesive properties. As an example of permanent deformation of dry adhesive fibers with single loading, Fig. 11.5 shows semi-rigid polyurethane (a) TC-8740 and (b) TC-9462

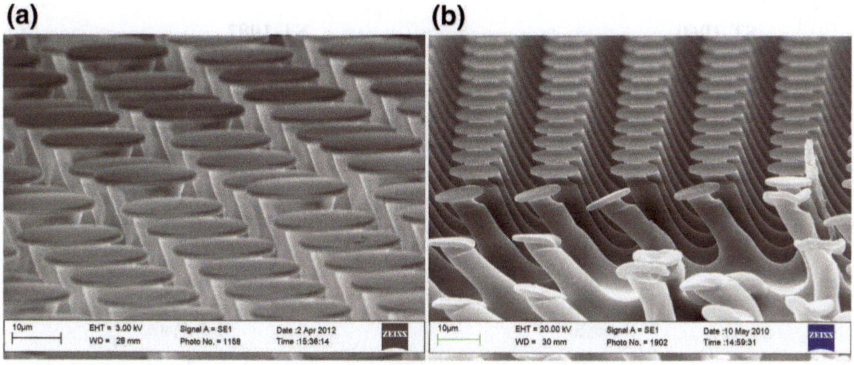

Fig. 11.5 Demonstration of plastic deformation with semi-rigid polyurethanes (>Shore D 40). **a** Axial plastic deformation of TC-8740 polyurethane after adhesion to a flat surface, **b** plastic deformation of TC-9462 polyurethane edge after cutting with scissors

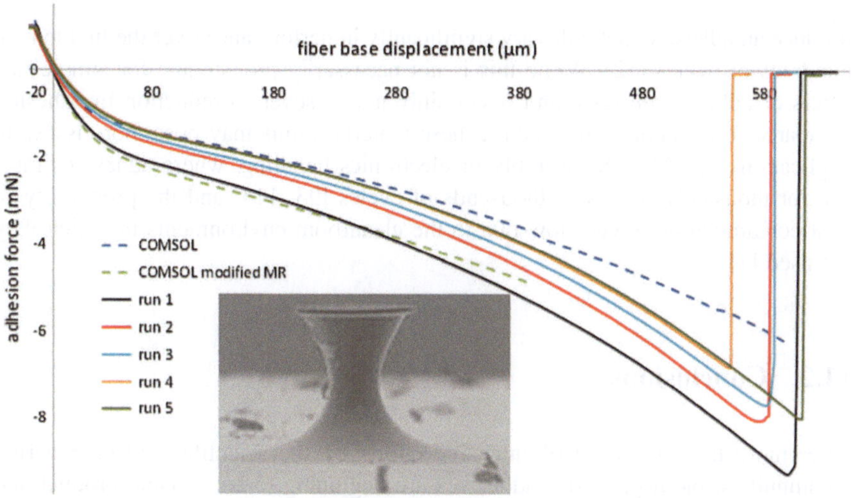

Fig. 11.6 Force versus displacement for a single Kraton G1657 fiber (tip diameter = 100 μm) displaced axially at 5 μm/s. Hyperelastic behavior is clearly seen, with parameters changing after each test due to material changes. A Mooney-Rivlin fit based on macroscale measurements and fiber geometry is plotted as dashed lines. Figure is adapted with permission from Bschaden 2014 [73]

with plastic deformation failures caused by axial adhesive force, and compressive loading during cutting. Even for elastomer materials that don't show obvious plastic deformation after use, the repeated cycling of tensile loads during operation may alter the mechanical properties as shown in Figs. 11.6 and 11.7.

As adhesives are further optimized from the perspective of maximum adhesion forces per fiber, it is feasible for designers to exceed material strength limits, and

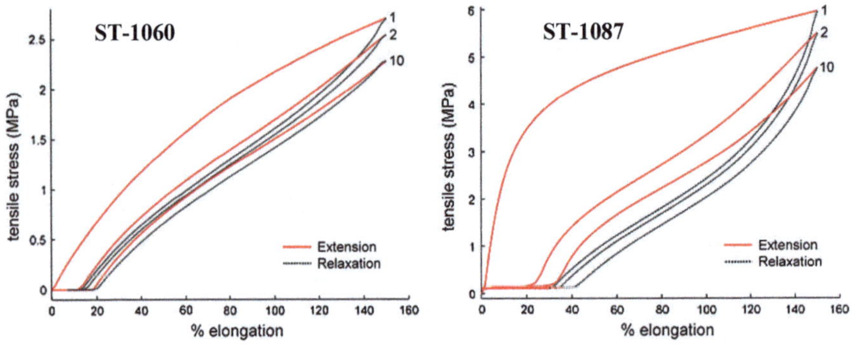

Fig. 11.7 Macroscale stress-strain curves for two polyurethanes commonly used for dry adhesive manufacturing [14] for the first, second and tenth cycles to 150% elongation. Significant plastic deformation and cycle dependent behavior is seen for both materials but is most severe for the ST-1087 polyurethane which has the highest initial modulus and durometer. Figure is adapted with permission from Bschaden [73]

produce an adhesive that will vary significantly in performance over the first tens or hundreds of cycles [52]. While this is not necessarily problematic for simple fasteners or adhesive designs, and is certainly not as severe a reduction in adhesion compared to contaminated systems, these material limits may pose more issues in applications like MEMS assembly or electronics handling, where adhesives must last for thousands or tens of thousands of cycles [93, 114] and the probability of dust contamination is very low due to the cleanroom environments in which they are used [71].

11.2 Conclusions

After more than a decade of intensive efforts by the scientific and engineering communities, bioinspired dry adhesives are beginning to be a viable product and may yet take their place alongside hook and loop fasteners as a successful adhesion mechanism adapted from the natural world. For long-term applications and products that are expected to last for years, silicone rubbers appear to be the best material choice, owning to their exceptionally low surface energies, solvent and temperature resistance, simple cleaning procedures and lower tendency to creep or yield than many other elastomers. Unfortunately their longer curing times, and difficulty of integrating silicone with other materials will likely mean their costs remain high for the foreseeable future. Polyurethanes and other non-silicone based curable elastomers will have a broader range of material modulus, elongation, and tear resistance than silicones, as well as easier integration with other materials, but will suffer in comparison in terms of chemical and thermal resistance, and in general will be harder to clean of contaminants because of their typically higher surface energies. With the exception of UV curing variations, these elastomers are also limited in production rates by their relatively long cure times. Thermoplastic elastomers may yet be a primary material choice for introducing bioinspired adhesives into everyday products, where reversible/reusable adhesion may be desirable for dozens or hundreds of cycles, but long term performance would be much more degraded through yielding and fatigue. Not all variations of these elastomers are appropriate for the production of fine features sizes with comparable quality to silicone and polyurethane rubbers, but several versions have shown successful production of complex microfibers and may yet be scalable to large production rates. There is no comparable structural material as easily adapted to the scalable manufacturing techniques of industrial thermoplastics and so these thermoplastic elastomers may be considered the prime industrial material of choice of large scale commercialization of dry adhesives.

Acknowledgements I would like to acknowledge the exceptional efforts of many students who have worked with me on the development of dry adhesives over the years, including Yasong Li, Juan Pablo Diaz-Tellez, Jeff Krahn, Helia Sharif, Brendan Ferguson, Ben Bschaden, Walid Khaled, Abdul Wasay and Babak Soltannia, as well as my post-doctoral supervisor Dr. Carlo Menon who provided me my first opportunity to work in this field. Financial support from NSERC, the European Space Agency and Micralyne Inc. has all contributed to my past or current work with dry adhesives.

References

1. M. Sitti, R.S. Fearing, J. Adhes. Sci. Technol. **17**, 1055 (2003)
2. D. Sameoto, C. Menon, Smart Mater. Struct. **19**, 103001 (2010)
3. M.K. Kwak, C. Pang, H.E. Jeong, H.N. Kim, H. Yoon, H.S. Jung, K.Y. Suh, Adv. Funct. Mater. **21**, 3606 (2011)
4. L. Heepe, A. Kovalev, M. Varenberg, J. Tuma, S. Gorb, Theor. Appl. Mech. Lett. **2**, 014008 (2012)
5. A. Geim, S. Dubonos, I. Grigorieva, K. Novoselov, A. Zhukov, S.Y. Shapoval, Nat. Mater. **2**, 461 (2003)
6. N. Glassmaker, A. Jagota, C.-Y. Hui, J. Kim, J. R. Soc. Interface **1**, 23 (2004)
7. C.-Y. Hui, N. Glassmaker, T. Tang, A. Jagota, J. R. Soc. Interface **1**, 35 (2004)
8. A. Peressadko, S.N. Gorb, J. Adhes. **80**, 247 (2004)
9. M.T. Northen, K.L. Turner, Nanotechnology **16**, 1159 (2005)
10. B. Yurdumakan, N.R. Raravikar, P.M. Ajayan, A. Dhinojwala, Chem. Commun. (30), 3799 (2005)
11. S. Kim, M. Sitti, Appl. Phys. Lett. **89**, 261911 (2006)
12. A. Del Campo, C. Greiner, E. Arzt, Langmuir **23**, 10235 (2007)
13. S. Gorb, M. Varenberg, A. Peressadko, J. Tuma, J. R. Soc. Interface **4**, 271 (2007)
14. M.P. Murphy, B. Aksak, M. Sitti, J. Adhes. Sci. Technol. **21**, 1281 (2007)
15. B. Aksak, M.P. Murphy, M. Sitti, in *IEEE International Conference on Robotics and Automation, 2008. ICRA 2008* (IEEE, 2008), p. 3058
16. J. Davies, S. Haq, T. Hawke, J. Sargent, Int. J. Adhes. Adhes. **29**, 380 (2009)
17. D. Sameoto, C. Menon, J. Micromech. Microeng. **19**, 115002 (2009)
18. D. Sameoto, C. Menon, J. Micromech. Microeng. **19**, 115026 (2009)
19. D. Sameoto, C. Menon, J. Micromech. Microeng. **20**, 115037 (2010)

20. H.E. Jeong, K.-Y. Suh, Soft Matter **8**, 5375 (2012)
21. J. Krahn, C. Menon, Macromol. React. Eng. **7**, 632 (2013)
22. Y. Wang, H. Hu, J. Shao, Y. Ding, ACS Appl. Mater. Interfaces **6**, 2213 (2014)
23. P.Y. Isla, E. Kroner, Adv. Funct. Mater. **25**, 2444 (2015)
24. H. Marvi, S. Song, M. Sitti, Langmuir **31**, 10119 (2015)
25. K.A. Daltorio, S. Gorb, A. Peressadko, A.D. Horchler, R.E. Ritzmann, R.D. Quinn, *Climbing and Walking Robots* (Springer, Berlin, 2006), p. 131
26. D. Sameoto, Y. Li, C. Menon, J. Bionic Eng. **5**, 189 (2008)
27. D. Sameoto, Y.S. Li, C. Menon, *Advances in Science and Technology* (Trans Tech Publ, Zürich, 2009), p. 439
28. A.G. Gillies, R.S. Fearing, Langmuir **27**, 11278 (2011)
29. D.H. Lee, Y. Kim, R.S. Fearing, R. Maboudian, Langmuir **27**, 11008 (2011)
30. D.Y. Lee, D.H. Lee, S.G. Lee, K. Cho, Soft Matter **8**, 4905 (2012)
31. D.-J. Guo, H. Zhang, J.-B. Li, S.-M. Fang, Z.-D. Dai, W. Tan, J. Mater. Chem. B **1**, 379 (2013)
32. Y.-C. Lin, Y.-N. Hsu, Y.-C. Chung, RSC Adv. **4**, 22931 (2014)
33. Y. Tian, Z. Zhao, G. Zaghi, Y. Kim, D. Zhang, R. Maboudian, ACS Appl. Mater. Interface **7**, 13232 (2015)
34. B. Aksak, M.P. Murphy, M. Sitti, Langmuir **23**, 3322 (2007)
35. H.E. Jeong, J.-K. Lee, H.N. Kim, S.H. Moon, K.Y. Suh, Proc. Natl. Acad. Sci. USA **106**, 5639 (2009)
36. T.I. Kim, H.E. Jeong, K.Y. Suh, H.H. Lee, Adv. Mater **21**, 2276 (2009)
37. M.P. Murphy, B. Aksak, M. Sitti, Small **5**, 170 (2009)
38. A. Parness, D. Soto, N. Esparza, N. Gravish, M. Wilkinson, K. Autumn, M. Cutkosky, J. R. Soc. Interface (2009). doi:10.1098/rsif.2009.0048
39. M.K. Kwak, H.E. Jeong, W.G. Bae, H.S. Jung, K.Y. Suh, Small **7**, 2296 (2011)
40. K. Jin, Y. Tian, J.S. Erickson, J. Puthoff, K. Autumn, N.S. Pesika, Langmuir **28**, 5737 (2012)
41. J. Tamelier, S. Chary, K.L. Turner, Langmuir **28**, 8746 (2012)
42. P. Day, E.V. Eason, N. Esparza, D. Christensen, M. Cutkosky, J. Micro Nano-Manuf. **1**, 011001 (2013)
43. K. Jin, J.C. Cremaldi, J.S. Erickson, Y. Tian, J.N. Israelachvili, N.S. Pesika, Adv. Funct. Mater. **24**, 574 (2014)
44. D. Ruffatto, A. Parness, M. Spenko, J. R. Soc. Interface **11**, 20131089 (2014)
45. Z. Wang, P. Gu, X. Wu, Soft Matter **10**, 3301 (2014)
46. B. Yoo, S. Cho, S. Seo, J. Lee, ACS Appl. Mater Interface **6**, 19247 (2014)
47. H. Li, Y. Fan, D. Conchouso, I.G. Foulds, Microsyst. Technol. **21**, 445 (2015)
48. H. Marvi, Y. Han, M. Sitti, Appl. Phys. Lett. **106**, 051602 (2015)
49. H. Yoon, H.E. Jeong, T.-I. Kim, T.J. Kang, D. Tahk, K. Char, K.Y. Suh, Nano Today **4**, 385 (2009)
50. Y. Rahmawan, H. Yoon, M.-W. Moon, H.-G. Jeong, K.-Y. Suh, J. Adhes. Sci. Technol. **28**, 367 (2014)
51. W.B. Khaled, D. Sameoto, Bioinspir Biomim **8**, 044002 (2013)
52. W.B. Khaled, D. Sameoto, ACS Appl. Mater. Interface **6**, 6806 (2014)
53. D. Sameoto, H. Sharif, J.P. Díaz Téllez, B. Ferguson, C. Menon, J. Adhes. Sci. Technol. **28**, 354 (2014)
54. S. Seo, J. Lee, K.-S. Kim, K.H. Ko, J.H. Lee, J. Lee, ACS Appl. Mater. Interfaces **6**, 1345 (2014)
55. L. Xue, J. Iturri, M. Kappl, H.-J. Butt, A. del Campo, Langmuir **30**, 11175 (2014)
56. Gecko Nanoplast Product Information. http://www.binder-gmbh.com/en/products/geckonanoplast/. Accessed 11 Jan 2016
57. nanoGriptech Inc. Homepage. www.nanogriptech.com. Accessed 11 Jan 2016
58. Akron Ascent Innovations. http://akronascentinnovations.weebly.com/. Accessed 11 Jan 2016

59. H. Lee, B.P. Lee, P.B. Messersmith, Nature **448**, 338 (2007)
60. M. Röhrig, M. Thiel, M. Worgull, H. Hölscher, Small **8**, 3009 (2012)
61. Y. Li, H.W. Ng, B.D. Gates, C. Menon, Nanotechnology **25**, 285303 (2014)
62. Y. Li, D. Sameoto, C. Menon, J. Bionic Eng. **7**, 228 (2010)
63. H. Kasem, A. Tsipenyuk, M. Varenberg, Soft Matter **11**, 2909 (2015)
64. M.K. Kwak, H.E. Jeong, K.Y. Suh, Adv. Mater. **23**, 3949 (2011)
65. S. Kim, M. Sitti, J.-H. Jang, E.L. Thomas, in *8th IEEE Conference on Nanotechnology, 2008. NANO'08* (IEEE, 2008), p. 780
66. J.D. Téllez, D. Sameoto, C. Menon, Sci. China Technol. Sci. **53**, 2942 (2010)
67. J. Krahn, D. Sameoto, C. Menon, Smart Mater. Struct. **20**, 015014 (2011)
68. D. Sameoto, H. Sharif, C. Menon, J. Adhes. Sci. Technol. **26**, 2641 (2012)
69. Y. Wang, J. Shao, Y. Ding, X. Li, H. Tian, H. Hu, *SPIE Smart Structures and Materials + Nondestructive Evaluation and Health Monitoring* (International Society for Optics and Photonics, Bellingham, 2015), p. 94291E
70. M. Varenberg, S.N. Gorb, Adv. Mater. **21**, 483 (2009)
71. D. Sameoto, ECS Trans. **45**, 515 (2012)
72. B. Bschaden, A. Baldelli, A. Amirfazli, and D. Sameoto, in *Proceeding of the 24th CANCAM* (Saskatoon, Canada, 2013)
73. B. Bschaden. *Developing Design Guidelines for Improved Gecko Inspired Dry Adhesive Performance*. (MSc Thesis, University of Alberta, 2014). doi:10.7939/R3GM4X
74. D. Sameoto, B. Ferguson, J. Adhes. Sci. Technol. **28**, 337 (2014)
75. B. Soltannia, D. Sameoto, ACS Appl. Mater. Interfaces **6**, 21995 (2014)
76. J.D. Eisenhaure, T. Xie, S. Varghese, S. Kim, ACS Appl. Mater. Interfaces **5**, 7714 (2013)
77. A. Wasay, D. Sameoto, Lab Chip **15**, 2749 (2015)
78. M.P. Murphy, S. Kim, M. Sitti, ACS Appl. Mater. Interfaces **1**, 849 (2009)
79. W.G. Bae, D. Kim, M.K. Kwak, L. Ha, S.M. Kang, K.Y. Suh, Adv. Health Mater. **2**, 109 (2013)
80. L.-F. Wang, J.-Q. Liu, B. Yang, X. Chen, C.-S. Yang, Microsyst. Technol. **21**, 1093 (2015)
81. W.-G. Bae, M.K. Kwak, H.E. Jeong, C. Pang, H. Jeong, K.-Y. Suh, Soft Matter **9**, 1422 (2013)
82. S. Akerboom, J. Appel, D. Labonte, W. Federle, J. Sprakel, M. Kamperman, J. R. Soc. Interface **12**, 20141061 (2015)
83. D. Brodoceanu, R. Elnathan, B. Prieto-Simón, B. Delalat, T. Guinan, E. Kroner, N.H. Voelcker, T. Kraus, ACS Appl. Mater. Interfaces **7**, 1160 (2015)
84. J.F. Najem, S.-C. Wong, G. Ji, Langmuir **30**, 10410 (2014)
85. C. Zhang, J.H.-W. Zhou, D. Sameoto, X. Zhang, Y. Li, H.W. Ng, C. Menon, B.D. Gates, J. Adhes. Sci. Technol. **28**, 320 (2014)
86. D.S. Kim, H.S. Lee, J. Lee, S. Kim, K.-H. Lee, W. Moon, T.H. Kwon, Microsyst. Technol. **13**, 601 (2007)
87. K. Liu, J. Du, J. Wu, L. Jiang, Nanoscale **4**, 768 (2012)
88. H. Hu, H. Tian, X. Li, J. Shao, Y. Ding, H. Liu, N. An, ACS Appl. Mater. Interfaces **6**, 14167 (2014)
89. ST-1060 Datasheet. http://www.bjbenterprises.com/media/wysiwyg/pdfs/Shore-A-High-Performance/ST-1060.pdf. Accessed 11 Jan 2016
90. D. Sameoto, A. Wasay, *SPIE BiOS* (International Society for Optics and Photonics, Bellingham, 2015), p. 932001
91. Y. Xia, G.M. Whitesides, Annu. Rev. Mater. Sci. **28**, 153 (1998)
92. Sylgard 184 Datasheet. http://www.dowcorning.com/DataFiles/090276fe80190b08.pdf. Accessed 11 Jan 2016
93. N. Gravish et al., J. R. Soc. Interface **7**, 259 (2010)
94. G. Castellanos, E. Arzt, M. Kamperman, Langmuir **27**, 7752 (2011)
95. N. Lakhera, A. Graucob, A.S. Schneider, E. Kroner, E. Arzt, C.M. Yakacki, C.P. Frick, Int. J. Adhes. Adhes. **44**, 184 (2013)
96. H. Yi et al., ACS Appl. Mater. Interfaces **6**, 14590 (2014)

97. M.D. Bartlett, A.B. Croll, D.R. King, B.M. Paret, D.J. Irschick, A.J. Crosby, Adv. Mater. **24**, 1078 (2012)
98. H. Minsky, K. Turner, Appl. Phys. Lett. **106**, 201604 (2015)
99. J. Krahn, C. Menon, Langmuir **28**, 5438 (2012)
100. J. Krahn, C. Menon, Smart Mater. Struct. **22**, 085010 (2013)
101. J. Eisenhaure, S. Kim, Polymers **6**, 2274 (2014)
102. A.G. Gillies, J. Kwak, R.S. Fearing, Adv. Funct. Mater. **23**, 3256 (2013)
103. D.M. Drotlef, P. Blümler, A. del Campo, Adv. Mater. **26**, 775 (2014)
104. J. Krahn, E. Bovero, C. Menon, ACS Appl. Mater. Interfaces **7**, 2214 (2015)
105. B. Ferguson. *Improved Gecko Inspired Dry Adhesives Applied to the Packaging of MEMS.* (MSc Thesis, University of Alberta, 2013). doi:10.7939/R3559J
106. D.-M. Drotlef, P. Blümler, P. Papadopoulos, A. del Campo, ACS Appl. Mater. Interfaces **6**, 8702 (2014)
107. M. Frensemeier, J.S. Kaiser, C.P. Frick, A.S. Schneider, E. Arzt, R.S. Fertig, E. Kroner, Adv. Funct. Mater. **25**, 3013 (2015)
108. H. Shahsavan, B. Zhao, Langmuir **27**, 7732 (2011)
109. G. Carbone, E. Pierro, Soft Matter **8**, 7904 (2012)
110. A.G. Gillies, J. Puthoff, M.J. Cohen, K. Autumn, R.S. Fearing, ACS Appl. Mater. Interfaces **5**, 6081 (2013)
111. J. Lee, R.S. Fearing, Langmuir **24**, 10587 (2008)
112. S. Kim, E. Cheung, M. Sitti, Langmuir **25**, 7196 (2009)
113. W. Hansen, K. Autumn, Proc. Natl. Acad. Sci. USA **102**, 385 (2005)
114. S. Chary, J. Tamelier, K. Turner, Smart Mater. Struct. **22**, 025013 (2013)

Chapter 12
Contact Mechanics of Mushroom-Shaped Adhesive Structures

Giuseppe Carbone and Elena Pierro

Abstract Very recently, both experimental and theoretical investigations have shown that micro-structured surfaces covered with mushroom shaped micropillars present strongly enhanced adhesive properties if compared to surfaces covered with cylindrical micropillars made of the same material. However, different geometries lead to different adhesive performance, and finding the optimal solution has become of utmost importance. In this chapter we summarize the main detachment mechanisms of flat-topped and mushroom-topped soft micro pillars and show how the geometry of the pillars should be designed in order to obtain the best adhesive performances. We also discuss the effect of air entrapment at the interface between the pillar and the substrate and investigate the influence of the non-uniform pillar height and thermal fluctuations on pull-off force. Comparisons with experiments are shown, to assess the theoretical findings, and the influence of the effect of tilted pull-off on the adhesion of individual mushroom shaped pillar is evaluated.

12.1 Introduction

Biomimetics has always played a crucial role in suggesting, conceiving and developing breakthrough solutions in mechanical, material and civil engineering, e.g.: (i) Velcro has been inspired by the hooks of burs, (ii) the shape of lumberjack blades has been inspired by wood-burrowing beetle, (iii) cat's eye reflectors have been inspired by the cats' system of reflecting cells, known as tapetum lucidum, (iv) 'morphing aircraft wings' that change shape according to the speed have been inspired by different bird species, (v) some paints and window surfaces have been

G. Carbone (✉)
TriboLab, Department of Mechanical and Management Engineering,
Politecnico di Bari, Viale Japigia 182, 70126 Bari, Italy
e-mail: carbone@poliba.it

E. Pierro
Scuola di Ingegneria, Università degli Studi della Basilicata,
Via dell'Ateneo Lucano 10, 85100 Potenza, Italy
e-mail: elena.pierro@unibas.it

L. Heepe et al. (eds.), *Bio-inspired Structured Adhesives*,
Biologically-Inspired Systems 9, DOI 10.1007/978-3-319-59114-8_12

engineered to be self-cleaning as in the case of lotus leaf, (vi) the cooling system of the Eastgate Centre building, in Harare, has been inspired by a termite mound [1]. In the last years, bio-inspired adhesive structures have drawn great attention from the research community [2–13], due to their extremely high adhesive performance. In nature, many biological attachment systems (e.g. Gecko foot pad [14]) are made of hairs which are constituted mainly of a relatively stiff material, the β-keratin (elastic modulus $E \approx 1$ GPa). In this case, as suggested by some experimental findings and theoretical investigations [15–26], the compliant fibrillar hierarchical geometry of such structures provides them with amazing adhesive properties. However, examples exist in Nature, e.g. the attachment pad of the males of some beetle species from the family Chrysomelidae, which do not present a hierarchical geometry as Gecko does, but are simply surfaces covered with mushroom-shaped microstructures [27–29]. In such cases [7], the shape of the terminal plate is crucial for the achievement of high adhesive strength values of the bio-inspired artificial surfaces [30]. In particular, experimental observations [7] have shown that mushroom shaped microstructures (Fig. 12.1a–c) strongly outperform, in terms of pull-off force, surfaces covered with miniaturized flat-punches made of the same material [polyvinylsiloxane (PVS), Young's modulus $E = 3$ MPa].

This property has been very recently exploited to develop bio-inspired band-aids [31, 32]. However, different mushroom-pillar geometries have led to different experimental results [21, 33] and to some differences if compared to theoretical predictions [30]. In this chapter we summarize the main findings, that explain the origin of the superior performance of mushroom shaped micropillars. The chapter is organized as follow. In Sect. 12.2 we present the mechanisms which govern the detachment of flat-topped cylindrical micro-pillars. In Sect. 12.3 we show that the presence of the plate in mushroom shaped pillar inhibits one of the detaching modes, leading to a strong increase of the pull-off force. In Sect. 12.4 we show how the geometry of the mushroom-shaped pillar may be optimized to increase the adhesive performance. In Sect. 12.5 we discuss the effect of entrapment of air at the interface on pull-off force. In Sect. 12.6 we show how the non-uniform micro-pillar height distribution affects the adhesive performance of the system. In Sect. 12.7 we present a different detaching mode which may occur on perfectly clean surfaces as a consequence of stress aided thermally activated defect nucleation. In Sect. 12.8 we show some comparisons with experiments to assess the theoretical findings, and the influence of the effect of tilted pull-off on the adhesion of individual mushroom shaped pillar is evaluated. In Sect. 12.9 we provide concluding remarks and outline future developments.

12.2 The Cylindrical Micropillar

Let us first consider in detail the case of a cylindrical micropillar in contact with a perfectly flat substrate (Fig. 12.1d). Let us assume the micro-pillar is loaded with a tractive force P.

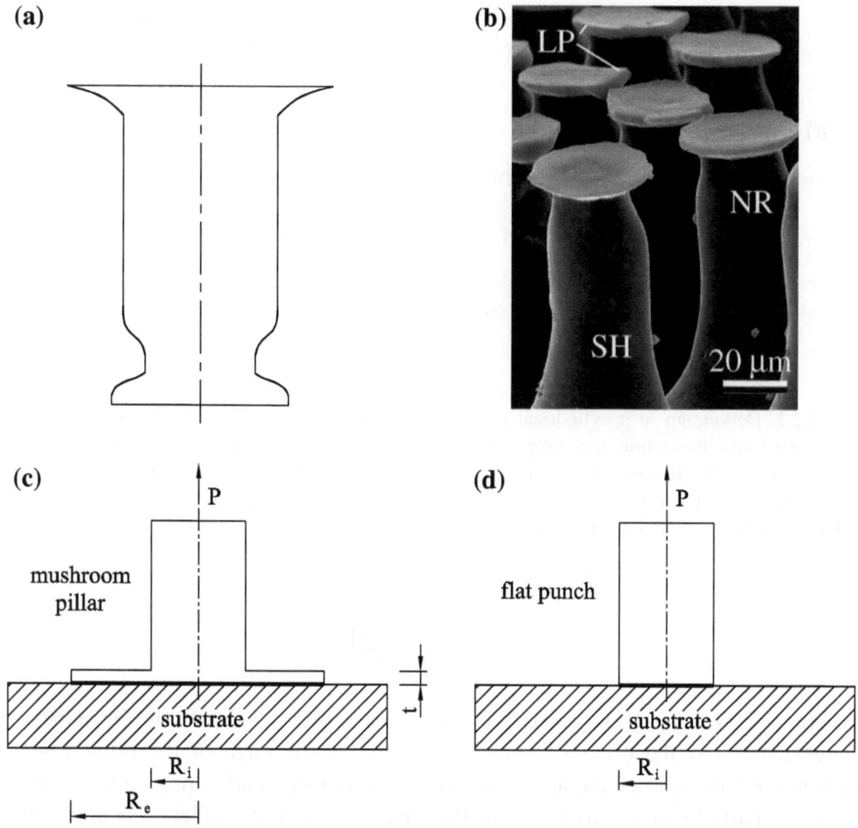

Fig. 12.1 A schematic drawing of a single mushroom shaped pillar (**a**); a SEM image of microfabricated PVS mushroom shaped pillars (courtesy of prof. Gorb [33]) (**b**); the mushroom shaped pillar (**c**) and the cylindrical pillar (**d**) in contact with a rigid substrate. The terminal plate in the mushroom shape (**c**) is the origin of the enhanced adhesive performance

In principle the detachment of the pillar from the substrate may be governed by three different mechanisms depending on the work of adhesion (also referred to as the Duprè energy of adhesion) $\Delta\gamma = \gamma_1 + \gamma_2 - \gamma_{12}$ (where γ_1 is the surface energy of solid 1, γ_2 is the surface energy of solid 2 and γ_{12} is the surface interaction energy when the two surfaces are in direct contact), on the geometry of the pillar, and Young's modulus E of the material: (I) crack propagation from the pillar edge (Fig. 12.2a), (II) propagation of interfacial defects from the inner side (Fig. 12.2b), (III) decohesion due to the achievement of theoretical contact strength $\sigma_{III} = \Delta\gamma/\rho$ where $\rho \approx 1 - 10$ nm is the typical range of van der Waals forces (Fig. 12.2c). Dimensional arguments for mode I debonding of micro-pillars of radius R in contact with the substrate, allow to assert that crack propagation from the edge initiates when the average tractive stress at the interface exceeds the value

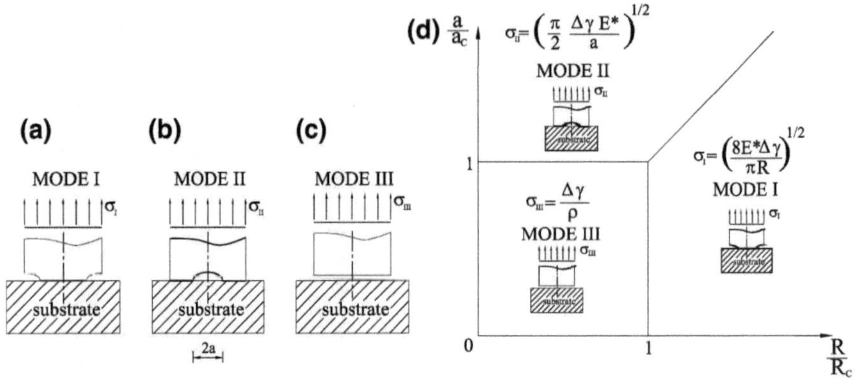

Fig. 12.2 Debonding of a cylindrical pillar may occur because of crack propagation from the pillar edge (**a**), nucleation and propagation of interfacial defects (**b**), decohesion due to the achievement of the theoretical contact strength (**c**). Map of debonding mechanisms of a cylindrical pillar (**d**). The quantities $a_C = (\pi/2)E^* \rho^2/\Delta\gamma$ and $R_C = (8/\pi)\,E^* \rho^2/\Delta\gamma$ are respectively a reference defect size and a reference pillar radius

$$\sigma_I = CE^* \left(\frac{8}{\pi}\frac{\delta}{R}\right)^{\lambda} \tag{12.1}$$

where $E^* = E/(1-\mu^2)$, $\mu = 0.5$ is Poisson's ratio, $\delta = \Delta\gamma/E^*$ is the adhesion length, $\lambda > 0$ is the order of stress singularity at the edge of the pillar [34]. λ depends on the corner angle θ and on the boundary conditions at the interface (no-slip, partial-slip, no friction). In the case of cylindrical pillars $\theta = \pi/2$, thus, assuming sticking friction at the interface, one has $\lambda = 0.4$ for incompressible materials [34]. The dimensionless quantity C is a constant factor which mainly depends on the geometry and boundary conditions at the interface. In the case of a flat punch in adhesive frictionless contact with an elastic half-space ($\theta = \pi$) one has $\lambda = 1/2$ and $C = 1$.

Equation (12.1) can be easily derived by observing that at the edge of the contact the stress $\sigma_{ij} \approx r^{-\lambda}$ where r is the distance from the edge [34]. Therefore also the critical stress σ_I for mode I debonding is $\sigma_I \approx R^{-\lambda}$. However σ_I also depends on E^* and $\Delta\gamma$, hence, invoking Buckingham's theorem [35, 36] one can write $\sigma_I/E^* \approx (\delta/R)^{\lambda}$, from which (12.1) follows. The strength of stress singularity at the edge is identified by the stress intensity factor K, which is defined through the relation $\sigma_{ij} = K\,(2\pi r)^{-\lambda} f_{ij}(\varphi)$, where $f_{ij}(\varphi)$ is a non-dimensional function of material parameters, corner angle, λ, and polar co-ordinate φ. Dimensional arguments allow to assert that $K \approx \sigma_0(\pi R)^{\lambda}$ where the σ_0 is the far field applied stress. Therefore mode I detachment will occur when K reaches the critical value

$$K_{\mathbf{c}} = \frac{1}{2}\alpha\sigma_I(\pi R)^{\lambda} = \frac{1}{2}\alpha CE^*(8\delta)^{\lambda} = \frac{1}{2}\alpha CE^{*1-\lambda}(8\Delta\gamma)^{\lambda} \tag{12.2}$$

where α is a constant of order unity. Equation (12.2) allows us to define an effective energy release rate G as

$$G = \frac{2^{1/\lambda}}{4\alpha^{1/\lambda}C^{1/\lambda}}\frac{K^{1/\lambda}}{2E^{*1/\lambda-1}} \tag{12.3}$$

so that the critical condition for crack propagation from the edge of the pillar can be written in the usual Griffith form $G = \Delta\gamma$. Of course in the case of a rigid flat punch in frictionless adhesive contact with an elastic half-space ($\lambda = 1/2$) one obtains the usual definition $G = K_I^2/(2E^*)$, with $\alpha = C = 1$.

Note that (12.1) is confirmed by more accurate calculations [37], which have been carried out within the framework of the Dugdale-Barenblatt cohesive zone. These calculations show that, for elastic cylindrical pillars in sticking contact with a flat rigid substrate, the critical mode I debonding stress is

$$\sigma_I \approx B\left(\frac{\sigma_{III}}{E^*}\right)^{1-2\lambda} E^*\left(\frac{\delta}{R}\right)^{\lambda} \tag{12.4}$$

which is the same as (12.1) except for a factor $(\sigma_{III}/E^*)^{1-2\lambda}$ of order unity.

As already observed, the exponent λ depends on the real geometry and on the real boundary conditions (e.g. no-slip, partial slip, no friction) at the interface, which usually differs from the nominal values. It follows that the real value of λ is not predictable with sufficient accuracy. Therefore, in what follows we will assume $\lambda \approx 0.5$. This assumption limits nowise the general validity of our treatment and main conclusions. The mode I pull-off stress, needed to separate an elastic cylindrical pillar in adhesive contact with a rigid substrate, then becomes

$$\sigma_I \approx E^*\left(\frac{8}{\pi}\frac{\delta}{R}\right)^{1/2} = \left(\frac{8E^*\Delta\gamma}{\pi R}\right)^{1/2} \tag{12.5}$$

Now let us estimate the mode II debonging stress, i.e. the stress needed to propagate existing interfacial defect (Fig. 12.2b). The presence of a circular defect of radius a determines a variation of adhesion energy $\Delta U_{ad} = \pi a^2\Delta\gamma$ and a corresponding variation of the elastic energy $\Delta U_{el} \approx -(4\sigma^2 a^3)/(3E^*)$. Therefore, the total energy variation is

$$\Delta U_{tot} \approx -\frac{4}{3}\frac{\sigma^2 a^3}{E^*} + \pi a^2\Delta\gamma \tag{12.6}$$

Enforcing the condition $\partial(\Delta U_{tot})/\partial a = 0$ one can calculate for any given defect size a the critical stress for mode II debonding as

$$\sigma_{II} \approx \left(\frac{\pi}{2}\frac{\Delta\gamma E^*}{a}\right)^{1/2} \tag{12.7}$$

Among the three different failure mechanisms (mode I, mode II and mode III) only the one corresponding to the minimum value of the critical stresses σ_I, σ_{II} and $\sigma_{III} = \Delta\gamma/\rho$ could actually take place. Therefore, a map of the debonding mechanisms can be proposed, as shown in Fig. 12.2d, where the quantities $a_C = (\pi/2)E^*\rho^2/\Delta\gamma$ and $R_C = (8/\pi)\, E^*\rho^2/\Delta\gamma$ are respectively a reference defect size, and a reference pillar radius. Let us observe that for soft micro flat-punch shaped pillars (e.g. PVS pillar [28]) assuming $\Delta\gamma \approx 16\,\text{mJ/m}^2$, $\mu = 0.5$, E = 3 MPa, and recalling that the range of van der Waals forces is about $\rho \approx 1\,\text{nm}$ one obtains $a_C = 0.39\,\text{nm}$ and $R_c = 0.63\,\text{nm}$, so that mode III debonding cannot occur in real cases where the pillar radius is of order several micrometers. Figure 12.2d also shows that cylindrical micropillars will detach by following the mode II mechanism only in the very seldom case of very large defects or impurities at the interface. In fact only when $a/a_C > R/R_C$, i.e. only if $a/R > a_C/R_C = \pi^2/16 \approx 0.62$, the stress σ_{II} is smaller than σ_I. Hence, soft cylindrical micropillars must necessarily detach by following mode I debonding. Assuming as in [28] that the pillar has a diameter $2R \approx 15\,\text{mm}$ one obtains a pull-off stress $\sigma_I \approx 4.6\,\text{kPa}$, and consequently a pull-off force $F_{out} = 0.8\,\text{N}$ in perfect agreement with experimental findings [28].

12.3 The Mushroom Shaped Pillar

Some studies [7, 28] have shown that if a very compliant annular plate is added to the base of the flat-punch pillar the pull-off stress is unexpectedly strongly increased. This suggests that a different debonding mechanism should occur. Let us

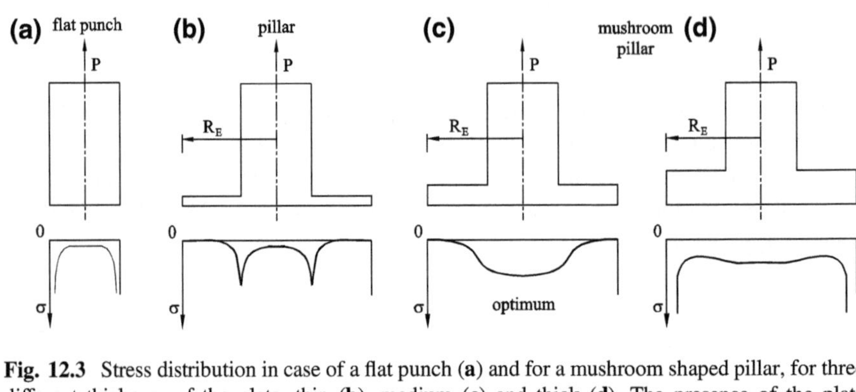

Fig. 12.3 Stress distribution in case of a flat punch (**a**) and for a mushroom shaped pillar, for three different thickness of the plate, thin (**b**), medium (**c**) and thick (**d**). The presence of the plate eliminates the stress singularity of the flat punch at r ≈ R. Stress peak in the mushroom pillar at r ≈ R will gradually vanish as the plate thickness t is increased up to its optimal value (**c**)

first observe that, the presence of the plate, if optimally designed (see also Sect. 12.4), may eliminate the square root stress singularity at $r \approx R$ (Fig. 12.3a).

However, for too thin plates a large stress peak will still be present at the interface (see Fig. 12.3b), which may facilitate the formation and propagation of cracks from the edge $r \approx R$ of the central pillar. The stress peak at $r \approx R$ will gradually vanish if the plate thickness t is increased up to an optimal value (see Fig. 12.3c), at which the stress distribution becomes almost uniform below the pillar and smoothly decreases until it vanishes at the plate perimeter $r = R_e$ [38]. In this case the uniform stress acts on a region of radius $R_\sigma \approx 1.1R$ [38]. If the thickness of the plate is increased further (see Fig. 12.3d) then the stress singularity may appear again at $r = R_e$ and the mushroom pillar will behave as a bigger and less performing flat punch. To confirm this behaviour, we have carried out a Finite Element Analysis of the optimized mushroom shaped pillar, using the commercial code ANSYS. The results are shown in Fig. 12.4, in terms of dimensionless interfacial normal stress distribution underneath the plate $\tilde{\sigma} = \sigma/E^*$, plotted against the ratio between the distance from the pillar center r and the internal radius R.

As expected, the stress singularity at the edge of the plate is almost completely disappeared, even more the interfacial normal stress almost completely vanishes as the external contact perimeter is approached. Hence, the main role of the plate is to prevent mode I debonding by canceling the stress singularity at the pillar edge. Therefore, in the case of optimally designed mushroom shaped micro-pillars, the most critical conditions are actually established under the central pillar at $r < R$, and detachment must follow mode II or mode III mechanisms, depending on which one of them is energetically more favorable. The debonding map is then modified as shown in Fig. 12.5, from which it follows that in the specific case of PVS samples under investigation ($\Delta\gamma \approx 16\,\mathrm{mJ/m^2}$, $E^* = 4\,\mathrm{MPa}$, $a_C = 0.39\,\mathrm{nm}$), mode II should be the real debonding mechanism as indeed confirmed experimentally in [21].

This actually means that since mode II debonding is independent of the pillar size but depends on the defect size (see 12.7), the pull-off force should be exactly the same both for thin and wide pillars, provided that the defect size is also the same in the two cases. This is indeed in perfect agreement with the experimental results

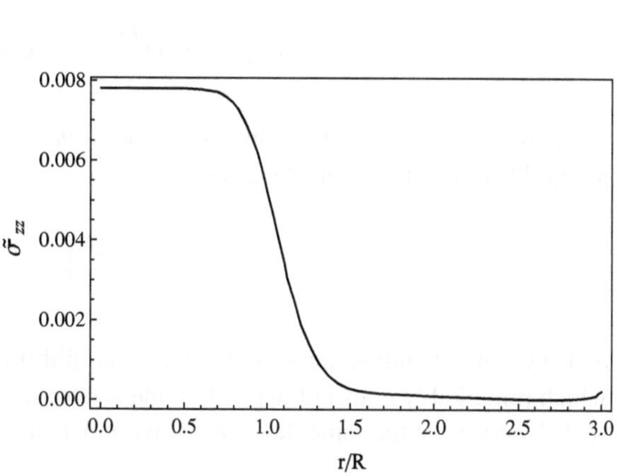

Fig. 12.4 The dimensionless normal stress distribution $\tilde{\sigma}_{zz}$ as a funtion of the ratio r/R, for an optimal mushroom shaped pillar

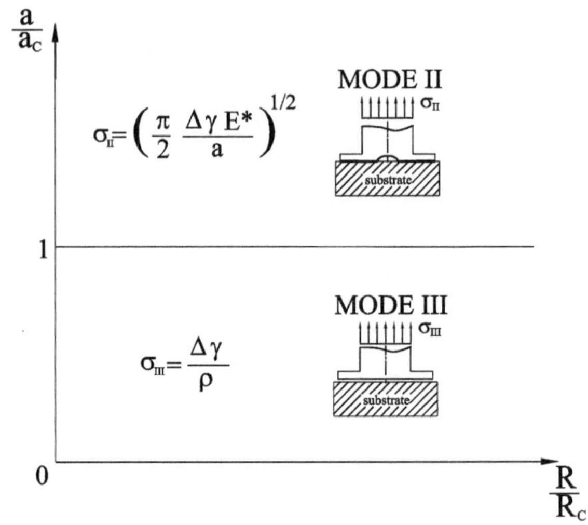

Fig. 12.5 Map of debonding mechanisms for an optimally designed mushroom shaped pillar. Detachments in this case must be governed by mode II or mode III debonding, depending on which one of them is energetically more favorable

presented in [28, 39]. A last consideration about the defect shape must be done. Some experimental observations, indeed, have shown that defects at the interface may assume the shape of simply connected closed regions with a size much smaller than the radius R of the pillar (see Fig. 12.6).

Therefore let us assume that a circular particle is found under the plate (see Fig. 12.6), it produces a circular defect of radius a and height h. Let us analyse the stability of the defect. This time the propagation of the defect will occur at constant height h since the size of the external particle does not change during propagation, hence the only contribution to the change of total energy will come from the change of elastic and adhesion energies. To calculate the change of elastic energy, consider that the external particle exerts on the plate a detaching force F given by $F = 16\pi Dh/a^2$ [40, 41], where $D = E^* t^3/12$. This gives $\Delta U_{el} = (1/2)Fh$, whereas the adhesion energy is $\Delta U_{ad} = \pi a^2 \Delta\gamma$. Thus, the change of total energy is

$$\Delta U_{tot} = 8\pi D\left(\frac{h}{a}\right)^2 + \Delta\gamma\pi a^2 \tag{12.8}$$

As before, enforcing the condition $\partial\Delta U_{tot}/\partial a = 0$ the size of the detached area at equilibrium can be calculated as

$$a_{eq} = \left(\frac{8Dh^2}{\Delta\gamma}\right)^{1/4} \tag{12.9}$$

and the corresponding detaching force at equilibrium becomes $F_{eq} = 4\pi\sqrt{2D\Delta\gamma}$, which remarkably does not depend on the size of the defect, i.e. external impurities will always exert the same detaching force independent of their size. Observing that

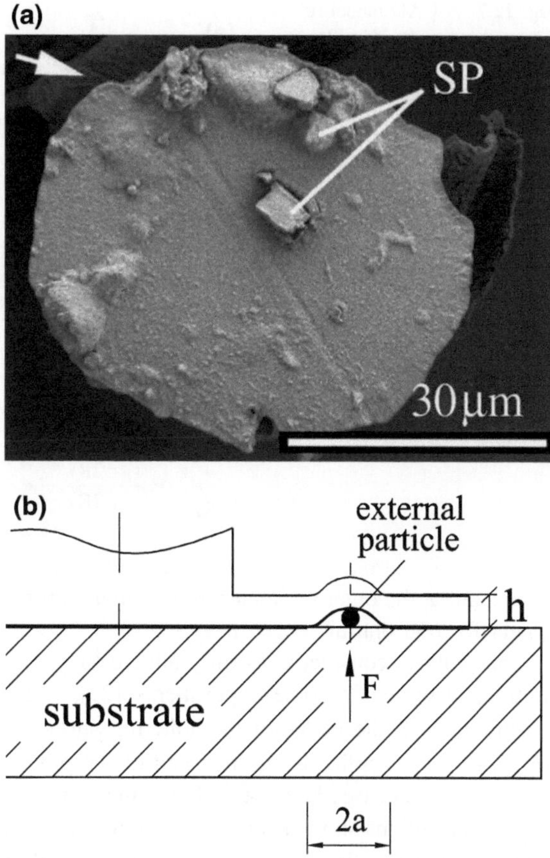

Fig. 12.6 Small dirt particles of about 3–5 μm in size contaminate the interface between the mushroom-shaped pillar and the substrate [33] (**a**). The presence of impurities at the interface determines a decrement of plate-substrate contact area, but, as our calculations show, does not compromise the plate stability, i.e. cannot cause defect propagation (**b**)

$\partial^2 \Delta U_{tot}/\partial a^2$ is always positive it follows that the presence of the plate stabilizes defects at the plate-substrate interface, i.e. defects under the plate cannot propagate but stay there. Therefore, the presence of the particle at the interface does not prevent the plate from adhering to the substrate. We also observe that this result is in agreement with Glassmaker's work [42], who showed that the presence of a terminal film connecting the ends of the pillars reduces the amount of energy available to propagate the interfacial defects.

12.4 Shape Optimization

Let us consider the pillar shape [33] shown in Fig. 12.1a, b. FEM calculations have been carried out for the simplified geometry represented in Fig. 12.7a. The grid (Fig. 12.7b) has been specifically generated in order to have higher elements density both at the inner radius R_i of the pillar (where interfacial stress peaks may

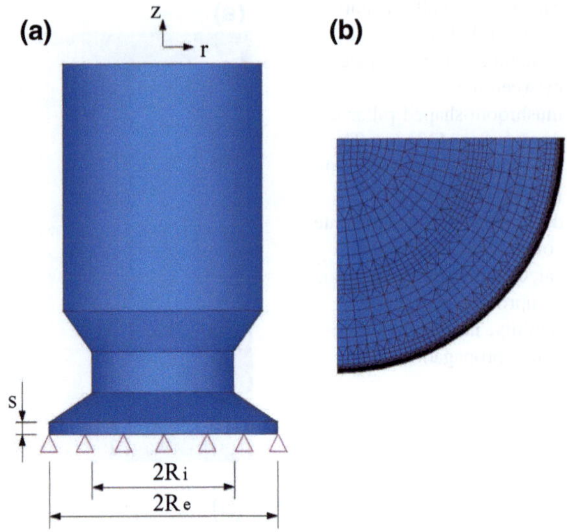

Fig. 12.7 A CAD model of the mushroom shaped pillar shown in Fig. 12.1 (**a**); and the FEM grid (**b**)

occur) and at the outer perimeter of the pillar (where stress square-root singularities appear for non optimized geometries).

The pillar is considered to fully adhere to the flat rigid substrate, and sticking friction, observed in some experiments [21], is taken into account by properly constraining all the nodes of the adhering surface. An external normal load $P = \pi R_i^2 \sigma_0$ is, then, applied to the free end of the pillar, and the interfacial normal σ_{zz} and shear σ_{zr} stresses have been determined by means of FEM calculations. Both stress distributions show a singular behaviour close to the external perimeter of the contact. The strength of such singularities is quantified by the corresponding stress intensity factors [43] K^I and K^{II}. The pull-off stress is calculated by enforcing the generalized Griffith condition [43] $G = \Delta\gamma$, i.e. by requiring that G is equal to the work of adhesion $\Delta\gamma$. The stress intensity factors K^I and K^{II} can be calculated by recalling that they are proportional to the externally applied stress σ_0 and depend, given the material properties, only on the shape of the pillar [44]. Therefore, once fixed the external applied stress σ_0, a single Finite Element analysis is sufficient [44] to calculate the reduced normal and tangential stress distributions $(\sigma_{zz}/\sigma_0, \sigma_{zr}/\sigma_0)$ at the interface and the stress intensity factors. The FE analysis has been carried out with the aid of the commercial software ANSYS [45].

As a first example, we have carried out calculation for the micropillar shown in Fig. 12.1b whose geometry has been taken from [33]. In particular the external diameter of the thin plate is $D_e = 2R_e \approx 40\,\mu\text{m}$, the internal diameter is $D_i = 2R_i \approx 25\,\mu\text{m}$ and the thickness s of the plate is $s = 2\,\mu\text{m}$. Calculations have shown [44] that this type of geometry is suboptimal. Indeed, the reduced normal and shear interfacial stress distributions σ_{zz}/σ_0 and σ_{zr}/σ_0, as functions of the ratio r/R_i, present an unwanted singular behavior at the external edge of the thin plate. The critical stress σ_I needed to activate mode I debonding mechanism [30] is, then, determined by enforcing the condition for crack propagation $G = K_{eq}^2/(2E^*) = \Delta\gamma$ which yields $\sigma_I = CE^*[8\delta/(\pi R_i)]^\lambda$, i.e.

$$\sigma_I = \frac{1}{\tilde{K}_{eq}} E^* \left(\frac{\Delta\gamma}{\pi R_i E^*}\right)^\lambda \tag{12.10}$$

where $(\tilde{K}_{eq}) \approx 0.2$ is a shape factor related to the pillar geometry. Recalling that $E = 3\,\text{MPa}$, $\nu = 0.5$ and $\Delta\gamma \approx 16\,\text{mJ/m}^2$, the calculated critical stress is $\sigma_I \approx 0.33\,\text{MPa}$. We may compare these results with the measurements reported on the same microstructures (see Fig. 12.1b) by Gorb et al. [33], where a surface with diameter $d = 2.9\,\text{mm}$ is covered with the mushroom-shaped pillars and presents a coverage factor $\alpha = 40\%$ (Fig. 12.1b). They measured on clean surfaces a pull-off force $F_{out} \approx 350\,\text{mN}$, whereas our estimated pull-off force is $F_{out} = \alpha(d^2/4R_e^2)\,\pi\sigma_I R_i^2 \approx 340\,\text{mN}$ in very good agreement with the experimental value. Of course the presence of impurities or defects at the interface may change the scenario and, in particular, may switch the debonding mechanism from mode I to mode II [30] and determine a reduction of the critical debonding stress.

The above results (see 12.10) show that the critical stress for mode I debonding rapidly increases as the shape factore $\tilde{K}_{eq}$ decreases. In principle σ_I should diverge when the shape factor $\tilde{K}_{eq}$ goes to zero, i.e. when the stress singularity at the pillar edge vanishes. In this case, for soft materials [30], the mode II mechanism should become the dominant one. Noticing that, on smooth and clean surfaces, the mode II debonding mechanism is activated by stress aided thermal fluctuations (see Sect. 12.7 and [30]), this should lead to very high adhesive forces. However the presence of stress peaks at the interface weakens the adhesion of the pillar to the substrate and should be avoided. Therefore, an optimally designed mushroom pillar should prevent the formation of stress intensification at the interface. This can be achieved, by properly engineering the geometry of the terminal plate and, in particular, by choosing the dimensionless quantities s/R_i and R_e/R_i to guarantee an almost uniform stress distribution in the central part of the pillar $0<r<R_i$ and a vanishing stress at the edge of the contact.

The FE analysis hereafter presented, has been carried out to investigate the influence of the two parameters R_e/R_i and s/R_i on the stress distribution under the plate.

Figure 12.8 shows the reduced normal stress distribution σ_{zz}/σ_0 at the interface between the mushroom shaped pillar and the rigid substrate, as a function of r/R_i, for three different values of R_e/R_i, i.e. $R_e/R_i = 2$ (a), $R_e/R_i = 3$ (b) and $R_e/R_i = 4$ (c). Results confirm what has been qualitatively discussed in Sect. 12.3. Indeed calculations show that, thick plates present a square root flat-punches at the edge of the contact. On the other hand, for very thin plates a stress peak appears at $r = R_i$. The physical reason behind this is very simple. When the thickness of the plate is very thin, the presence of the plate itself negligibly modify the stress distribution which would be observed if the plate were completely absent, i.e. in the case of a

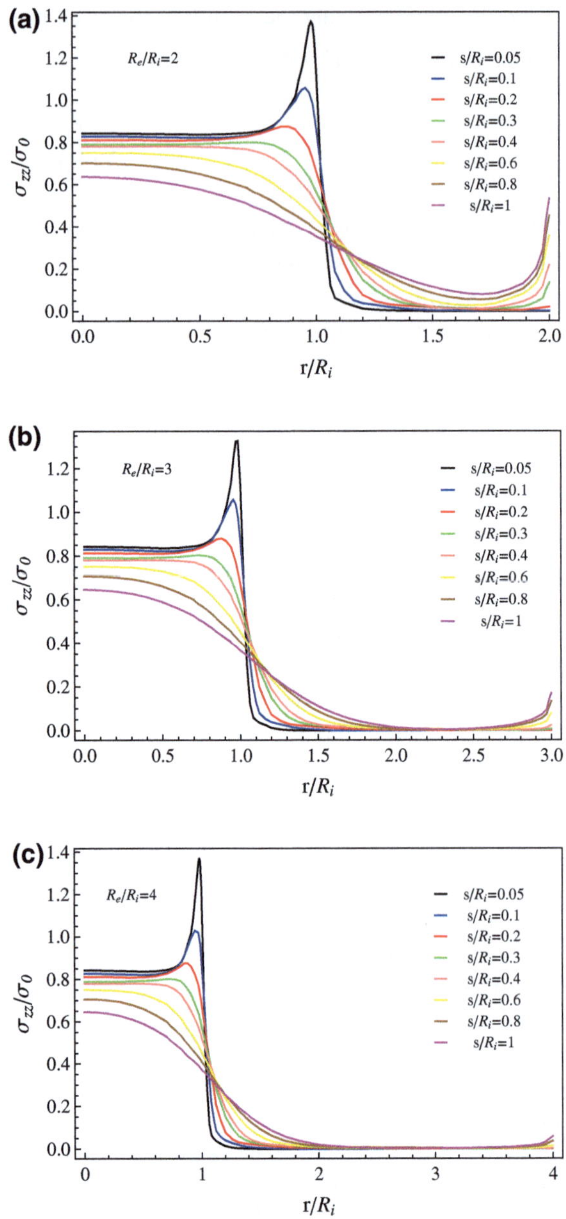

Fig. 12.8 The reduced normal stress distribution σ_{zz}/σ_0 as a function of r/R_i, for $R_e/R_i = 2$ (**a**); $R_e/R_i = 3$ (**b**); and $R_e/R_i = 4$ (**c**). For $R_e/R_i = 2$ (or smaller) and $0.2 < s/R_i < 1$ the stress increases at the external perimeter of the plate. For thinner plates ($s/R_i < 0.2$), on the contrary, a stress peak appears at the internal radius R_i. At larger values of R_e/R_i it is possible to find optimal configurations with $s/R_i \approx 0.2 - 0.3$. In this case the singularity at the $r = R_e$ almost disappears as well as the stress peak at $r = R_i$

flat punch micropillar. But, flat punch pillars present a square root singularity of stress distribution at $r = R_i$.

Thus, thin plates will be only able to slightly smoothen the stress singularity, without making the peak disappear. On the other hand, very thick plates behave as flat punch pillars of radius R_e and the stress singularity will then appear at $r = R_e$.

The plate thickness is therefore the very crucial quantity that must be controlled to optimize the adhesive performance of the system. However, for $R_e/R_i = 2$ (Fig. 12.8a) it is not very easy to find the optimal value of s/R_i that guarantees zero stress at $r = R_e$ and avoids the stress peak at $r = R_i$, a good value is $s/R_i = 0.2$. For larger values of R_e/R_i (Fig. 12.8b, c) this optimal choice can be identified with $s/R_i \approx 0.3$. In conclusion the optimized geometry should be characterized by $s/R_i \approx 0.2 - 0.3$ and values of $R_e/R_i \geq 2$ and preferably close to 3. Values of R_e/R_i much larger than 3 should be avoided since they might reduce the coverage factor of the surface and worsen the performance of the whole adhesive (see below).

Recalling that for a cylindrical micropillar of radius R_i the mode I debonding stress is $(\sigma_I)_{FP}= \left[1/\left(\tilde{K}_{eq}\right)_{FP}\right] E^*[\Delta\gamma/(\pi E^* R_i)]^{\lambda}$, we can easily calculate for the case of a mushroom shaped pillar with internal radius R_i the stress enhancement factor

$$\frac{\sigma_I}{(\sigma_I)_{FP}} = \frac{\left(\tilde{K}_{eq}\right)_{FP}}{\tilde{K}_{eq}} \tag{12.11}$$

which provides a measure of how much the presence of the terminal plate is beneficial in enhancing the adhesive performance. Figure 12.9 shows, under the assumption $\lambda \approx 0.5$, the debonding maps for different type of micropillars: (1) flat punch shaped pillar, (2) the non-optimized mushroom shaped pillar, and (3) the optimized mushroom shaped pillar. We recall that the mode III debonding mechanism shown in Fig. 12.9 is due to the achievement of the theoretical contact strength $\sigma_{III} = \Delta\gamma/\rho$ where $\rho \approx 1$ nm is the typical range of van der Waals forces (for more details see [30]). In Fig. 12.9, the quantities $a_C = (1/2)\pi E^* \rho^2/\Delta\gamma$ and $R_C = E^* \rho^2/[\pi\Delta\gamma(\tilde{K}_{eq})^2_{FP}]$ are respectively a reference defect size, and a reference pillar radius. Assuming, as before, $\Delta\gamma \approx 16\,\text{mJ/m}^2$, $v = 0.5$, $E = 3\,\text{MPa}$, and recalling that $\rho \approx 1$ nm we calculate $a_C = 0.39$ nm and $R_C = 2$ nm. For the mushroom shaped pillar (see Fig. 12.9b), the real debonding mechanism, given the defect size, depends on ratio $\sigma_I/(\sigma_I)_{FP}$.

In particular, for a possible optimal pillar shape $R_e/R_i \approx 3$ and $s/R_i \approx 0.3$, we calculate $\sigma_I/(\sigma_I)_{FP} \approx 0.32 \times 10^3$, so that the mode I debonding will occur for pillars of radius $R_i \geq \left[\sigma_I/(\sigma_I)_{FP}\right]^2 \times R_C = 0.2$ mm, i.e. for optimally designed mushroom shaped micro-pillars with $R_i \approx 10\,\mu\text{m}$ and $R_e/R_i \approx 3$ mode I debonding mechanism cannot take place (see Fig. 12.9c). Our predictions are confirmed by some experimental results [21].

Therefore we may propose a design rule of mushroom shaped micro-pillars. Indeed, on basis of our analysis the optimal pillar should be fabricated in such a way to fulfill the following geometrical constraints: $2 \leq R_e/R_i \leq 3$ and $s/R_i \approx 0.2 - 0.3$. Recently, Aksak et al. [46] have investigated the adhesion mechanism of bio-inspired mushroom shaped pillars by implementing the Dugdale-Barenblatt cohesive zone model into FE calculations. In their geometry, they introduce a wedge angle θ in the mushroom plate zone in contact with the substrate, which is considered as an optimization variable, together with the ratio

Fig. 12.9 The map of debonding mechanisms for a flat punch shaped pillar (**a**); the modified map for a non-optimized mushroom shaped pillar (**b**); the debonding map of an optimized mushroom shaped pillar (**c**). The quantities $a_C = (1/2)\pi E^* \rho^2/\Delta\gamma$ and $R_C = E^* \rho^2/[\pi\Delta\gamma(\tilde{K}_{eq})^2_{FP}]$ are a reference defect size and a reference pillar radius respectively

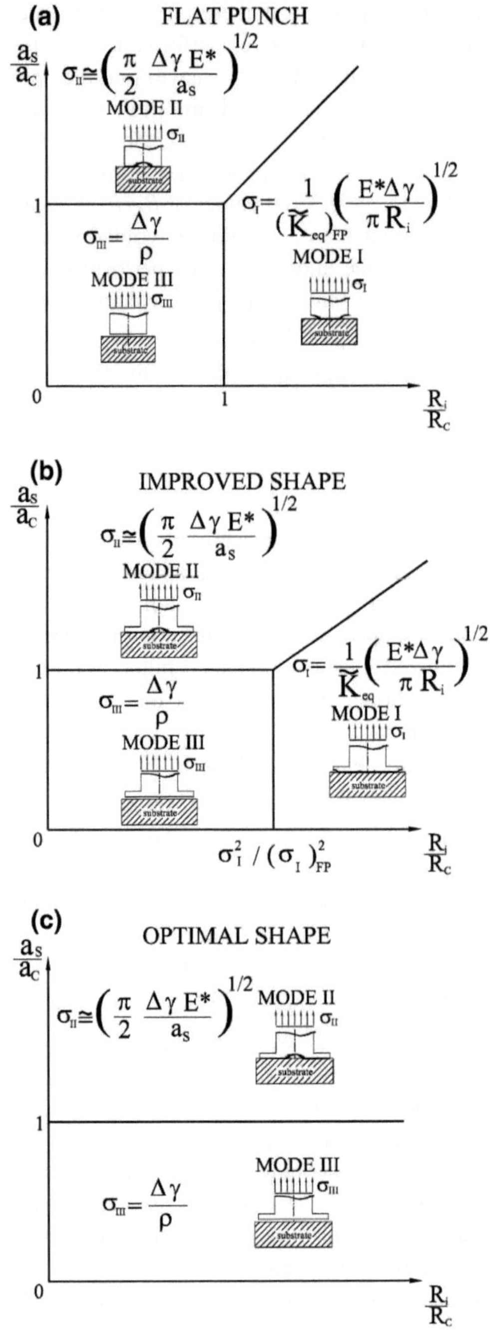

R_e/R_i. They find the optimal mushroom shaped pillar configuration with $R_e/R_i = 1.1 - 1.2$ and $\theta = 45°$, which is different by our results, i.e. $2 \leq R_e/R_i \leq 3$. However, as also stressed by Aksak et al. [46], the two models present some differences in the model geometry (we only consider the case $\theta = 90^{\circ}$), optimization function (their optimization function is the pull-off stress per unit contact area of a single fiber) and adhesion modeling (they neglect fracture Mode II).

12.5 Interfacial Entrapped Air

Often during the approach of the adhesive to the substrate, air bubbles may remain entrapped at the interface. This may lead to a strong reduction of the adhesive performance, which need to be investigated. In order to carry out the analysis the total energy change of the system when a bubble of air is present at the interface must be precisely calculated. We assume that the bubble of air is much smaller than the diameter and height of the pillar so that one can treat the pillar as an elastic half-space in contact with a rigid flat surface. Assuming isothermal conditions and a constant uniform asymptotic far field tractive stress σ, the equilibrium of the system can be sought by requiring that the total free energy at the interface (i.e. the interfacial Gibbs energy) is stationary. Given the defect size and assuming the material linear elastic, the calculation of the energy change of the system must consider four different contributions (see also [47] for a different derivation): (1) the contribution to the interfacial elastic energy due to the asymptotic applied uniform tractive stress σ, (2) the contribution to the interfacial elastic energy due the air pressure p, (3) the internal energy of the air bubble, (4) the variation of surface energies due to the presence of van der Waals forces. Let us consider the system shown in Fig. 12.10 where the bottom (initially flat) surface of an elastic half-space is glued to a rigid plate except on a circular region of radius a. Let us displace the rigid plate of a quantity u_0 (see Fig. 12.10) so that a small void is formed at the interface. Assume that the air pressure in the void is p.

To calculate the elastic energy of the system, let us first observe that the contact problem may have an equivalent formulation in terms of interfacial elastic energy, i.e. in terms of the amount of elastic energy stored at the interface as a consequence of local interfacial deformations [48]. Accordingly, the elastic interfacial energy is [48]

$$\mathrm{E} = \frac{1}{2}\int d^2x\sigma_{zz}(\mathbf{x})[u_z(\mathbf{x}) - \langle u_z(\mathbf{x})\rangle] \tag{12.12}$$

where $\mathbf{x}$ is the in-plane position vector, $\sigma_{zz}(\mathbf{x})$ is the non-uniform normal interfacial stress, $u_z(\mathbf{x})$ is the local normal displacement of the surface, and $\langle u(\mathbf{x})\rangle$ is the average displacement at the interface (the symbol $\langle\cdot\rangle$ is the average operator). Considering that because of force balance the uniform stress σ at infinity is $\sigma = \langle\sigma_{zz}(\mathbf{x})\rangle$ one can rephrase (12.12) as

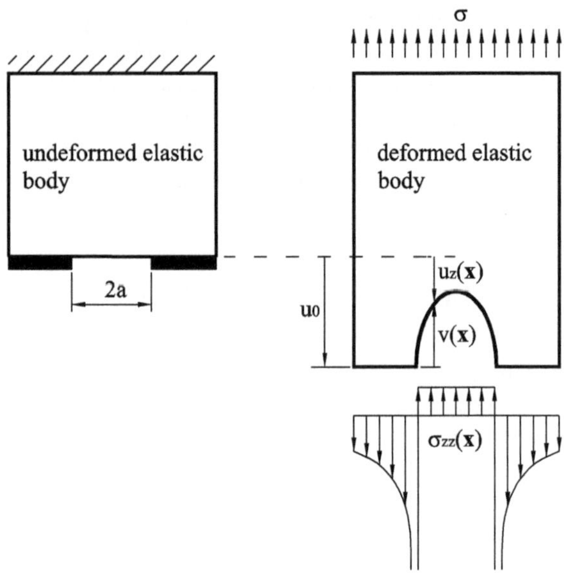

Fig. 12.10 The displacement, gap and stress distributions involved in the calculation of the free interfacial energy

$$\mathtt{E} = \frac{1}{2}\int d^2x[\sigma_{zz}(\mathbf{x}) - \sigma]u_z(\mathbf{x}) \tag{12.13}$$

Now let us define (see Fig. 12.10) the gap distribution $v(\mathbf{x})$ as $v(\mathbf{x}) = u_0 - u_z(\mathbf{x})$. Of course $v(\mathbf{x}) \neq 0$ on the circular region of radius a, whereas it vanishes elsewhere. Using $v(\mathbf{x})$ and considering that $\sigma_{zz}(\mathbf{x}) = -p$ for $|\mathbf{x}| < a$, (12.13) becomes

$$\mathtt{E} = \frac{1}{2}(p + \sigma)\, V \tag{12.14}$$

where $V = \int d^2xv(\mathbf{x})$ is the volume of the air bubble. The total Helmholtz free interfacial energy F is then the sum of the elastic interfacial energy, the free internal energy $U_A = -p_0V_0 \ln(V/V_0)$ of the entrapped air, and the surface energy, i.e.

$$\mathtt{F}(V, a) = \mathtt{E}(V, a) + U_A(V) + \pi a^2\Delta\gamma \tag{12.15}$$

From thermodynamics one concludes that under constant bubble volume V the equilibrium of the system corresponds to the stationary values of the energy F. However, in our analysis we, instead, keep constant the asymptotic load $\langle\sigma(\mathbf{x})\rangle = \sigma$. In this case the equilibrium of the system corresponds to the stationary values of the interfacial Gibbs energy $\mathtt{G}(\sigma, a)$. Following the standard approach of thermodynamics [49], we obtain $\mathtt{G}(\sigma, a)$ by enforcing a Legendre transformation, i.e.

$$\mathtt{G}(\sigma,\ \mathtt{a}) = \mathtt{F}(\mathtt{V},\ \mathtt{a}) - \left.\frac{\partial \mathtt{F}}{\partial \mathtt{V}}\right|_{\mathtt{a}} \mathtt{V} \tag{12.16}$$

Observing that $\partial \mathtt{E}/\partial V|_a = p + \sigma$ and $\partial U_A/\partial V|_a = -p$, one yields the required expression for the interfacial Gibbs energy $\mathtt{G}$, i.e.

$$\mathtt{G}(\sigma, a) = \Delta U_{tot} = \frac{1}{2}(p - \sigma)\, V + U_A + \pi a^2 \Delta\gamma \tag{12.17}$$

Beside (12.17) the two additional equations are needed

$$pV = p_0 V_0 \tag{12.18}$$

and

$$V = \frac{8a^3}{3E^*}(\sigma + p) \tag{12.19}$$

Equations (12.18) and (12.19) allow to calculate the quantity V and p as a function of the radius a of the non contact circular area. Therefore, the total energy change ΔU_{tot} given by (12.17) finally depends only on the applied constant stress σ and the size of the voids a. By following a similar approach as in JKR theory [50], requiring that $\partial \Delta U_{tot}/\partial a|_\sigma = 0$ allows to calculate the values of a at equilibrium, given the applied uniform stress σ. The critical pull-off stress σ_{cr}, which destabilize the defect and causes the detachment of the pillar from the rigid flat substrate, is determined by requiring that at equilibrium the relation $\partial^2 \Delta U_{tot}/\partial a^2|_\sigma = 0$ is also satisfied.

The above equations can be rephrased in a dimensionless form. To this end let us define the adhesion length $\delta = \Delta\gamma/E^*$ and the dimensionless quantities $\tilde{\sigma} = \sigma/E^*$, $\tilde{p} = p/E^*$, $\tilde{a} = a/\delta$, $\tilde{V} = V/\delta^3$. The dimensionless total energy change of the system is therefore $\Delta\tilde{U}_{tot} = \Delta U_{tot}/(\delta^3 E^*)$, where $\tilde{V} = (8/3)(\tilde{\sigma} + \tilde{p})\,\tilde{a}^3$ and $\tilde{p}\tilde{V} = \tilde{p}_0 \tilde{V}_0$.

We assume that the environment pressure is 1 bar, so that the initial pressure of the entrapped air bubble is also $p_0 = 0.1\,\text{MPa}$. We also notice that the asymptotic applied stress σ is $\sigma = \sigma_0 - p_0$, where $\sigma_0 = P/A$ is the external applied average stress, P the applied load and A the cross section area of the pillar. Figure 12.11 shows the total dimensionless energy change $\Delta\tilde{U}_{tot}$ as a function of the dimensionless radius $\tilde{a}$, for different values of the initial dimensionless size $\tilde{a}_0$ of the void. In our calculations we have used $\sigma = 0.2\,\text{MPa}$ (i.e. $\sigma_0 = 0.3\,\text{MPa}$), $\Delta\gamma \approx 16\,\text{mJ/m}^2$, $E = 3\,\text{MPa}$ and $\nu = 0.5$. The figure shows that for any value of $\tilde{a}_0$, two equilibrium conditions exist, i.e. the stable state (energy minimum) and the unstable state (energy maximum). However, as expected, when the asymptotic applied stress is

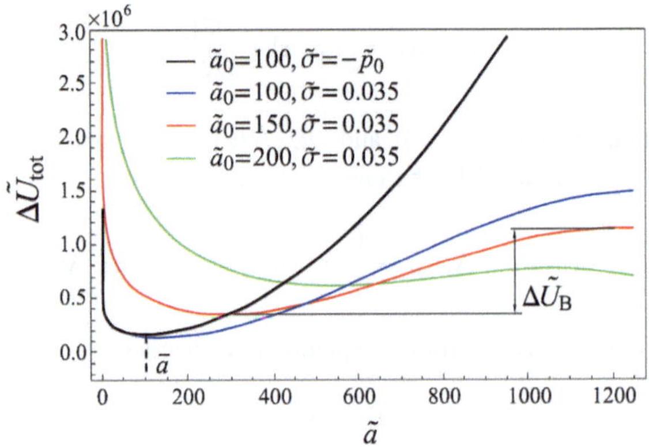

Fig. 12.11 The dimensionless total energy $\Delta\tilde{U}_{tot}$ as a function of the radius of the detached area $\tilde{a}$, for three different values of the initial radius $\tilde{a}_0$, given the same value of dimensionless stress $\tilde{\sigma} = 0.035$ (*blue*, *red* and *green curves*). Increasing $\tilde{a}_0$ determines a decrease of the energy barrier $\Delta\tilde{U}_{tot}$ between the stable and unstable equilibrium states. The *solid black curve* represents the total energy as a function of a when the pillar is subjected to the environment pressure only, i.e. $\tilde{\sigma} = -\tilde{p}_0$, and for $\tilde{a}_0 = 100$. Notice that in this case there is only one equilibrium condition at $\tilde{a}_0 = \bar{a}$ which is necessarily stable

zero or even negative only a stable equilibrium state must be present (see the black line in Fig. 12.11 with $\tilde{a}_0 = 100$ and $\tilde{\sigma} = -\tilde{p}_0$, i.e. $\tilde{\sigma}_0 = 0$). When an external stress $\sigma > 0$ is applied, an energy barrier must be exceeded in order to destabilize the system.

The energy barrier $\Delta\tilde{U}_B$ in this case is defined as the difference between the energy value of the unstable equilibrium state and the energy value of the stable equilibrium state. From Fig. 12.11, given the same applied stress, one observes that the energy barrier $\Delta\tilde{U}_B$ decreases as the initial radius $\tilde{a}_0$ of the bubble (i.e. its initial volume) is increased. When $\Delta\tilde{U}_B = 0$, i.e. when $\partial\Delta U_{tot}/\partial a|_\sigma = 0$ and $\partial^2\Delta U_{tot}/\partial a^2|_\sigma = 0$, the critical defect size $(\tilde{a}_0)_{cr}$ is found which prevents the pillar from adhering to the substrate.

Given the initial defect size $\tilde{a}_0$, one may also analyze what happens when the applied stress σ is increased. In particular, Fig. 12.12 shows that, for a fixed value of the radius $\tilde{a}_0$ (we have considered $a_0 = 0.6\ \mu\text{m}$, i.e. $\tilde{a}_0 = 150$), when the applied stress σ increases an unstable equilibrium state appears, which is again separated from the corresponding stable equilibrium by an energy barrier $\Delta\tilde{U}_B$. As the stress $\tilde{\sigma}$ is further increased, the energy barrier $\Delta\tilde{U}_B$ decreases and vanishes at a certain stress level $\tilde{\sigma}_{cr}$ (the so called critical pull-off stress) at which the air bubble of initial size $\tilde{a}_0$ is destabilized and the pillar detaches from the substrate.

It is interesting now to compare the critical stress in case of solid defects of size a_S (dust particles, impurities, etc.) $\sigma_{II} = [\pi\Delta\gamma E^*/(2a_S)]^{1/2}$ with the critical pull-off stress σ_{cr} obtained in the case a bubble of air entrapped at the interface. The comparison must be carried out assuming that, at $\tilde{\sigma} = -\tilde{p}_0$ (i.e. $\tilde{\sigma}_0 = 0$), the

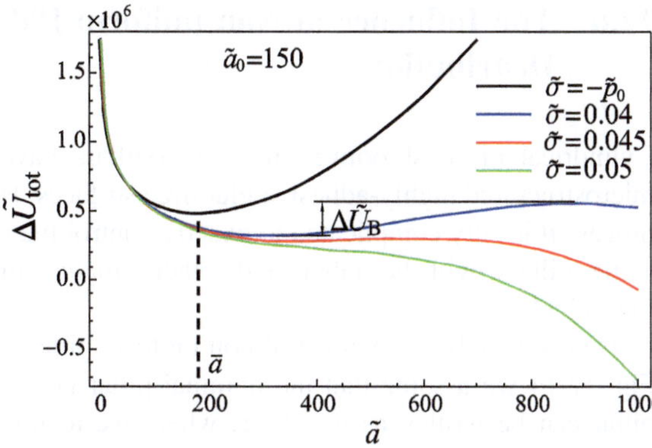

Fig. 12.12 The dimensionless total energy $\Delta\tilde{U}_{tot}$ as a function of the radius of the detached area $\tilde{a}$, for four different values of the applied stress $\tilde{\sigma}$, and for $\tilde{a}_0 = 150$. Increasing $\tilde{\sigma}$ determines a decrease of the energy barrier $\Delta\tilde{U}_{tot}$ between the stable and unstable equilibrium states until it vanishes and the air bubble is destabilized

(dimensionless) size $\bar{a}$ of the air bubble at equilibrium is identical to the (dimensionless) solid defect size, i.e. $\bar{a} = \tilde{a}_S$. Figure 12.13 compares the critical stress $\tilde{\sigma}_{0cr} = \tilde{\sigma}_{cr} + \tilde{p}_0$ in the two cases as a function of the radius $\bar{a}$.

We observe that, in the case of air bubble, the debonding stress $\tilde{\sigma}_{0cr}$ is always significantly smaller than the one obtained in the case of solid defect nucleation process s with a reduction of about 35–40% over the entire range of defect size considered in the calculation, i.e. $\bar{a} = \tilde{a}_S = 0.4 - 4\ \mu\text{m}$. Indeed, micro-air bubbles weaken the adhesive link between the pillar and the rigid substrate more than the presence of external particles, since their gas pressure exerts an additional debonding force and reduce the suction effect which contributes to keep the pillar in contact with the substrate. This represents a practical problem during fast attaching-detaching of this kind of microstructure, since in this case the entrapment of air can hardly be avoided.

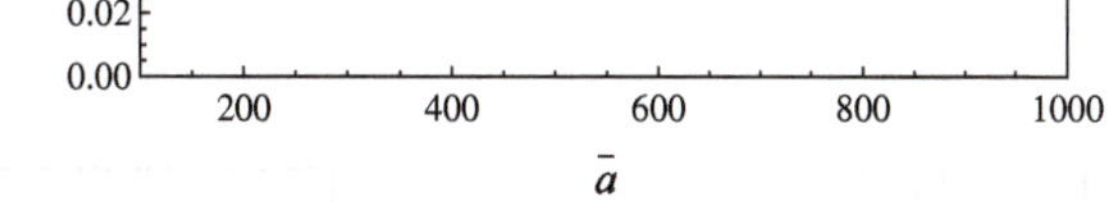

Fig. 12.13 The dimensionless external critical stress $\tilde{\sigma}_{0cr} = \tilde{\sigma}_{cr} + \tilde{p}_0$ as a function of the air bubble or solid particle size a (see text for more details). The *blue curve* refer to the air bubble case, the *black curve* to the interfacial solid particle case

12.6 The Influence of Non-uniform Pillar Height Distribution

Cylindrical or mushroom-shaped micropillars have been employed to fabricate microstructured highly-adhesive glue-free surfaces. However, during the fabrication process it is very complicated to exactly control the height of each single pillar, i.e. some pillars will be taller and other smaller than the nominal height (see Fig. 12.14).

Here we analyse the effect of non uniform pillar height distribution. Referring to Fig. 12.14 we assume that the nominal pillar height is l, then the height h of the pillar can be written as $h = l + z$, where we assume that the distances z are distributed according to a Gaussian probability density distribution

$$p(z) = \frac{1}{\sqrt{2\pi\langle z^2\rangle}} \exp\left(-\frac{z^2}{2\langle z^2\rangle}\right) \tag{12.20}$$

where $\langle\ \rangle$ is the statistical average operator, and $\langle z^2\rangle$ is the mean square distance. Now assume that, after preloading, all pillars come into contact with the substrate. Therefore, assuming that the substrate is located at a distance s from the mean plane $z = 0$, we can calculate the force needed to elongate the fibre as

$$F = \pi R^2 E \frac{s - z}{l + z} \tag{12.21}$$

where R is the circular radius of the pillar cross section. Observe that the force F increases with $s - z$ up to the limiting pull-off value $F_{out} = \pi R_\sigma^2 \sigma_{II}$ at which the pillar will detach from the substrate. Also observe that if the quantity $s - z$ is less than zero then a compressive (negative) force will act on the pillar, and this compressive force will increase as $s - z$ is decreased down to a limiting (negative) load F_B, which causes the elastic buckling of the pillar

$$F_B = -\pi^2 EJ/[2(l + z)]^2 \tag{12.22}$$

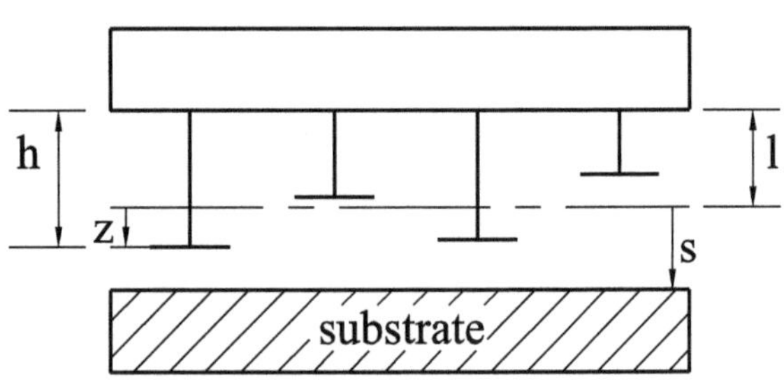

Fig. 12.14 A microstructured surface covered with mushroom-shaped pillars. Some pillars are taller and others smaller than the nominal height. The nominal pillar height is l, the actual height h of the pillar can be written as $h = l + z$

where $J = \pi R^4/4$. We assume that below this value the pillar is not able to support the load. Thus, we can calculate, as a function of s, the range of distances z, that will contribute to the total force acting on the microstructured surface. By requiring that $F_B < F < F_{out}$, one obtains $z_{\min}(s) < z < z_{\max}(s)$ where

$$z_{\min}(s) = l\left(\frac{s}{l} - \frac{R_\sigma^2}{R^2}\frac{\sigma_{II}}{E}\right)\left(1 + \frac{R_\sigma^2}{R^2}\frac{\sigma_{II}}{E}\right)^{-1} \tag{12.23}$$

$$z_{\max}(s) = \frac{l}{2}\left[\sqrt{\left(1+\frac{s}{l}\right)^2 + \pi\frac{J}{l^2R^2}} - \left(1-\frac{s}{l}\right)\right] \tag{12.24}$$

and the average stress $\langle\sigma(s)\rangle$ in units of E at equilibrium is

$$\frac{\langle\sigma(s)\rangle}{E} = \int_{z_{\min}(s)}^{z_{\max}(s)} p(z)\frac{s-z}{l+z}dz \tag{12.25}$$

Let us consider, as before, the case of a microstructured surface made of PVS ($\Delta\gamma \approx 16\,\mathrm{mJ/m^2}$, $E^* = 4\,\mathrm{MPa}$) covered with mushroom-shaped pillars with internal radius $R = 12.5\,\mu\mathrm{m}$ and nominal pillar height $l = 100\,\mu\mathrm{m}$ [33]. Assuming a limiting pull-off stress $\sigma_{II} \approx 0.26\,\mathrm{MPa}$ (see Sect. 12.3), we obtain the average stress $\langle\sigma(s)\rangle$ in units of E as a function of the dimensionless separation s/l for different values of the reduced root mean square (rms) distance $\langle z^2\rangle^{1/2}/l$ as shown in Fig. 12.15.

Figure 12.15 shows the strong reduction of the average stress $\langle\sigma(s)\rangle$, which occurs as the quantity $\langle z^2\rangle^{1/2}/l$ is increased. In particular the maximum dimensionless pull-off stress $\langle\sigma\rangle_{\max}/E$ can be plot as a function of $\langle z^2\rangle^{1/2}/l$. Figure 12.16

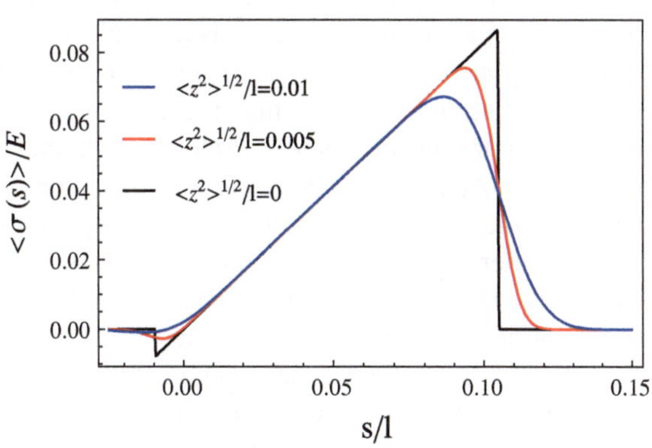

Fig. 12.15 The mean stress $\langle\sigma(s)\rangle$ in units of the elastic modulus E of the material, as a function of the dimensionless separation s/l for different values of the quantity reduced root mean square (rms) distance $\langle z^2\rangle^{1/2}/l$

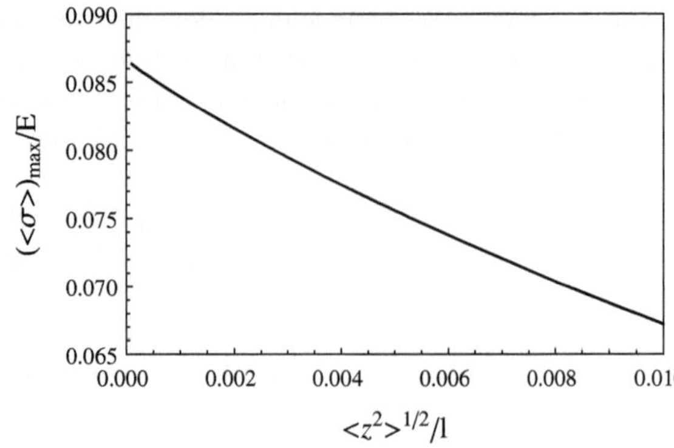

Fig. 12.16 The maximum average stress $\langle\sigma\rangle_{\max}$ (in units of the elastic modulus E) as a function of $\langle z^2\rangle^{1/2}/l$

shows, indeed, that a value of the rms distance $\langle z^2\rangle^{1/2}$ equal to $\approx 1\%$ of the nominal height l is already enough to produce a decrement of the pull-off stress of about $\approx 20\%$.

12.7 Stress Aided Thermally Activated Defect Nucleation

The above calculations have been carried out for cases where defects are already present at the interface between the micro-pillars and the substrate. However, in some conditions the surfaces may be very clean and very smooth. When this happens one may be tempted to conclude that, not depending on the type of pillar we are considering, mode II debonding mechanism can never occur. In particular one would expect that flat-punch pillars would detach by following mode I or mode III mechanisms, whereas mushroom-shaped pillars would detach by following mode III only. However, in this section we show that this is not true and that mode II remains a possible debonding mechanism since defects at the interface may be formed as a consequence of a stress aided thermally activated nucleation process. Indeed, thermal fluctuations are always able to overcome the stress dependent energy barrier $\Delta U_B = \pi^3 \Delta\gamma^3 E^{*2}/(12\sigma^4)$ (see [30]) and nucleate the defect. Statistical mechanics shows that the rate w at which these fluctuations occur depends on the energy barrier ΔU_B through the Maxwell-Boltzmann equation $w = \nu \exp(-\Delta U_B/k_B T)$, where T is the temperature of the system, $\nu \approx 10^{14}\ \mathrm{s}^{-1}$ is a very large prefactor related to the high entropy associated with placing the nucleus in many different places on the contact area [10] and k_B is the Boltzmann constant. We observe that, increasing the applied stress determines a strong reduction of the energy barrier ΔU_B and therefore an increase of the nucleation rate. Of course, if the applied stress is small one must wait a long time before the first nucleus is formed and pillar detachment can be observed.

However, as the applied stress is increased, defect nucleation and pillar detachment will occur on shorter time intervals and will become observable when these time intervals will reach values of order 1 s or less, i.e. when the nucleation rate w is equal to or smaller than $w_0 \approx 1\,\text{Hz}$. When this happens the stress-dependent energy barrier takes the value $\Delta U_B = k_B T\ \ln(v/w_0)$. At room temperature, i.e. $T = 300$ K, one obtains $\Delta U_B \approx 1\,\text{eV}$.

This means that mode II debonding mechanism can be observed, even in case of perfectly smooth and clean surfaces, when the applied stress is large enough to reduce the energy barrier against nucleation at values equal to or smaller that $\Delta U_B \approx 1\,\text{eV}$. From ΔU_B one can calculate the critical external stress σ_{II} which activates the stress aided thermally activated debonding mechanism of type II:

$$\sigma_{II} = \left[\frac{\pi^3}{12}\frac{\Delta\gamma^3 E^{*2}}{\Delta U_B}\right]^{1/4} \tag{12.26}$$

For the case of PVS micropillars under consideration one obtains $\sigma_{II} = 5\,\text{MPa}$. This value is relatively large and may be even larger than the rupture stress of the fibre. This of course means that soft micropillars might be, in principle, ripped off before mode II debonding can take place. Now, neglecting other forms of failure, and concentrating only on interfacial detachment, one can produce a new map of debonding mechanisms for the case of flat-punch shaped micropillars in contact with clean and smooth surfaces.

This new map is shown in Fig. 12.17, where

$$E_C^* = \left(\frac{12}{\pi^3}\frac{\Delta\gamma\Delta U_B}{\rho^4}\right)^{1/2} \tag{12.27}$$

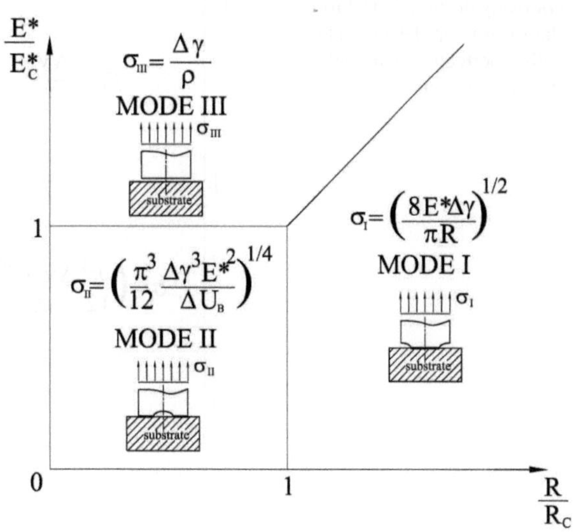

Fig. 12.17 Map of debonding mechanisms for a flat-punch shaped micropillar in contact with perfectly clean and smooth substrates

Recalling that $\Delta U_B \approx 1\,\text{eV}$, and that for PVS micropillars $\Delta\gamma \approx 0.016\,\text{J/m}^2$ and $\rho \approx 1\,\text{nm}$ one obtains that the stress-aided thermally activated nucleation at the interface is energetically more favorable than uniform decohesion (mode III) when $E^* < E_C^* = 31\,\text{MPa}$. Hence, in case of PVS ($E^* = 4\,\text{MPa}$) flat-punch shaped micro-pillars we expect that only mode I and mode II debonding mechanisms can be activated. To find which one is the real mechanism it is enough to observe that mode I mechanism occurs when $\sigma_I < \sigma_{II}$, which gives

$$R > R_C = \sqrt{\frac{768}{\pi^5}\frac{\Delta U_B}{\Delta\gamma}} \tag{12.28}$$

Being $\Delta\gamma \approx 0.016\,\text{J/m}^2$, we obtain $R_C \approx 5\,\text{nm}$ which simply means that, even in case of contact with perfectly smooth and clean surfaces, man-made flat-punch shaped PVS micropillars (size of order 10 μm [33]) always detach from the substrate by following the mode I debonding mechanism.

However the presence of an optimally designed annular plate inhibits mode I failure mechanism, thus, in the case of mushroom-shaped pillar in contact with perfectly smooth and clean surface, the map of debonding mechanisms changes as shown in Fig. 12.18, from which one concludes that relatively soft mushroom-shaped pillars detach at much higher stress because of stress aided thermally activated defect nucleation (mode II mechanism) even in case of clean and smooth surfaces.

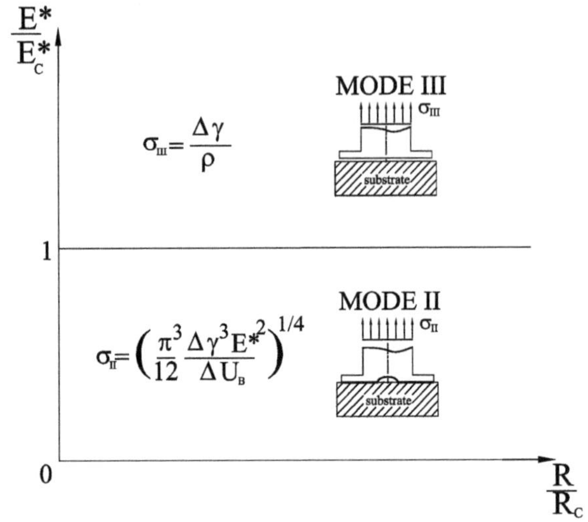

Fig. 12.18 Map of debonding mechanisms for an optimally designed mushroom shaped micropillar in contact with a perfectly clean and smooth substrate

12.8 Adhesion Tilt-Tolerancy

Mushroom-shaped contact elements have been independently developed at the macro-, micro-, and nano-scale in the evolution of many organisms from different lineages (animals, plants, fungi, and bacteria) living in both terrestrial and aquatic environments [28, 51]. This specific geometry is able to stay passively adhered without external effort and is mainly related to long-term and permanent adhesion of organisms [28]. Thus, one may assume that mushroom-shaped contact elements adhering to a substrate should be tolerant to changing and varying load conditions as, e.g., life in wave-swept seashores (e.g. sea anemones) and the lond-term pairing process in some organisms.

However, in biological systems, the load conditions can strongly vary in different environmental and behavioural situations. That is why one may hypothesize that this particular contact geometry is not only optimized to normal adhesion, but is also tolerant to the angle of applied pull off force.

In order to test the hypothesis that this contact geometry is tolerant to varying load conditions, in [52] adhesion experiments of individual mushroom-shaped adhesive microstructures (MSAMSs) have been considered, see Fig. 12.19a, pulled under different tilt angles.

In the experiments four individual MSAMSs, denoted by sample **1**–**4**, were detached from a smooth glass slide under different tilt angles $\alpha = \{-45°, -10°, 0°, 10°, 45°\}$ simultaneously recording the failure dynamics.

Individual MSAMSs were cut off from the Gecko-Tape (Gottlieb Binder GmbH & Co. KG, Holzgerlingen, Germany) with a thickness of the supporting poylmer

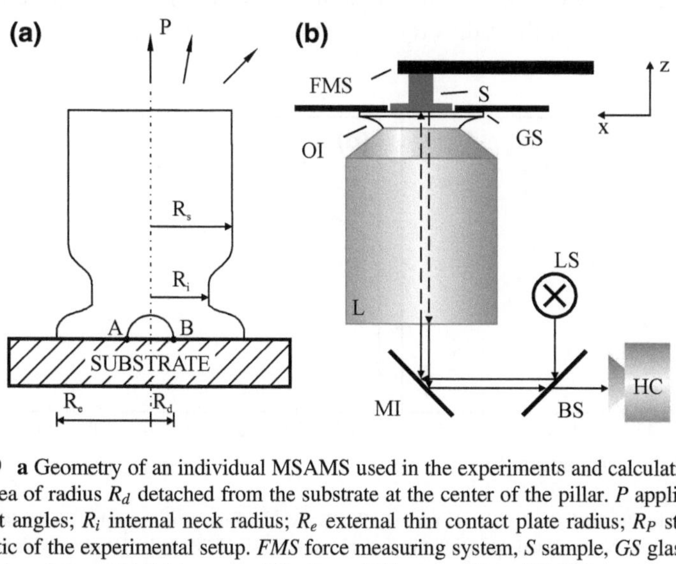

Fig. 12.19 **a** Geometry of an individual MSAMS used in the experiments and calculations with a circular area of radius R_d detached from the substrate at the center of the pillar. *P* applied force at various tilt angles; R_i internal neck radius; R_e external thin contact plate radius; R_P stalk radius. **b** Schematic of the experimental setup. *FMS* force measuring system, *S* sample, *GS* glass slide, *OI* oil immersion, *L* lens, *LS* light source, *MI* mirror, *BS* beam splitter, *HC* high-speed camera

film of about 900 μm. Pull-off forces P_z [z direction only, see Fig. 12.19b] were measured using a force measuring system (FMS) consisting of a tensometric force transducers FORT-10 (World Precision Instruments Inc., Sarasota, Florida) fixed on a three-axis micromanipulator F-131.3SS (Physik Instrumente GmbH & Co. KG, Karlsruhe, Germany). The FMS was installed on an inverse microscope Observer. A1 (Carl Zeiss MicroImaging GmbH, Göttingen, Germany) observing the detachment behavior with an attached high-speed camera Photron Fastcam SA1.1 (VKT Video Kommunikation GmbH, Pfullingen, Germany). Figure 12.19b shows the schematic of the experimental setup. To be able to repeatedly attach and detach samples, individual MSAMSs were glued to the force transducer. To ensure parallel alignment between samples and the glass slide, individual MSAMSs were first attached to the glass slide manually using tweezers observing the proper contact via the microscope. Then, attached to the glass slide samples were withdrawn at a retraction velociy of $v_z = 10\,\mu\text{m/s}$ normal to the glass slide (z direction). In order to pull under different tilt angles a velocity $v_x = \pm v_z \tan\alpha$ in the x direction (see Fig. 12.19b) was superimposed on v_z.

Figure 12.20a shows the experimentally obtained pull-off forces P_z for the different tilt angles α normalized by $P_0 \equiv P_z(\alpha = 0°)$. Except for an outlier (sample **2**, $\alpha = -45°$), normalized pull-off forces scatter surprisingly only by about $\pm$ 10% around P_0 for all angles measured without certain trend.

We have carried out a Finite Element analysis of an individual MSAMS shown in Fig. 12.19a with the aim of calculating the critical stress at which the MSAMS is pulled off from the substrate, due to the presence of a circular detached area at the interface of radius R_d, located close to the pillar axis (Fig. 12.19a). We have employed a mesh made of SOLID45 elements (penta6 and hex8 types have been used) with very high density close to those regions of the interface where singular stress behavior is expected, i.e. mainly at the perimeter of the circular defect [45]. A numerical value of the radius $R_d = 3\,\mu\text{m}$ has been estimated from experimental

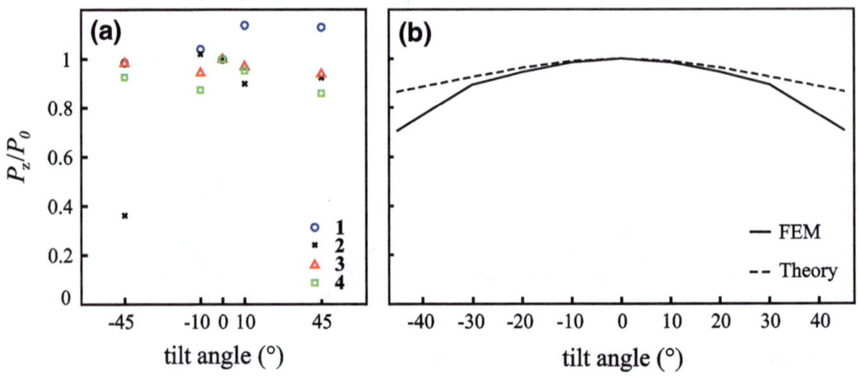

Fig. 12.20 **a** Normalized pull-off forces P_z/P_0 for different tilt angles α obtained from experiments. **b** Normalized pull-off forces obtained by FEM calculations (*solid line*) and theoretical prediction (*dashed line*)

observations of typical defects size [53]. The pull-off load and stress intensity factors K^I and K^{II} are calculated following the procedure described in Sect. 12.4.

Observe that, this time, because of load tilting (see Fig. 12.19a) K^I and K^{II}, are not uniform on the perimeter of the detached area (Fig. 12.19a]. The most critical condition is therefore reached at point A in Fig. 12.19a. In Fig. 12.20b we show the results of our FE simulations, in terms of normalized pull-off force P_z/P_0 (solid line), for different tilt angles of the external applied load, i.e. $\alpha = \{-45°, -10°, 0°, 10°, 45°\}$.

The FEM calculation have been carried out under the assumptions of no slip at the interface. However, experiments in [53] and high-speed video observations of the contact behavior in [52] show that slip actually occurs at the interface (5 μm) with a certain amount of friction. Therefore we also considered a different approach to calculate the pull-off force. To this end, first observe that, in absence of any defect, an optimized MSAMS [44] (i.e. the interfacial stress at the interface does not show any singular behaviour or any strong peak), presents an almost uniform normal stress distribution σ_{zz} in the central part of the contact. However, because of tilting, additional stress components will appear at the interface. In fact, tilting produces three types of stresses as shown in Fig. 12.21: (i) the almost uniform normal stress σ_{zz}, (ii) the tangential stress σ_{zx}, and (iii) a triangular stress

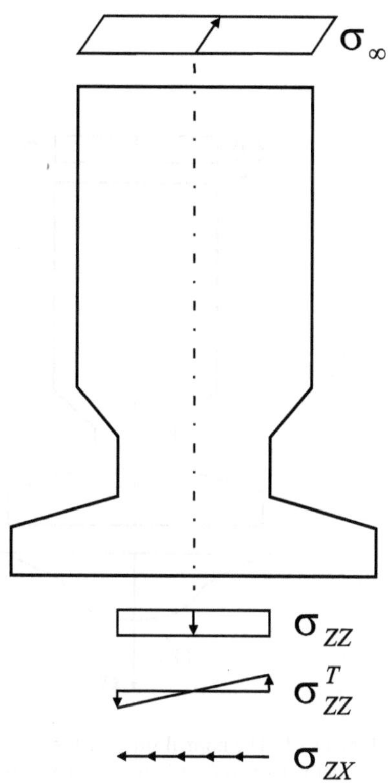

Fig. 12.21 The three types of stress distributions under the plate due to the external tilted force F applied to the pillar: the uniform normal stress σ_{Mz}, the tangential stress σ_{Mx}, and the triangular stress distribution σ^T_{Mz}

distribution σ_{zz}^T. Notice that if the defect is located relatively close to the center of the pillar the triangular stress distribution σ_{zz}^T, caused by bending, is very negligible compared to σ_{zx} and σ_{zz}. Therefore, we expect that, in presence of central defect, the dominant contribution to K_{eq} will come from the stresses σ_{zz} and σ_{zx}. We notice that the stress σ_{zx} is not uniformly distributed at the interface however a rough estimation K^I and K^{II} can be done assuming that σ_{zx} is constant at the interface.

In this case the pull-off force P_z can be estimated observing that under tilted-load conditions the central defect resembles the case of a penny-shaped crack subjected to the combination of two remote loads: (i) a pure traction [54], and (ii) a pure shear load [55]. One then can use the superposition principle to combine the two stress states. The stress intensity factors relevant for pull-off calculations are: (i) $K_\perp = \max K^I$ for remote traction load, and (ii) $K_{II} = \sqrt{(K^{II})^2 + (K^{III})^2}$ for shear mode. $K_\perp$ and K_{II} can be estimated as

$$K_\perp = 2\sqrt{\frac{R_d}{\pi}}\sigma_T \cos^2 \alpha \tag{12.29}$$

$$K_{II} = \frac{2}{(2-\nu)}\sqrt{\frac{R_d}{\pi}}\sigma_T \sin(2\alpha) \tag{12.30}$$

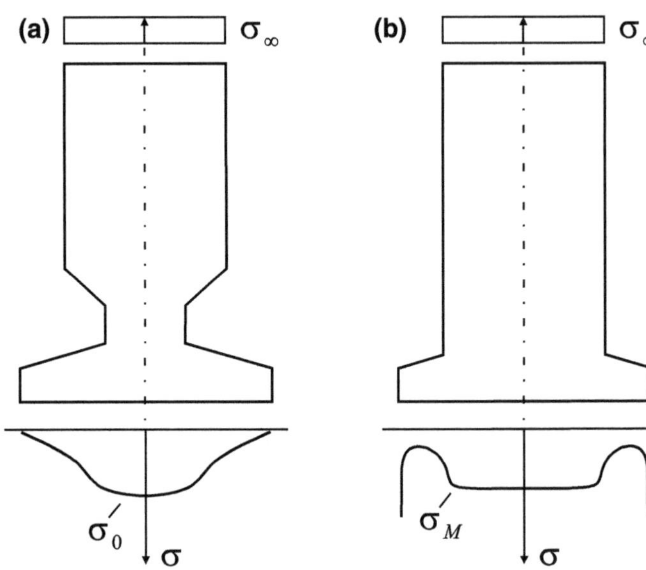

Fig. 12.22 The normal stress distribution σ_0 under the mushroom shaped pillar with the neck and the normal stress distribution σ_M under the modified pillar

where v is the Poisson's ratio, $\sigma_T = f\sigma_\infty$ is the equivalent remote total stress at the interface, $f \approx (R_P/R_i)^2 > 1$ is a factor which takes into accout that σ_T and σ_∞ cover different areas (see Fig. 12.22 where R_P is the pillar radius and R_i is the neck radius], and α is the tilting angle.

The accurate value of f has been calculated from the FE analysis for $\alpha = 0^\circ$. Incidentally observe that K_{II} is uniform on the perimeter of the defect [55]. The equivalent stress intensity factor is in this case $K_{eq} = \sqrt{K_\perp^2 + K_{II}^2}$ and enforcing the Griffith condition $G = K_{eq}^2/2E^* = \Delta\gamma$ allows the calculation of the z -component P_z of the pull-off force for different tilting angles α through equation.

$$\frac{P_z}{P_0} = \sqrt{\left[\cos^4\alpha + \left(\frac{\sin 2\alpha}{2-v}\right)^2\right]^{-1}} \cos\alpha \tag{12.31}$$

Figure 12.20b (dashed curve) shows the results of (12.31) compared to FE predictions. We observe that for tilting angles α ranging within ± 20° FE calculations and (12.31) are in very good agreement. However, the difference is no longer negligible at high tilting angles meaning that the effect of the two different assumptions (no slip for FE calculations and uniform frictional stress for analytical calculations) becomes relevant. Comparison with experiments shows that the analytical approach (12.4) is more effective that FE model to capture the slight dependence of P_z on α.

We observe that, the marginal influence of tilting on pull-off force, proved by experiments and numerical/theoretical calculations, is strictly a consequence of the optimal geometry of the pillar which prevent the formation of singular stress distribution at the edge of the contact area, i.e. at the perimeter of the plate, see Fig. 12.22a. Indeed, non-optimal geometries as the one shown in Fig. 12.22b will lead to the occurrence of stress singularity at the edge of the plate, that will detach following the mode I behavior (i.e. crack propagation from the edge) described in [30]. This of course will very significantly increase the sensitivity of pull-off force on tilting because of the appearance of triangular normal stress distribution at the interface, which will take its maximum values at the edge of plate and will strongly enhance the maximum value of the stress intensity factor at the perimeter of the plate.

Another important question concerns the effect of the neck on tolerance to tilting of mushroom shaped pillars. Indeed, the crucial role of the neck is that it strongly reduces the bending stiffness of the pillar itself. This makes the pillar much less sensitive to the force tilting. In fact, when an external tilted force is applied, a certain amount of tilting will also occur (Fig. 12.23), leading to a reduction from α to α' of the angle between the applied force and pillar axis, i.e. to a reduction of the lever arm. For high bending stiffness (i.e. pillar without neck) the change $\Delta\alpha = \alpha - \alpha' \ll \alpha$ and a strongly detrimental triangular stress distribution will appear at the interface and may cause (even for optimized geometry) the occurrence of mode I debonding. On the other hand, the presence of the neck, may strongly reduce the

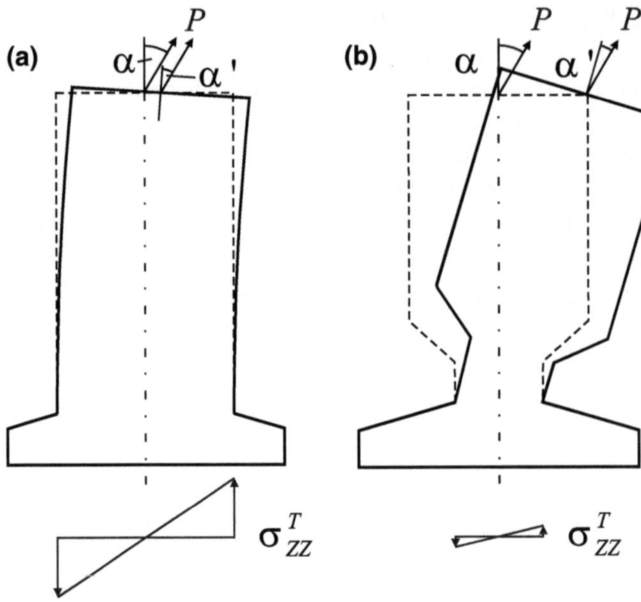

Fig. 12.23 Due to the presence of the neck, the bending stiffness of the pillar decreases. Therefore, when a tilted force F is applied to the pillar, a small tilting of the pillar can be expected

stiffness and, in the limiting case of very high bending compliance, $\Delta\alpha = \alpha - \alpha' \approx \alpha$, i.e. $\alpha' = 0$. This causes an almost complete vanishing of the triangular stress distribution at the interface thus keeping the high values of pull-off force of mode II debonding mechanism.

So far we have assumed a crack to appear in the center of the contact interface between MSAMS and the substrate. This assumption may not always hold in real experiments [53]. However, in light of above mentioned, a reduction of or a vanishing triangular stress distribution would also allow MSAMS with eccentric cracks to detach by the high values of pull-off force of mode II. Thus, the presence of a neck (or joint-like feature) provides tilt-tolerant adhesion also under experimental conditions.

Moreover, a neck may also offer further advantages. Structures with a neck not only provide tilt-tolerant detachment, but could also assist attachment. The reduced bending stiffness would allow for a better adaptability to uneven and non-parallel substrates compared to MSAMS without neck or a flat punch. In addition, recent experiments suggest that a neck makes MSAMSs also tolerant to high compressive loads. In situ scanning electron microscopy (SEM) obersavtions of the contact behavior of MSAMSs indicated that the thin contact plate remained in contact, although MSAMS were buckled [21]. By contrast, flat punch pillars were shown to lose adhesive contact to the substrate at a critical load [56].

12.9 Conclusions and Outlook

We have reviewed the main detachment mechanisms of microstructured adhesives made of regular arrays of micropillars and shown how mushroom shaped micropillar can be optimally designed to strongly enhance the performance of this type of biomimetic adhesives. Although these optimized systems present pull-off forces which may be even larger than those obtained in the case of biological systems (e.g. geckos and beetles), they present direction-independent adhesion, i.e. the strength of adhesion is the same not depending on the direction of the applied load. This is an important drawback when the goal is related to applications in the field of locomotion and object manipulation. We may say that, notwithstanding the great advances in the field of microstructured adhesives, our attempt to copy Nature is incomplete. Indeed, a lot of research is still going on and will be needed in the future to propose and fabricate bio-mimetic surfaces with controlled direction dependent adhesion.

References

1. http://en.wikipedia.org/wiki/Bionics
2. A.K. Geim et al., Nat. Mater. **2**(7), 461 (2003)
3. T. Kim, H.E. Jeong, K.Y. Suh, H.H. Lee, Adv. Mater. **21**, 2276 (2009)
4. H. Lee, B.P. Lee, P.B. Messersmith, Nature **448**, 338 (2007)
5. C. Majidi, R.E. Groff, Y. Maeno, B. Schubert, S. Baek, B. Bush, R. Maboudian, N. Gravish, M. Wilkinson, K. Autumn, R.S. Fearing, Phys. Rev. Lett. **97**, 076103 (2006)
6. R. Spolenak, S. Gorb, E. Artz, Acta Biomater. **1**, 5 (2005)
7. A. del Campo, C. Greiner, E. Arzt, Langmuir **23**, 10235 (2007)
8. C. Greiner, R. Spolenak, E. Arzt, Acta Biomater. **5**, 597 (2009)
9. C. Greiner, A. del Campo, E. Arzt, Langmuir **23**, 3495 (2007)
10. B.N.J. Persson, A.I. Volokitin, E. Tosatti, Eur. Phys. J. E **11**, 409 (2003)
11. B.N.J. Persson, Wear **254**, 832 (2003)
12. B.N.J. Persson, S. Gorb, J. Chem. Phys. **119**, 11437 (2003)
13. G. Carbone, B.N.J. Persson, J. Chem. Phys. **121**, 2246 (2004)
14. K. Autumn, Y.A. Liang, S.T. Hsleh, W. Zesch, W.P. Chan, T.W. Kenny, R. Fearilng, R. J. Full, Nature **405**, 681 (2000)
15. G. Carbone, L. Mangialardi, B.N.J. Persson, Phys. Rev. B **70**, 125407 (2004)
16. G. Carbone, P. Decuzzi, J. Appl. Phys. **95**, 4476 (2004)
17. B.N.J. Persson, O. Albohr, U. Tartaglino, A.I. Volokitin, E. Tosatti, J. Phys. Condens. Matter **17**, R1 (2005)
18. M. Scherge, S. Gorb, *Biological micro- and nanotribology* (Springer, Berlin, 2001)
19. S. Kim, M. Sitti, Appl. Phys. Lett. **89**, 261911 (2006)
20. M. Varenberg, N.M. Pugno, S. Gorb, Soft Matter **6**, 3269 (2010)
21. M. Varenberg, S. Gorb, J. R. Soc. Interface **5**, 785 (2008)
22. M.K. Kwak, H.E. Jeong, G. BaeW, H.S. Jung, K.Y. Suh, Small **7**, 2296 (2011)
23. W.L. Noderer, L. Shen, S. Vajpayee, N.J. Glassmaker, A. Jagota, C.Y. Hui, Proc. R Soc. A **463**, 2631 (2007)
24. N.J. Glassmaker, A. Jagota, C.Y. Hui, W.L. Noderer, M.K. Chaudhury, Proc. Natl. Acad. Sci. U.S.A. **104**, 10786 (2007)

25. L. Afferrante, G. Carbone, J. R. Soc. Interface **9**, 77 (2012)
26. M.P. Murphy, S. Kim, M. Sitti, Appl. Mater. Interfaces **1**(4), 849 (2009)
27. K.A. Daltorio, S. Gorb, A. Peressadko, A.D. Horchler, R.E. Ritzmann, R.D. Quinn, in *International Conference on Intelligent Robots and Systems* (2005), pp. 131–138
28. S.N. Gorb, M. Varenberg, J. Adhes. Sci. Technol. **21**, 1175 (2007)
29. M. Varenberg, S. Gorb, J. R. Soc. Interface **4**, 721 (2007)
30. G. Carbone, E. Pierro, S. Gorb, Soft Matter **7**, 5545 (2011)
31. J.M. Karp, R. Langer, Nature **477**, 42 (2011)
32. M.K. Kwak, H.E. Jeong, K.Y. Suh, Adv. Mater. **23**, 34 (2011)
33. S. Gorb, M. Varenberg, A. Peressadko, J. Tuma, J. R. Soc. Interface **4**, 271 (2007)
34. D.B. Bogy, J. Appl. Mech. **38**(2), 377 (1971)
35. E. Buckingham, Phys. Rev. **4**, 345 (1914)
36. E. Buckingham, Nature **96**, 396 (1915)
37. T. Tang, C.Y. Hui, N.J. Glassmaker, J. R. Soc. Interface **2**(5), 505 (2005)
38. A.V. Spuskanyuk, R.M. McMeeking, V.S. Deshpande, E. Arzt, Acta Biomater. **4**, 1669 (2008)
39. S. Gorb (Private communication)
40. S.P. Timoshenko, S. Woinowsky-Kreiger, *Theory of Plates and Shells*, 2nd edn. Engineering mechanics series (McGraw-Hill, London, 1959). ISBN 0-07-085820-9
41. L.D. Landau, E.M. Lifshitz, *Theory of Elasticity* (Pergamon, London, 1959)
42. N.J. Glassmaker, A. Jagota, C.Y. Hui, W.L. Noderer, M.K. Chaudhury, Proc. Natl. Acad. Sci. U.S.A. **104**(26), 10786 (2007)
43. D. Maugis, *Contact, Adhesion and Rupture of Elastic Solids*, Springer series in solid state sciences (Springer, Berlin, 1999)
44. G. Carbone, E. Pierro, Small **8**(9), 1449 (2012)
45. ANSYS, User's manual. Version 10.0
46. B. Aksak, K. Sahin, M. Sitti, Beilstein J. Nanotechnol. **5**, 630 (2014)
47. G. Carbone, E. Pierro, Soft Matter **8**, 7904 (2012)
48. G. Carbone, L. Mangialardi, J. Mech. Phys. Solids **56**(2), 684 (2008)
49. H.B. Callen, *Thermodynamics and an Introduction to Thermostatistics* (Wiley, USA, 1985). ISBN 0-471-86256-8
50. K.L. Johnson, K. Kendall, A.D. Roberts, Proc. R. Soc. Lond. A **324**, 301 (1971)
51. S.N. Gorb, *Attachment Devices of Insect Cuticle* (Springer, New York, 2001)
52. L. Heepe, G. Carbone, E. Pierro, A.E. Kovalev, S.N. Gorb, Appl. Phys. Lett. **104**(1), 011906 (2014)
53. L. Heepe, A.E. Kovalev, A.E. Filippov, S.N. Gorb, Phys. Rev. Lett. **111**, 104301 (2013)
54. I. Sneddon, Proc. R Soc. Lond. A Math. Phys. Sci. **187**, 229 (1946)
55. H. Tada, P.C. Paris, G.R. Irwin, H. Tada, *The Stress Analysis of Cracks Handbook*, vol. 130 (ASME Press, New York, 2000)
56. D. Paretkar, M. Kamperman, A.S. Schneider, D. Martina, C. Creton, E. Arzt, Mater. Sci. Eng. C **31**, 1152 (2011)

Chapter 13
Bioinspired Mushroom-Like Fiber Adhesives

The Effect of Cap Dimensions on Adherence

Serdar Gorumlu and Burak Aksak

Abstract Mushroom-like synthetic microfibers have been shown to generate strong adhesion to smooth surfaces, resulting in adhesive performance on par with the natural adhesive pads of the gecko. Theoretical and numerical investigations regarding the pull-off mechanism of this family of fibers suggest that adherence is primarily controlled by the geometry of the cap. Here, we designed an experimental study to determine the effect of tip size and edge angle on pull-off stress for comparison with existing theoretical/numerical results. An adhesion measurement system was designed to measure pull-off force of an individual fiber in an array. A fabrication technique was developed to fabricate microfibers with varying cap diameters and constant edge angle. Pull-off experiments performed with fibers of two different edge angles with varying tip sizes showed that pull-off stress, which reached as high as 2.08 MPa, increased with decreasing tip diameter and are in agreement with theoretical expectations. Repeatability experiments performed with a fiber employing a small cap (tip-to-stalk ratio = 1.3) showed that the fiber lost only about 10% of its initial adherence after six consecutive pull-off measurements. Conversely, a microfiber with a larger cap diameter (tip-to-stalk ratio = 2) lost 66% of its initial adherence after the same number of runs. Our findings suggest that fibers with smaller caps are better adhesives due to their enhanced durability and pull-off stress.

13.1 Introduction

A relatively new generation of adhesives, inspired by the clinging abilities of the gecko lizard, exhibit unusual properties such as self-cleaning [1, 2], repeatable adhesion [3], and superhyrophobicity [4, 5]. Realization of the so-called gecko-inspired synthetic adhesives that feature the unique properties of its biological equivalent could be transformative in many industries ranging from robotics

S. Gorumlu · B. Aksak (✉)
Department of Mechanical Engineering, Texas Tech University, Lubbock, USA
e-mail: burak.aksak@ttu.edu

L. Heepe et al. (eds.), *Bio-inspired Structured Adhesives*,
Biologically-Inspired Systems 9, DOI 10.1007/978-3-319-59114-8_13

to household adhesives. It not only enables applications that require reusable adhesives, but also brings an environment friendly alternative to existing adhesives that are discarded after one-time use. In short, gecko footpads employ densely packed microfibers that branch into sub-micron fibers with specialized saucer-shaped tips [6]. This fibrillar construction provides the footpads with necessary compliance to form full contact with the opposing surface where the adhesive interaction of each contacting fiber leads to a significant adhesive force. It is hypothesized that the saucer shaped tips factor in the high adhesion strength of an individual contacting fiber.

Several research groups have looked into mimicking the saucer tip shape of the gecko fiber mostly from softer polymers such as silicone rubber [7], polydimethysiloxane (PDMS) [8] and polyurethanes [9]. Referred to in literature as mushroom, these types of structures can achieve adhesion strengths as high as 180 N/m^2 [9], on par with the natural gecko adhesive. Recently Carbone and Pierro [10] studied the mechanism of adhesion for mushroom-like fibers, specifically for the geometry they obtain as a result of the fabrication method shown in [7], and offered an optimized shape for maximum adhesion stress, defined as the pull-off force per the actual contact area of a single mushroom-like fiber. They argued that optimized fiber geometry should employ a tip diameter 2–3 times the diameter of the fiber stalk with the tip/cap thickness 0.2–0.3 times the stalk diameter. Once the geometry yielding the highest adhesion stress is determined, the fibrillar adhesive can be designed in the array form to maximize actual contact area by maximizing the fiber packing density. Recently, Wang et al. [11] carried out adhesion experiments on arrays of mushroom-like fibers with varying cap sizes and found that the pull-off force increases with increasing cap size for a constant packing density as expected. They report adhesion stress based on the area of the indenter but offer no conclusions regarding the adhesion stress based on the actual contact area.

13.1.1 Results from the Numerical Study on the Effect of Tip Shape

Our recent work suggests that both the ratio of the tip-to-base diameter, β and the tip edge angle, θ (i.e. cone angle of the 2D tip projection, see Fig. 13.1) together determine the adhesion stress for mushroom-like microfibers [12]. In that study, we carried out a numerical/analytical study on the effect of tip size on adhesion strength for the fiber shape shown in Fig. 13.1. In this Finite Elements (FE) based study, the traction-separation relationship at the interface between a mushroom-like fiber and a rigid, smooth adhering surface was defined using the Dugdale–Barenblatt (DB) cohesive zone model [13, 14]. According to this model, the interface separates when the normal interfacial stress reaches the theoretical strength of the interface σ_o and it continues to separate at this stress up to a critical distance δ_c. Beyond this critical distance the attractive stress drops to null, resulting in a crack to initiate.

The work of adhesion is given by $w_{\mathrm{adh}} = \sigma_{\mathrm{o}}\delta_{\mathrm{c}}$. In this study, we numerically analyzed the effect of tip shape on the adhesion stress defined as the pull-off force per unit tip area. We performed a parametric sweep for varying β and θ for a fixed non-dimensional value χ, a measure of the ratio of the cohesive length to the radius of the fiber defined as [12, 15].

$$\chi \equiv \frac{\sigma_{\mathrm{o}}^2 a(1-\nu^2)}{2\pi E w_{\mathrm{adh}}} = \frac{\sigma_{\mathrm{o}} a(1-\nu^2)}{2\pi E \delta_{\mathrm{c}}}, \tag{13.1}$$

where σ_{o} is the intrinsic adhesion stress, a is the fiber radius (fiber stalk radius for mushroom-like fibers), ν is the Poisson's ratio, and E is the elastic modulus of the fiber. Additionally, the adhesion stress of a mushroom-like fiber can be expressed using the equation below as a function of β, θ, χ and the location of crack initiation:

$$\Phi = \frac{\sigma_s}{\sigma_o} = \begin{cases} B_e[\beta\chi]^{-\alpha} & \text{for edge crack,} \quad \text{(a)} \\ B_c[\beta\chi]^{-0.5} + \Gamma_c & \text{for center crack.} \quad \text{(b)} \end{cases} \tag{13.2}$$

Here,Φ is the normalized adhesion stress and is simply a measure of what fraction of the intrinsic adhesion stress can be obtained on average for a given adhesive interface. The exponent α is a function of edge angle θ as well as the elastic moduli and the Poisson's ratio of both the adhering surface and the fiber material and can be calculated using the classical work by Bogy [16]. For a crack initiating at the center, there is a square root singularity and thus, $\alpha = 0.5$ as shown in (13.2a). Numerical constants B_e and B_c are form factors, which are determined by fitting (13.2a) and (13.2b) to the simulation data for given β, θ and χ. Constant Γ_c is approximately the value of the pull-off stress when the crack initiation switches from the edge to the center of the fiber for $\theta > 45°$ and can also be extracted from the numerical simulations. For $\theta \leq 45°$ and a crack initiating at the center for all χ, Γ_c is the lower bound for normalized pull-off stress at the limit $\chi \rightarrow \infty$.

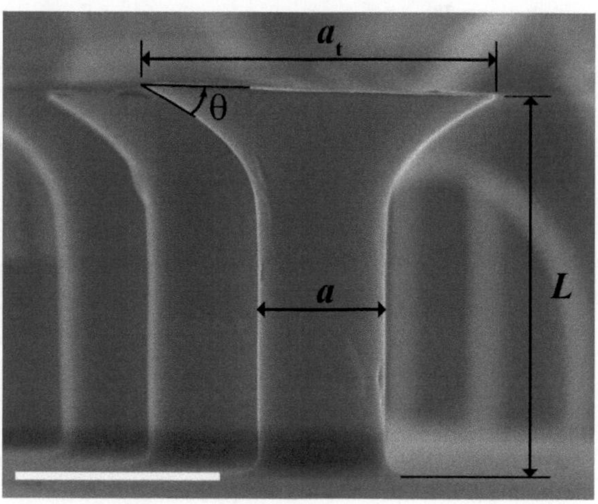

Fig. 13.1 Dimensional parameters shown on a side-view scanning electron microscope (SEM) image of a mushroom-like fiber. *Scale bar* 50 μm

Figure 13.2 shows the results of the parametric study for four different χ values. Note that for all the χ values shown in Fig. 13.2, pull-off stress is maximum for $\theta \approx 45°$ and $\beta \approx 1.1$–1.2. One can also observe that for edge angles between 30° and 50°, pull-off stress is quite sensitive to the change in β and exhibits a sharper peak. For angles larger than 50°, pull-off stress peak is broader indicating lower sensitivity to β. In summary, the maximum adhesion stress is obtained for a cap diameter 1.2 times the diameter of the fiber stalk ($\beta = 1.2$), it decreases monotonically for larger cap diameters, and the optimal edge angle is $\theta = 45°$ degrees, all of which are observable in Fig. 13.2.

Motivated by the results of this numerical study, Marvi et al. [17] have recently performed pull-off experiments with mushroom-like fibers with $\beta = 1.1$–1.5 and $\theta = 30°$–$90°$, manufactured via two-photon lithography for master template fabrication and casting using two different polyurethanes to obtain the mushroom-like fiber. Their results show a discrepancy between model predictions and experimental results, which as they suggested could be attributed to the blunt edges of the tip [12]. More specifically, they found that the maximum adhesive stress occurs at

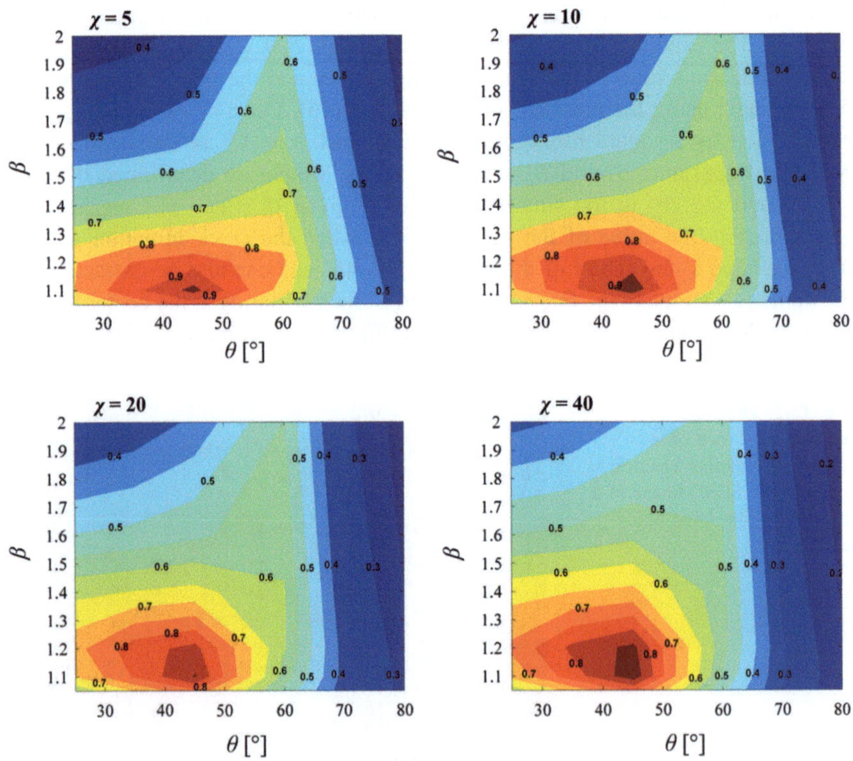

Fig. 13.2 Normalized pull-off stress Φ contour plots for $\chi = 5$ (*top left*), $\chi = 10$ (*top right*), $\chi = 20$ (*bottom left*), and $\chi = 40$ (*bottom right*) as a function of β and θ. The peak Φ for each case lies within $\beta = 1.1$–1.2 and $\theta = 45°$ [12]

$\theta = 30°$, and $\beta = 1.4$ and 1.2 for the stiffer and the softer polyurethane fibers, respectively. The dependency of the optimal values to the fiber material as well as the optimal angle being less than the predicted 45° both point to the influence of the blunt edges on pull-off stress. Furthermore, for a given β, they measured decreasing adhesion stress with increasing θ. This is another consequence of the blunt edges. Lower θ values lead to lower stress intensity factors at the edge of the fiber, and as a consequence pull-off loads and the corresponding average stresses increase. The thickness of the edge is in the order of 1 μm possibly due to the resolution limit of the device used for master template fabrication. Here, we designed an experimental study, which employs fibers with "sharp" tip edges, defined here as an edge with a radius of curvature much smaller than the flange size of the mushroom tip (i.e. $a_t - a$). Additionally, this study employs fibers with a larger β range to effectively test the validity of theoretical predictions [12]. Both criteria are met using a process based on a previously developed fabrication technique, which allows for fabrication of mushroom-like fibers with sharp edges in varying tip diameters and tip edges. The pull-off force of a single mushroom-like fiber employing various tip diameters and edge angles is measured using a custom adhesion measurement system. Experiments are performed on an optical microscope, which allows for the identification and measurement of an individual fiber to be tested, confirmation of contact and pull-off, and the visual observation of the entire experiment from contact with the fiber to pull-off.

Fabrication process of the mushroom-like fibers, designed to enable fibers with varying cap diameters and edge angles, is described in Sect. 13.2. The experimental setup is detailed in Sect. 13.3. Experimental results, discussion points regarding the agreement between theory and experiments, as well as the repeatability of adhesive performance are presented in Sect. 13.4. Finally, conclusions are given in Sect. 13.5.

13.2 Fabrication of Mushroom-Like Microfibers with Varying Cap Dimensions

Mushroom-like microfibers were fabricated using the technique detailed in previous work by the author and his coworkers [3, 18]. First, two separate arrays of cylindrical fibers 32 μm in diameter, 85 μm in height, and 200 μm in center-to-center spacing are fabricated using a two-part polyurethane (ST-1085, BJB Enterprises, USA). These fibers are then brought into contact with a thin layer of liquid ST-1085 that is uniformly coated on a polystyrene substrate using a spin coater (WS-650 MS Spinner, Laurell Technologies, USA), and removed immediately to retain some of the liquid polymer on their tips, a process called "dipping" [3, 6, 18] (Fig. 13.3, step 1). Using spin speeds ranging from 4000 to 12,000 rpm, the thickness of the liquid polymer, and thus the volume of liquid each microfiber retains upon contact were varied. The dipped cylindrical fibers are then placed on a smooth tip forming substrate for the liquid polymer to spread and morph into the desired tip shape (Fig. 13.3, step 2). Two factors determine the diameter and the edge angle of the tip. First is the amount of liquid an individual fiber collects after dipping: the thicker the liquid layer is, the more volume the fiber retains, the larger the fiber tip becomes. The second factor is the affinity of the liquid polymer to the tip forming substrate. For a given volume of retained liquid, higher affinity of the liquid polymer to the substrate causes the liquid to spread into a wider footprint with relatively sharper edges and vice versa. It is expected that the static contact angle of the polymer with the tip forming substrate determines the edge angle of the final tip shape regardless of the dipped volume. In order to obtain fibers with varying tip edge angles, two materials were selected as the smooth substrate: polydimethyl siloxane (PDMS), and glass. Microscope slides (McMaster Carr, USA) were used as the glass substrate. Glass substrates were cleaned with acetone and isopropyl alcohol, rinsed using di-water, and dried with clean dry air. PDMS substrate was prepared by spin coating a layer of PDMS (10:1 weight mixing ratio, Sylgard 184, Dow Corning, USA) onto a 4-inch silicon wafer followed by curing at 70 °C in an oven for 24 h.

After the fibers are placed on the tip forming substrates, they are cured for 16 h at room temperature followed by post-curing for 12 h at 70 °C in an oven. Upon curing, they are placed in a freezer for 3 h and the fiber array is peeled-off immediately from the substrate to obtain the mushroom-like fibers (Fig. 13.3, step 3). The freezing step is found to be crucial to prevent damage by limiting fiber and tip deformation during the removal of the fiber array from the substrate. All of the

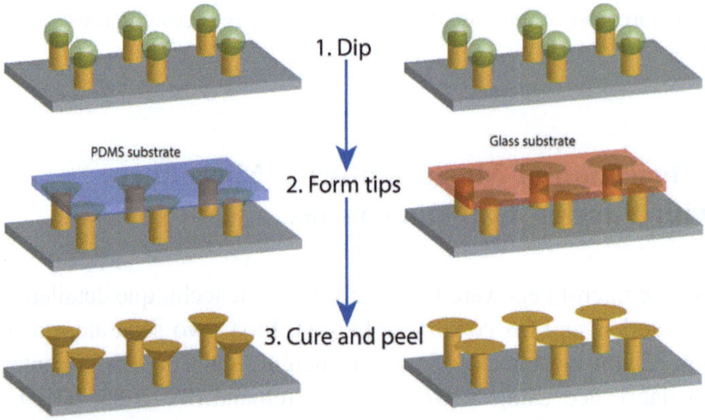

Fig. 13.3 Process flow for the fabrication of thick cap and thin cap mushroom-like fibers. *1 Dip* Two identical cylindrical fiber arrays are dipped into a thin layer of polyurethane. *2 Form tips* Dipped fiber arrays are placed on a PDMS and Glass substrates, respectively. The difference in the affinity of the liquid polyurethane to the substrate causes the polymer spread differently and subsequently results in differing tip diameters and edge angles. *3 Cure and peel* After the polymer is let cure, it is placed in a deep freezer and mechanically removed from the substrate

cylindrical fibers were fabricated at the same time from a single batch of polyurethane mixture to prevent sample-to-sample material property variations. The same batch of polyurethane was used for dipping and the mushroom-tip fabrication process, as well. Figure 13.4 shows SEM images of four mushroom-like fibers fabricated on glass and PDMS substrates, respectively. Measurements performed on the SEM images of the fabricated microfibers using Adobe Illustrator indicates that edge angles are in the range 30°–40° and 50°–60° when tips are formed on glass and PDMS substrates, respectively. These angles were found to be only substrate dependent, and irrespective of the tip size. The resulting β for both the thin cap and the thick cap fibers are in the range 1.3–3. While a lower bound of 1.1 would be ideal for this study to determine whether maximum pull-off stress happens at the predicted $\beta = 1.1$–1.2 range, it proved difficult to reliably produce and dimensionally characterize fibers with $\beta < 1.3$ fabricated using the technique employed in this work. For simplicity, we will refer to the fibers with tips formed on PDMS as *thick cap* (Fig. 13.4a, c) and on glass as *thin cap* (Fig. 13.4b, d).

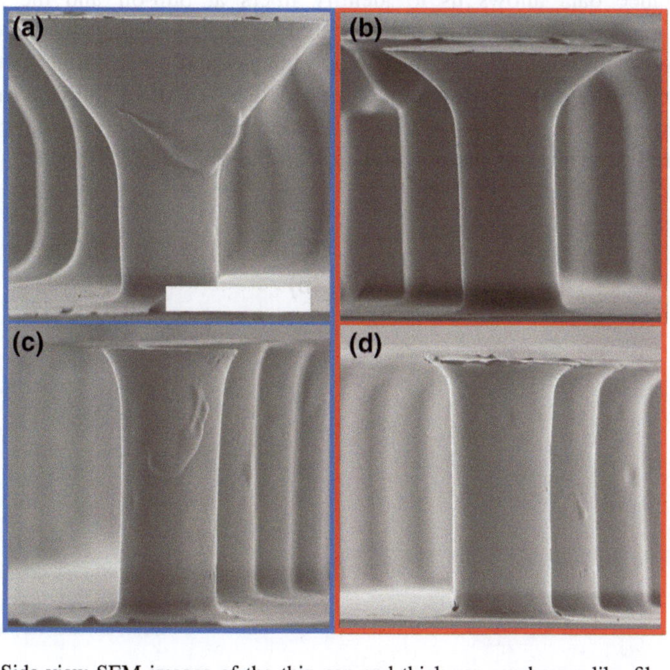

Fig. 13.4 Side-view SEM images of the thin cap and thick cap mushroom-like fibers. **a** Thick cap, large cap diameter; **b** thin cap, large cap diameter; **c** thick cap, small cap diameter; and **d** thin cap, small cap diameter. *Scale bar* 50 μm

13.3 Single Microfiber Pull-Off Experiments

Adhesion experiments are performed using an automated system comprised of a precision linear stage (MFA-CC, Newport Cooperation, USA), a 10-g load cell (GSO-10, Transducer Techniques, USA), a 6 mm diameter fused silica indenter (QU-HS-6, ISP Optics, USA) and a LabVIEW (National Instruments, USA) based custom software that controls preload, approach and retract speeds, and contact time (see Fig. 13.5a). The custom software displaces the silica hemisphere toward the sample at 1 μm/s until the hemisphere contacts a target fiber and applies a compressive load of 0.2 mN. After 1 s of contact time, the hemisphere is retracted from the sample until detachment. The force and the displacement are recorded during the process to obtain a force-distance curve as shown in Fig. 13.5b. Pull-off force is the absolute value of the negative peak in the force-distance curve. We elect to measure the pull-off load of a single fiber at a time in order to eliminate the effect of fiber-to-fiber variations in an array and the ease of observing a single fiber rather than many in contact. However, instances where more than one fiber contacts the hemisphere, the force-distance data includes adhesion peaks equal to the number of fibers in contact. Visual inspection using the microscope as well as the force-distance data allows us to identify fibers at pull-off and the associated adhesion peaks (see the inset in Fig. 13.5b). The tip diameter of the fiber is determined using the top view microscope images and circle fitting using Adobe Illustrator CS6 (Adobe Systems Incorporated, USA). All of the experiments are performed in a 72 h window so as to limit the effect of time dependent changes in the material properties of the polyurethane on the pull-off experiments [19].

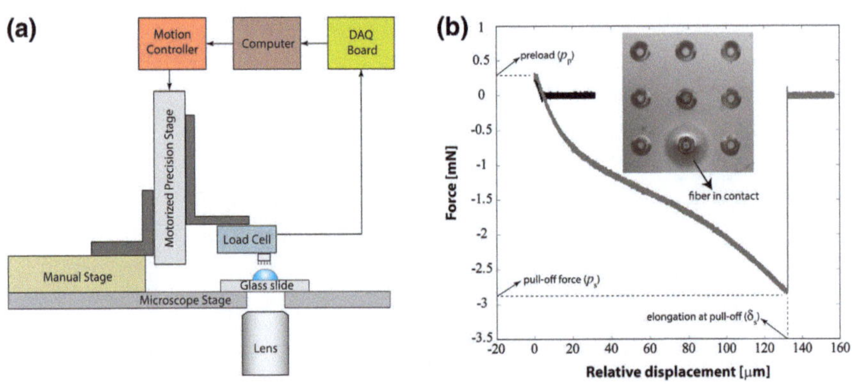

Fig. 13.5 **a** Schematics of the pull-off force measurement system. **b** Sample force-distance curve obtained from a single fiber. The positive peak corresponds to the preload, the absolute value of the negative peak is the pull-off force. The indenter is assumed to make first contact with the fiber (i.e. zero relative displacement) when the compressive load equals 0.1 mN between the indenter and the fiber. The out of focus fiber in the inset marks the fiber in contact with the hemispherical indenter

Each fiber is tested only once because pull-off force obtained from a fiber decreases with repetition in particular for larger tips as will be discussed in the next section. Additionally, each fiber in an array tends to be slightly different in tip diameter and therefore pull-off forces obtained from different fibers could not be averaged. Therefore, data obtained from the experiments is shown as a scatter plot where each datum corresponds to a single test from a single fiber.

13.4 Results and Discussion

13.4.1 Pull-Off Experiments

Figure 13.6 shows the scatter plot of the pull-off force and pull-off stress measurements for individual thin cap and thick cap mushroom-like fibers, each cap type with varying tip-to-base ratio, β ranging from 1.3 to 3. As expected, the pull-off force increases with β for both edge angles due to increase in the contact area for larger cap diameters. Adhesion stress σ_s on the other hand, an average value defined as pull-off force per unit tip area of the individual fiber ($\sigma_s = p_s/\pi a_t^2$), decreases with β for both the thin cap and thick cap mushroom-like fibers. Adhesion stress drops at a higher rate with β for thin cap fibers than for thick cap fibers. When the thin cap and thick cap fiber samples are compared to each other, thick cap fibers yield a higher adhesion stress in general for all the tip-to-base ratios tested. The highest adhesion stress is obtained for the smallest β for thin cap and a moderate β for the thick cap fibers. The associated pull-off stresses are 2.08 MPa for thick cap fibers with $\beta = 1.68$, and 1.58 MPa for thin cap fibers with $\beta = 1.39$. However,

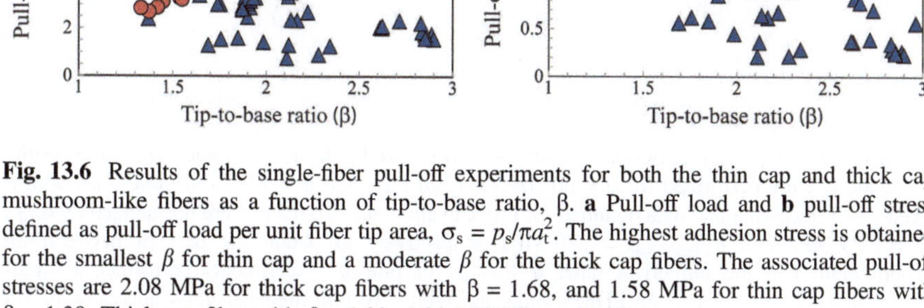

Fig. 13.6 Results of the single-fiber pull-off experiments for both the thin cap and thick cap mushroom-like fibers as a function of tip-to-base ratio, β. **a** Pull-off load and **b** pull-off stress defined as pull-off load per unit fiber tip area, $\sigma_s = p_s/\pi a_t^2$. The highest adhesion stress is obtained for the smallest β for thin cap and a moderate β for the thick cap fibers. The associated pull-off stresses are 2.08 MPa for thick cap fibers with $\beta = 1.68$, and 1.58 MPa for thin cap fibers with $\beta = 1.39$. Thick cap fiber with $\beta = 1.33$ yield 1.99 MPa pull-off stress

the smallest thick cap fiber with $\beta = 1.33$ yields a pull-off stress of 1.99 MPa, which is very similar to the highest adhesion stress of 2.08 MPa obtained with thick cap fibers.

13.4.2 Comparison Between Simulations and Experiments

Figure 13.7a. shows a comparison between simulations [12] and experiments, where the measured adhesion stress is normalized by an intrinsic adhesion stress, σ_o. Recall that the intrinsic adhesion stress is the maximum normal stress an interface can bear before starting to separate in normal direction as defined in the DB cohesive zone model. Here, σ_o is used as a fitting parameter and the measured adhesion stress was normalized by a value of $\sigma_{o,thin\_cap} = 2.25$ MPa for thin cap, and $\sigma_{o,thick\_cap} = 2.70$ MPa for thick cap, resulting in a ratio of 1.2 between the two cap types. The simulation results were obtained for $\chi = 5$. Details of simulations as well as definitions of σ_o and χ, and their significance in the adhesion stress of mushroom-like fibers are provided elsewhere in detail [12] and are briefly explained in Sect. 1.1.

One would expect that the σ_o values be independent of the cap type. Indeed, to ensure consistency in properties, they all are fabricated from the same batch of polyurethane mixture at the same time for both the cylinder fabrication and the subsequent steps to fabricate the mushroom-like fibers. However, the fibers were cured on two different substrates during tip formation, which could affect the surface properties of the tips, in particular the adhesive properties. To check this,

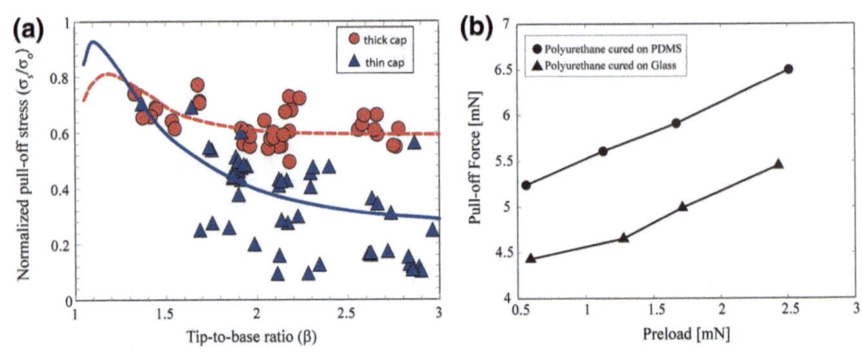

Fig. 13.7 **a** Comparison between experimental normalized pull-off stress and simulation results. Simulations were performed for $\theta = 60°$ (*dashed red line*) and $\theta = 40°$ (*solid blue line*) both at fixed $\chi = 5$. Experimental data obtained for thin cap and thick cap fibers were normalized by $\sigma_{o,thin\ cap} = 2.25$ MPa and $\sigma_{o,thick\ cap} = 2.70$ MPa, respectively to obtain the best fit between the experimental and the numerical data. **b** Results of pull-off experiments performed on bulk polyurethane samples cured on PDMS and glass with a 10 mm diameter glass hemisphere. The pull-off force measured for polyurethane sample cured on PDMS is on average ~1.2 times that of the sample cured on glass for all applied preloads

we performed a set of control experiments with two 1 cm × 1 cm × 0.5 cm bulk ST-1085 samples such that one was cured in contact with glass (thin cap forming substrate), the other with PDMS (thick cap forming substrate). We performed pull-off experiments with a 10 mm diameter glass hemisphere (QU-HS-6, ISP Optics, USA) and measured pull-off force as a function of applied preload at 1 μm/s approach/retract speeds. It was found that the pull-off force for the polyurethane sample cured on PDMS is on average ~1.2 times that of the sample cured on glass for all applied preloads (see Fig. 13.7b). In a test where a hemispherical indenter contacts a flat smooth surface (i.e. JKR test), such as the ones performed here, the pull-off force is equal to $(3/2)\pi R w_{adh}$, where R is the radius of curvature of the indenter and w_{adh} is the work of adhesion [20]. Therefore, we expect that the difference in the measured pull-off force is a direct consequence of the discrepancy in work of adhesion. Work of adhesion and intrinsic adhesion stress are directly related [21], which explains why different σ_o values were used to normalize the experimental data. While we recognize that the pull-off force may not be equal to $(3/2)\pi R w_{adh}$ [22–25] due to bulk and surface viscoelastic effects, if we assume that the viscoelastic behavior is only present at the vicinity of the crack, the JKR relationship is still valid with the use of an effective work of adhesion that includes the dissipative effects [26]. However, it is challenging to differentiate the bulk and surface viscoelastic effects in a JKR test, and even at this scale bulk effects may factor into the discrepancy between the samples in terms of both the bulk and the single fiber pull-off values [25]. While the speed was kept at 1 μm/s for both the bulk and the single fiber experiments, due to the differences in dimensions and indentation depths, we expect the effective work of adhesion to be different for each case. Consequently, σ_o values are not determined from the bulk pull-off experiments but rather from fitting the theory to the experimental data. On the other hand, the relative values of pull-off loads obtained in the bulk experiments for the samples cured on PDMS and glass should be approximately equal to the relative values of σ_o for the thick and thin capped fibers. We therefore use $\frac{\sigma_{o,\text{thick cap}}}{\sigma_{o,\text{thin cap}}} = 1.2$.

13.4.3 Repeatability of Adhesion Stress

Experimental results suggest that pull-off stress increases with smaller tip diameter. However, larger tips still experience larger pull-off force as shown in Fig. 13.8 and therefore experience larger elongations before pull-off. Recalling that the stalk diameters of all the fibers are identical, fibers that experience larger pull-off forces also experience larger stresses at the stalk of the fiber. Therefore, we hypothesize that small tipped fibers have the added advantage of durability when it comes to repeated use due to lower maximum normal stress. Indeed, we observed a sharp drop in pull-off force (66%) for a thick cap fiber with $\beta \approx 2.0$ for as low as six repeated adhesion experiment. On the other hand, a relatively negligible decrease of 10% was observed for a fiber with $\beta \approx 1.3$ after six experiments. These results

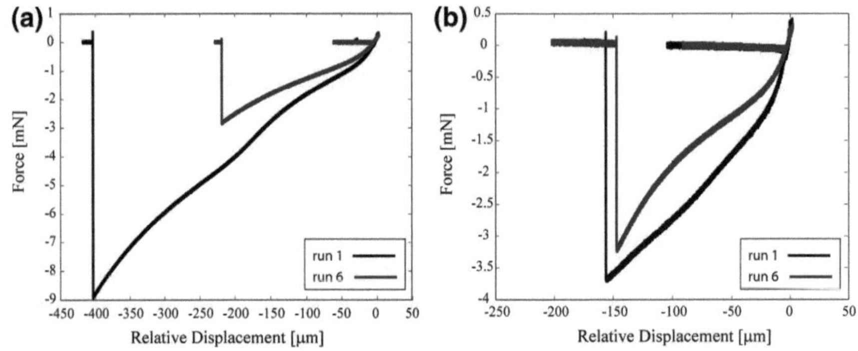

Fig. 13.8 Force-distance data for the first and the sixth run of six consecutive pull-off experiments for a thick cap fiber with **a** $\beta \approx 2.0$ and **b** $\beta \approx 1.3$. Fiber with the larger cap diameter, $\beta \approx 2.0$, loses 66% of its initial adherence after six runs. Conversely, the fiber with $\beta \approx 1.3$ loses only about 10% of its initial adherence after six runs. Both fibers exhibit different mechanical properties and behave softer in the consecutive pull-off experiments compared with the initial run

along with adhesion stress data in Fig. 13.6 suggest that fibers with smaller tips are advantageous both in terms of attainable pull-off stress and repeatability.

The force-distance data in Fig. 13.8 shows a discrepancy between the first and the sixth run for both fibers. It is observed that fibers behave stiffer in the first test compared to the subsequent tests. However, a fiber exhibits similar force-deformation characteristics in the subsequent run, with the only observed difference being the reduction in the attainable pull-off load for each test. The change in force-distance data could be attributed to the plastic deformation the fibers experience on the first run. The similarity of the force-distance curves for the subsequent tests suggest that plastic deformation takes place only on the first run.

Similar to our observations in this work, a significant reduction in pull-off force for subsequent experiments have been shown by Kroner et al. [27] for PDMS substrates. They suggested that this behavior is due to the steady coverage of the indenter with free oligomers. While a similar mechanism could be at play here, further experiments are necessary to determine the exact cause of the differences in the rate of change of pull-off with subsequent tests.

13.5 Conclusions

In this study, it was shown experimentally that the pull-off stress, independent of the edge angle, increases with smaller cap diameters. The experimental data also shows a reasonable agreement with the previously reported numerical study. Additionally, and perhaps as expected due to the discrepancy in stress the stalk of the fiber experiences, fibers with smaller cap diameters were shown to degrade much less compared to their counterparts with larger diameters. These findings suggest that a mushroom-like fiber array should be designed such that the adhesive patch is comprised of fibers with small cap diameters ($\beta \approx 1.2$) and they should be packed as closely as possible to increase the actual contact area of the fiber array. Fibers with larger cap diameters not only perform inferior in terms of adhesion stress, their durability and the consistency of pull-off stress are also compromised due to what we believe the relatively larger stresses experienced by the stalk of the fiber. It is worth noting that there are limitations on packing density due to the tendency of fibers to stick together permanently when packed closely, which should be taken into account in design. Additionally, results presented here are applicable when a fiber experiences normal loading exclusively. The expectation from a universal adhesive is that it resists not only normal separation, but also peeling moments and shear loads. This implies that under normal operation, the fibers that comprise the adhesive patch will experience normal and transverse loads and associated moments. It is therefore imperative that the dependency of pull-off to fiber geometry be investigated in the presence of transverse shear and a multi-objective optimization carried out to design fibrillar adhesives that satisfy performance requirements associated with peel, shear, and normal resistance.

References

1. W. Hansen, K. Autumn, Proc. Natl. Acad. Sci. USA **102**, 385 (2005)
2. Q. Xu, et al., Nat. Commun. **6**, 8949 (2015)
3. M.P. Murphy, B. Aksak, M. Sitti, Small **5**, 170 (2009)
4. K. Liu, J. Du, J. Wu, L. Jiang, Nanoscale **4**, 768 (2012)
5. A.Y. Stark, T.W. Sullivan, P.H. Niewiarowski, J. Exp. Biol. **215**, 3080 (2012)
6. K. Autumn, Y.A. Liang, S.T. Hsieh, W. Zesch, W.P. Chan, T.W. Kenny, R. Fearing, R. J. Full, Nature **405**, 681 (2000)
7. S. Gorb, M. Varenberg, A. Peressadko, J. Tuma, J. R. Soc. Interface **4**, 271 (2007)
8. A. Del Campo, C. Greiner, E. Arzt, Langmuir **23**, 10235 (2007)
9. S. Kim, M. Sitti, Appl. Phys. Lett. **89**, 261911 (2006)
10. G. Carbone, E. Pierro, Small **8**, 1449 (2012)
11. Y. Wang, J. Shao, Y. Ding, X. Li, H. Tian, H. Hu, *SPIE smart structures and materials + nondestructive evaluation and health monitoring* (International Society for Optics and Photonics, Bellingham, 2015), p. 94291E
12. B. Aksak, K. Sahin, M. Sitti, Beilstein J. Nanotechnol. **5**, 630 (2014)
13. D. Dugdale, J. Mech. Phys. Solids **8**, 100 (1960)
14. G.I. Barenblatt, Adv. Appl. Mech. **7**, 55 (1962)
15. T. Tang, C.-Y. Hui, N.J. Glassmaker, J. R. Soc. Interface **2**, 505 (2005)
16. D.B. Bogy, J. Appl. Mech. **38**, 377 (1971)
17. H. Marvi, S. Song, M. Sitti, Langmuir **31**, 10119 (2015)
18. M.P. Murphy, B. Aksak, M. Sitti, J. Adhes. Sci. Technol. **21**, 1281 (2007)
19. http://www.bjbenterprises.com/media/wysiwyg/pdfs/Shore-A-High-Performance/ST-1085.pdf
20. K.L. Johnson, K. Kendall, A.D. Roberts, Proc. R. Soc. Lond. A Math. Phys. Eng. Sci. **324**, 301 (1971)
21. D. Maugis, *Contact, adhesion and rupture of elastic solids* (Springer, New York, 2013)
22. G. Haiat, M.C.P. Huy, E. Barthel, J. Mech. Phys. Solids **51**, 69 (2003)

23. C.-Y. Hui, J. Baney, E. Kramer, Langmuir **14**, 6570 (1998)
24. M.K. Chaudhury, Mater. Sci. Eng. R Rep. **16**, 97 (1996)
25. B. Lorenz, B. Krick, N. Mulakaluri, M. Smolyakova, S. Dieluweit, W. Sawyer, B. Persson, J. Phys. Condens. Matter **25**, 225004 (2013)
26. K.R. Shull, Mater. Sci. Eng. R Rep. **36**, 1 (2002)
27. E. Kroner, R. Maboudian, E. Arzt, Adv. Eng. Mater. **12**, 398 (2010)

Chapter 14
Adhesion Enhancement of a Gel-Elastomer Interface by Shape Complementarity

Dadhichi Paretkar, Pooja Malhotra, C.-Y. Hui and Anand Jagota

Abstract Organogel samples molded against structured Polydimethylsiloxane (PDMS) surfaces consisting of periodic ridge-valley and hexagonally packed fibrillar array patterns were separated employing a wedge adhesion test. We found that the complementary gel-elastomer interface can be up to 5 times stronger than a control, non-structured, gel-elastomer interface. The higher the surface feature (ridge/fibril) and the smaller the spacing between the features, the more is the energy required to separate the interface. The principal mechanisms for enhanced adhesion are crack-trapping by the structured interface and friction between interpenetrating features.

14.1 Introduction

Structuring of polymeric surfaces with fibrillar (hair-like), wavy or pad-like and related structures which enable splitting of contact generates surfaces with higher effective compliance, increasing contact conformability with a given substrate, with enhanced and switchable properties [1, 2]. Nature demonstrates extensive use of this principle across the insect and animal kingdom with end contact features in the micro-to-nanometer scale, often with an advanced hierarchical arrangement as in the case of the gecko adhesive system [3–5]. Several basic principles have been identified and exploited to create biomimetic structures, including crack-trapping that enhances adhesion and friction by mechanical instabilities, contact splitting,

---

D. Paretkar · P. Malhotra · A. Jagota (✉)
Department of Chemical and Biomolecular Engineering,
Lehigh University, Bethlehem, USA
e-mail: anj6@lehigh.edu

C.-Y. Hui
Department of Mechanical and Aerospace Engineering,
Cornell University, Ithaca, USA
e-mail: CH45@cornell.edu

L. Heepe et al. (eds.), *Bio-inspired Structured Adhesives*,
Biologically-Inspired Systems 9, DOI 10.1007/978-3-319-59114-8_14

directional adhesion and friction, and so on [1, 2, 6–12]. A significant part of this work is for adhesion under dry conditions, but recent studies have looked at mechanisms for wet adhesion as well [13, 14].

The examples just cited are of attachment systems designed to work against a variety of surfaces, that is, to have surface mechanical properties such as friction and adhesion that are relatively insensitive to the choice of substrate. There are also many ways in which nature designs systems that aim to be highly selective in their choice of attachment surface. One of the ways in which such selectivity can be achieved in nature is by matching surface geometry such as structures that can mechanically interlock, of which *Velcro®* is a celebrated example [15]. Examples of such shape complementarity can be found at all length scales ranging, for example, from the extremely small molecular level for proteins and nucleotides [16] to the mesoscale attachment system of the dragonfly [4, 17].

Vajpayee et al. [18] have previously shown that highly selective adhesion can be achieved between two compliant elastomer surfaces by patterning them with ripples. In their study an elastomeric surface was patterned with ripples having periodicity in the tens of microns and amplitude in the range of a few microns. They found that such patterned surfaces had enhanced adhesion (up to a factor of two) against complementary surfaces and highly attenuated adhesion against non-complementary ones. Adhesion enhancement of complementary surfaces was shown to be due to a crack-trapping mechanism that develops because the interfacial crack is periodically forced to evolve along a direction other than the most favorable one. Mechanical interlocking of arrays of micropillar-hole complementary surface structures as well as periodic ridge-valley structures has been previously shown to enhance adhesion both at first separation (as molded) [19, 20] as well as repeated separation and pressing back in contact [21, 22].

In this work, we investigate the use of shape-complementary interfaces to enhance the adhesion between an organogel and a hydrophobic elastomer. This work is relevant to adhesion of soft structured surfaces such as our skin that pose significant challenges for an adhesive. Apart from the biological requirements such as non-toxicity and strong bonding to an organic substrate without tissue damage, the adhesive should also be able to perform in wet conditions e.g. in the presence of body fluids and under mechanical stress from a compliant surface. Many of the structured biomimetic structures have particular difficulty to attach strongly to compliant hydrophilic surfaces. In this work, we applied the idea of surface structuring to enhance adhesion between the hydrophobic PDMS and a much softer hydrophilic material such as a gel.

14.2 Experimental

14.2.1 Sample Preparation and Adhesion Testing

Pre-gel mixture was prepared by dissolving 10 wt% gelatin powder (type A from porcine skin, Sigma Aldrich) in a 70 wt% glycerol −30 wt% deionized water mixture at 85 °C for 4 h. The mixture was held at this temperature for an additional hour without stirring to allow air bubbles to escape from the pre-gel mixture. The pre-gel mixture was subsequently poured on the structured PDMS mold, held with the structured side up, inside a petri dish as shown schematically in Fig. 14.1.

The pre-gel mixture wetted the PDMS surface completely and displaced the air from the periodic ridge- or the hexagonally arranged fibrils on the PDMS mold [23]. The hot pre-gel mixture cooled down during pouring and cross-linked partially at room temperature within a period of 10–15 min. Subsequently, the gel was placed in a refrigerator at 4 °C overnight (~16.5 h). After this fixed gelation period the samples were removed from the refrigerator and allowed to equilibrate at room temperature for an hour prior to adhesion testing.

Patterned polydimethylsiloxane (PDMS, Sylgard 184) samples that served as a mold for the organogel were prepared by molding elastomer base/curing agent mixture with a ratio of 10 parts by weight of elastomeric base to 1 part by weight of curing agent into an etched silicon wafer coated with a thin layer of silane following a procedure described previously [22]. Samples were cured for 1 h at 80 °C, and then peeled carefully off the silicon wafer. Two types of structured PDMS molds were used: a periodic array of hexagonally packed fibrils and a periodic ridge-channel structure. The fibril diameter and the ridge width were both ~10 μm.

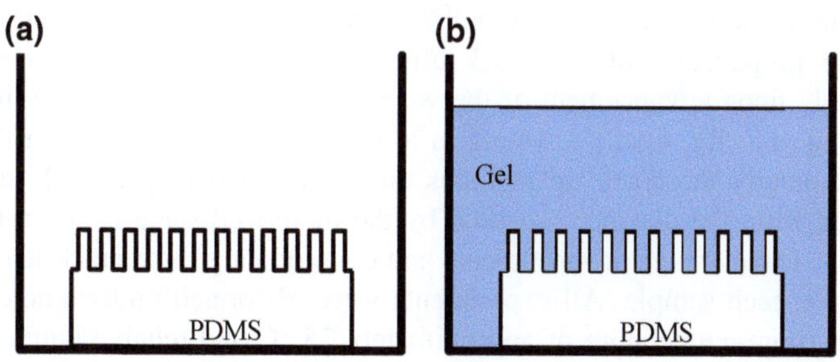

Fig. 14.1 Schematic drawing of sample fabrication. A PDMS strip structured with fibrils or ridges/channels is fabricated using standard microlithographic patterning of a silicon wafer followed by molding and curing of the PDMS (**a**). The PDMS is in turn used as a mold against which the liquid gel mixture is poured and allowed to gel into a solid (**b**)

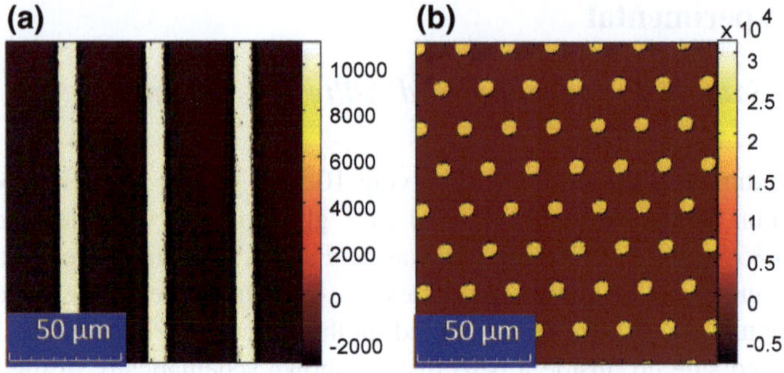

Fig. 14.2 Optical profilometry images of (**a**) ridge-channel structures, ~10 μm in height, and (**b**) hexagonally-packed fibrillar array (~30 μm in height)

The fibril/ridge height and center-center spacing between the fibrils/ridges were varied systematically. For the fibrillar samples different PDMS molds with fibril heights h of 10, 20 and 30 μm and center-center spacing S of 20, 30, 40 and 50 μm were used. Similarly, for the ridge/channel pattern, different ridge molds with heights of 10, 30 and 40 μm with ridge spacing S of 20, 35, 50, 65, 80, 95 and 110 μm were used.

Two optical profilometry images representative of the two types of samples (ridge-channel and fibrillar patterns) are shown in Fig. 14.2.

Adhesion is measured by using a wire (circular cross section with diameter of about 1.6 mm) as a wedge to separate the structured and shape-complementary gel-PDMS interface, as depicted in Fig. 14.3. The wire was sufficiently stiff that it showed no observable bending. (See [22] for a picture of the set-up and how the wire is held in place.)

The sample was placed on a motorized stage under an optical microscope to record the propagation of the crack after the wedge was advanced manually. Immediately upon advancement of the wedge the crack began to grow but with a diminishing velocity, finally arresting in less than 15 min. By bringing the wedge and subsequently the crack tip in focus under the microscope, crack length was determined from the distance travelled by the motorized stage between these two conditions. Three wedge advancements and consequently three crack lengths were recorded for each sample. All experiments were performed under ambient conditions with room temperature of approximately 25 °C and relative humidity in the range 45–50%.

14.2.2 Finite Element Model

We conducted finite element calculations (using the commercial finite element package, ABAQUS, version 6.13) to relate the measured crack length to energy

Fig. 14.3 Schematic drawing of a periodically patterned gel-PDMS interface separated using a wire as a wedge

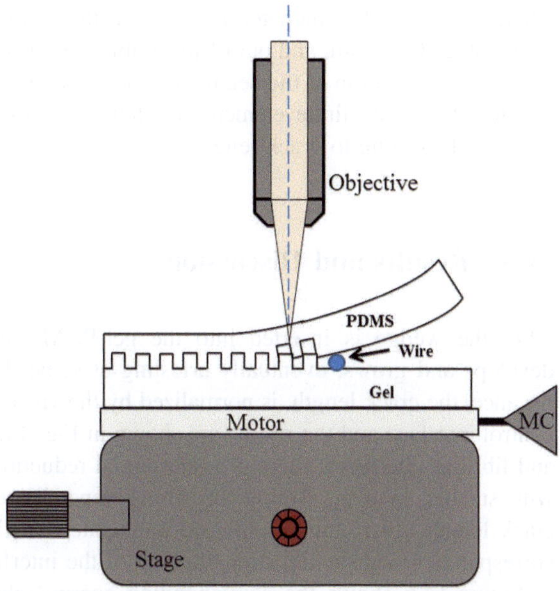

release rate over the entire experimental range of crack lengths. Both PDMS and gel were modelled as incompressible elastic solids with Young's moduli of the gel E_{gel} = 40 kPa and PDMS E_{PDMS} = 2.5 MPa and $\nu_{gel} = \nu_{PDMS} = 0.499$. The moduli were determined by separate indentation experiments [24] and verified by the beam bending method [23]. The stainless steel wire that is used as a wedge in the adhesion test was represented as a rigid cylinder and frictionless hard surface contact was assumed between the wedge-gel and wedge-PDMS surfaces.

The wedge adhesion test is simulated in two steps. In the first step, the crack is held fixed and the wedge is inserted between the PDMS beam and the gel base to a certain position such that crack length (distance between wedge and crack tip) is smaller than the minimum value observed experimentally. In the second step, constraints holding interfacial nodes in the PDMS beam to their counterpart nodes on the gel surface are released one by one. The aim is to obtain the total elastic energy of the system as a function of crack length. Nodes are released starting from a small crack length of 1.8 mm (lower than the minimum crack length observed experimentally) to a large crack length of 12.9 mm (much longer than highest crack length observed experimentally) while noting the total elastic energy of the system.

For crack lengths, a, significantly longer than the gel thickness, the energy release rate, G, should approach the result for a simple cantilever beam [25] adhered to a foundation,

$$G = \frac{9E_{PDMS}I\delta^2}{2a^4} \tag{14.1}$$

where δ is the wedge diameter, $I = t^3/12$ is the moment of inertia (t is the thickness of the plate, 0.55 mm, and out-of-plane thickness is one). For shorter crack lengths the large deformation of the beam and gel substrate cause significant deviation from this result and the finite element simulation results are therefore used to related energy release rate to crack length.

14.3 Results and Discussion

After the wedge is inserted into the gel-PDMS interface, an interfacial crack develops and grows, eventually arresting at some distance from the wedge. This distance, the crack length, is normalized by the value found for a flat, unstructured, control interface and the results are shown in Fig. 14.5. For both the ridge-channel and fibrillar interfaces, there is a substantial reduction in crack length for samples with smaller spacing. Noting the strong dependence of energy release rate with crack length (14.1, Fig. 14.4b), we anticipate that this reduction in crack length corresponds to substantial strengthening of the interface.

Figure 14.6 shows the corresponding energy release rate for the fibrillar and ridge-channel samples as a function of spacing and fibril/ridge depth, h. The energy release rate is normalized by the value found for a flat, unstructured, control interface. For both sets of samples, we find up to a factor of five enhancement in adhesion as measured by energy release rate and this occurs for the smallest spacing. For both sample sets, adhesion decreases with increasing periodic spacing,

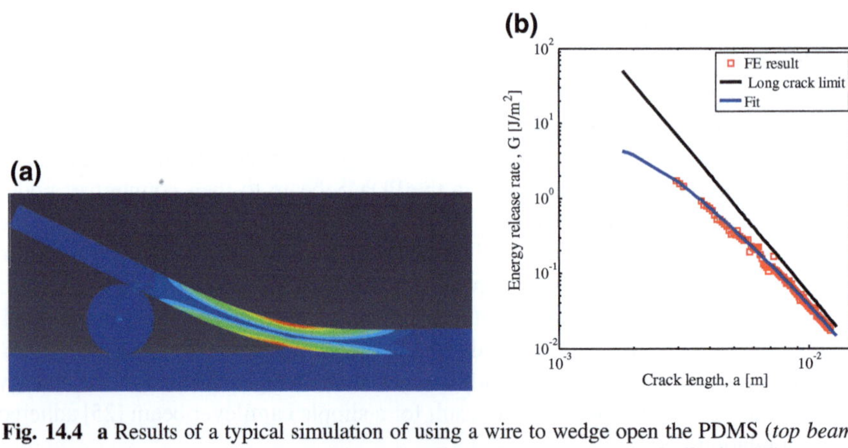

Fig. 14.4 **a** Results of a typical simulation of using a wire to wedge open the PDMS (*top beam*) and gel (*bottom*) interface. *Colors* represent (qualitatively) the intensity of effective Mises stress with *red* being the highest value and *blue* being stress-free. **b** Finite element simulation results for energy release rate as a function of crack length along with a fit used to relate experimentally measured crack length a to the corresponding energy release rate G. The expression for energy release rate valid in the long-crack limit (14.1) is shown for reference

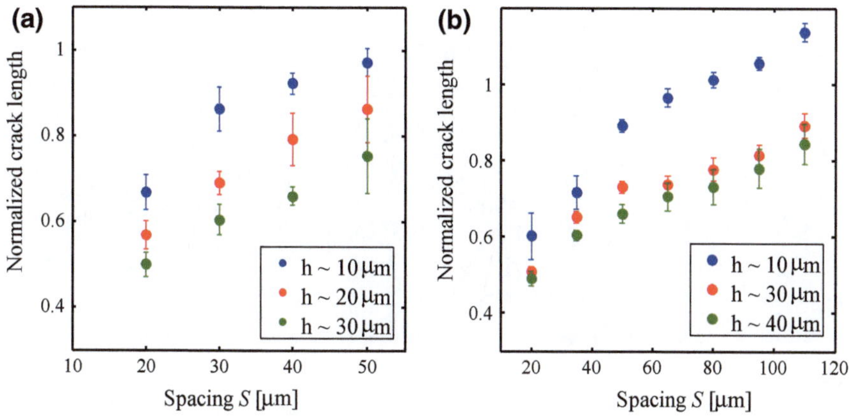

Fig. 14.5 Normalized crack-length as a function of center-center spacing and three different heights for (**a**) hexagonally arranged fibrils and (**b**) ridge-channel pattern

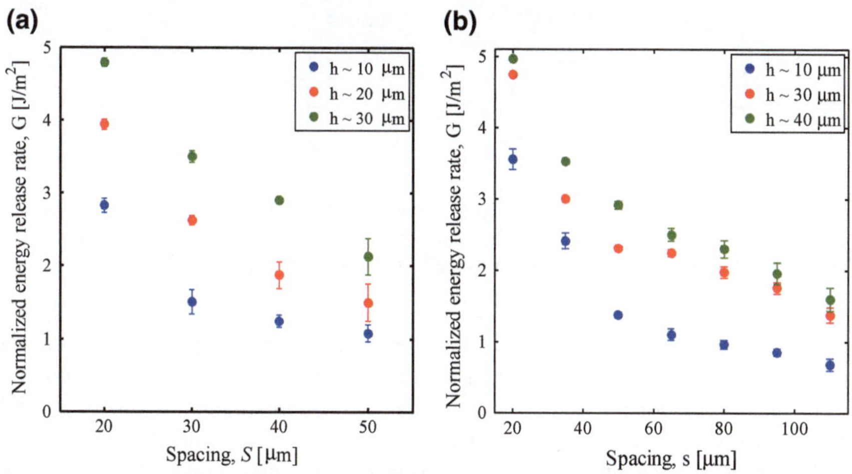

Fig. 14.6 Normalized energy release rate as a function of center-center spacing and three different heights for (**a**) hexagonally arranged fibrils and (**b**) ridge-channel pattern

in some instances even below that of the unstructured flat control. Finally, again for both sample sets, there is a systematic increase in adhesion with increasing depth, h.

Figure 14.7 shows series of snapshots from videos of an advancing crack front (left to right) for typical ridge (a) and fibrillar samples. These pictures reveal several interesting features of the mode of crack advance and a difference between ridge and fibrillar structures.

In the ridge-channel samples, crack advance is accomplished by nucleation of a debonded region, in all instances out of the immediate field of view. A likely initiation site, we propose, are the edges of the sample where the ridge-channel

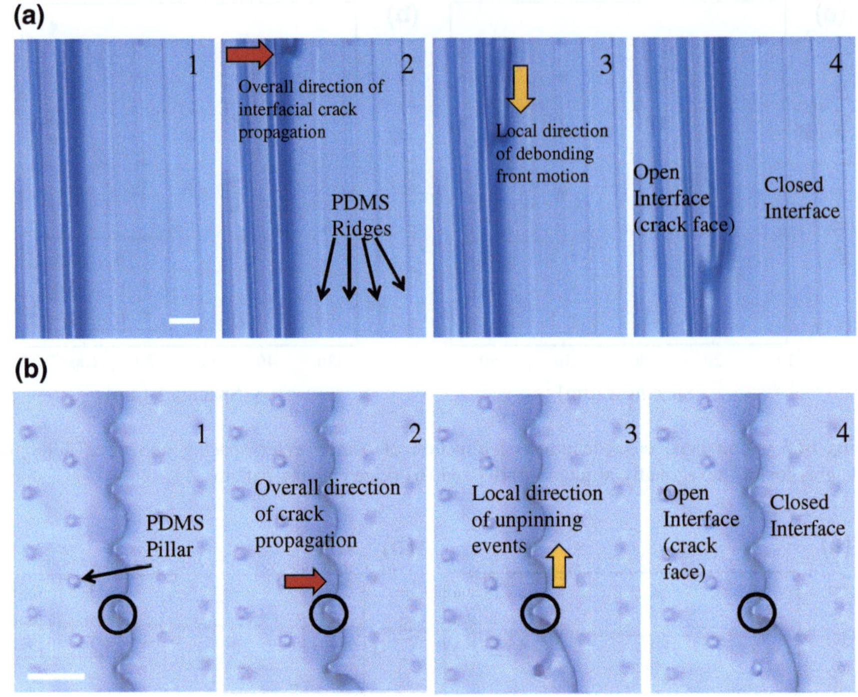

Fig. 14.7 Video snapshots of an advancing crack front (from *left* to *right*) at the PDMS-Gel interface for (**a**) periodic ridges and (**b**) hexagonally packed fibrils (*scale bar* 50 μm). Pillars and ridges are both 10 μm in width/diameter as well as 10 μm in height

meets the free surface. We observe that these debond fronts propagate along the ridge direction, and thus overall crack propagation is accomplished by unzippering events in an orthogonal direction. This mode of crack propagation is consistent with previous experiments on a closely related system [21, 22]. Similar crack growth by nucleation of a defect and its propagation orthogonal to main crack growth direction was also found for an undulating shape-complementary interface [18].

Figure 14.7b shows a typical set of micrographs of a propagating crack in the fibrillar sample. The first micrograph clearly reveals that the crack front is strongly pinned by the PDMS fibrils embedded in the gel. Crack front profiles we observe are similar to those found experimentally and theoretically for crack trapping of an interface crack by obstacles [26–28]. Similar to the ridge-channel case, crack propagation occurs by a series of unpinning events that propagate in a direction orthogonal to the overall direction of crack propagation. We observe two distinct events, debonding of the fibril/gel interface followed by extraction of the fibril from the gel in which it is embedded. This is illustrated by the four circles in Fig. 14.7b.

The circles in panels 1 and 2 show the PDMS pillar still fully embedded in the gel, pinning the crack. Between panels 2 and 3 there is a debonding event in which the interface between the PDMS pillar and the gel decohere but the pillar remains embedded in the gel. Between panels 3 and 4 the pillar is extracted from the gel, which requires work against friction between the two.

We consider two possible mechanisms that account for the strengthening of the interface, (a) Crack trapping, i.e., the resistance to crack propagation is effectively increased because the interfacial crack is forced to kink repeatedly [21, 22] (b) Work against friction to remove ridges/pillars from their complementary cavity in the gel. We have previously shown that crack trapping can explain quantitatively the enhancement of adhesion for two undulating complementary surfaces [18]. However, for a ridge-channel complementary surface (PDMS on both sides) that we previously studied, we found that adhesion enhancement due to crack-trapping is relatively small in comparison with frictional resistance [21, 22]. A third possibility is that the removal of fibrils is resisted by suction created between the bottom of the ridge/fibril and the channel or hole from which it is removed. We neglect this possibility because we measure crack length after the system has relaxed, i.e., allowing sufficient time for rate processes at the interface and diffusion of air into any cavities.

The current results support a similar interpretation, i.e., that frictional resistance to separation of complementary shapes is the main mechanism for adhesion enhancement, for a number of reasons:

(i) For the ridge-channel geometry, the crack-trapping mechanism depends primarily on the kink angle and not on the spacing between ridges. However, the experimental data in Fig. 14.7b clearly show a reduction in adhesion enhancement with spacing down to the value corresponding to an unstructured control, which is consistent with a frictional mechanism.

(ii) Similarly, the crack-trapping mechanism is nearly independent of ridge or pillar height h, whereas the experiments show a strong increase in adhesion enhancement with increase in h, which also favors a frictional mechanism.

(iii) Also for the ridge-channel geometry, the nucleation of debond sites, if it occurs without a barrier at the sample edges, obviates the crack-trapping mechanism, which relies on the entire crack front being trapped or pinned.

(iv) Similar considerations for the fibrillar geometry, for example, the decrease in adhesion with increase in separation between fibrils and decrease in h both indicate that frictional work is the dominant contributor to adhesion enhancement.

Adhesion enhancement in similar PDMS-PDMS shape complementary interfaces was considerably greater than found in the experiments we report here on a PDMS-gel interface. These we attribute to the fact that friction between gel and PDMS is significantly lower than between PDMS and PDMS.

14.4 Conclusion

We have studied two types of shape complementary interfaces between an organogel and structured PDMS, one in which the gel is molded around ridges in PDMS and the second in which it is molded around fibrils in PDMS. In both cases we show that this interdigitation results in very significant enhancement of adhesion, which increases with fibril or ridge height and decreases with increased spacing between fibrils or ridge spacing. We surmise that adhesion enhancement is primarily due to frictional sliding during extraction of the fibril or ridge from its complementary hole or channel, although crack trapping is also a contributing factor. In other work, we have shown that the ridge-channel structure can be used for repeated closure and opening [21, 22] but that the fibril-hole structure works only once. This suggests two different types of applications for such shape-recognition patterns, one in which repeated adhesion is desired and others where first-time adhesion is useful but subsequent adhesion is undesirable.

Acknowledgements This research was supported by the U.S. Department of Energy, Office of Basic energy Sciences, Division of Materials Sciences and Engineering under award DEFG02-07ER46463.

References

1. A. Jagota, C.-Y. Hui, Mater. Sci. Eng. Rep. **72**, 253 (2011)
2. B. Bhushan, Philos. Trans. R. Soc. A Math. Phys. Eng. Sci. **367**, 1445 (2009)
3. K. Autumn, Y.A. Liang, S.T. Hsieh, W. Zesch, W.P. Chan, T.W. Kenny, R. Fearing, R. J. Full, Nature **405**, 681 (2000)
4. M. Scherge, S.N. Gorb, *Biological Micro-and Nanotribology: Nature's Solutions* (Springer, New York, 2001)
5. E. Arzt, S. Gorb, R. Spolenak, Proc. Natl. Acad. Sci. **100**, 10603 (2003)
6. M. Kamperman, E. Kroner, A. del Campo, R.M. McMeeking, E. Arzt, Adv. Eng. Mater. **12**, 335 (2010)
7. S. Kim, B. Aksak, M. Sitti, Appl. Phys. Lett. **91**, 221913 (2007)
8. N.J. Glassmaker, A. Jagota, C.Y. Hui, W.L. Noderer, M.K. Chaudhury, Proc. Natl. Acad. Sci. USA **104**, 10786 (2007)
9. J.Y. Chung, M.K. Chaudhury, J. R. Soc. Interface **2**, 55 (2005)
10. A. Majumder, A. Ghatak, A. Sharma, Science **318**, 258 (2007)
11. L. Ge, S. Sethi, L. Ci, P.M. Ajayan, A. Dhinojwala, Proc. Natl. Acad. Sci. USA **104**, 10792 (2007)
12. L. Xue, J. Iturri, M. Kappl, H.-J.R. Butt, A.N. del Campo, Langmuir **30**, 11175 (2014)
13. D.M. Drotlef, L. Stepien, M. Kappl, W.J.P. Barnes, H.J. Butt, A. del Campo, Adv. Funct. Mater. **23**, 1137 (2013)
14. J. Iturri, L. Xue, M. Kappl, L. García-Fernández, W.J.P. Barnes, H.J. Butt, A. del Campo, Adv. Funct. Mater. **25**, 1498 (2015)
15. G. de Mestral, in www.uspto.gov, edited by USPTO (Velcro S.A. (Soulie, Nyon, CH), USA, 1973)
16. A.J. Bruce Alberts, J. Lewis, M. Raff, K. Roberts, P. Walter, *Molecular Biology of the Cell*, 4th edn. (Garland Science (Taylor & Francis Group), New York, 2002)

17. S.N. Gorb, Proc. R. Soc. Lond. Ser. B Biol. Sci. **266**, 525 (1999)
18. S. Vajpayee, K. Khare, S. Yang, C.-Y. Hui, A. Jagota, Adv. Funct. Mater. **21**, 547 (2011)
19. H. Shahsavan, B. Zhao, Langmuir **27**, 7732 (2011)
20. M. Lamblet, E. Verneuil, T. Vilmin, A. Buguin, P. Silberzan, L. Léger, Langmuir **23**, 6966 (2007)
21. C. Jin, A. Jagota, C.Y. Hui. "Structure and Energetics of Dislocations at Micro-Structured Complementary Interfaces Govern Adhesion." Adv. Funct. Mater. **23**, 27 (2013): 3453–3462.
22. A.K. Singh, Y. Bai, N. Nadermann, A. Jagota, C.-Y. Hui, Langmuir **28**, 4313 (2012)
23. D. Paretkar, X. Xu, C.-Y. Hui, A. Jagota, Soft Matter **10**, 4084 (2014)
24. S. Vajpayee, C.Y. Hui, A. Jagota, Langmuir **24**, 9401 (2008)
25. M. Kanninen, Int. J. Fract. **9**, 83 (1973)
26. H. Gao, J.R. Rice, J. Appl. Mech. **56**, 828 (1989)
27. A. Bower, M. Ortiz, J. Mech. Phys. Solids **39**, 815 (1991)
28. S. Xia, L. Ponson, G. Ravichandran, K. Bhattacharya, J. Mech. Phys. Solids **83**, 88 (2015)

Chapter 15
On the Bioadhesive Properties of Silicone-Based Coatings by Incorporation of Block Copolymers

Thị Chinh Ngo, Radostina Kalinova, Rosica Mincheva, Audrey Beaussart, Elise Hennebert, Patrick Flammang, Yves Dufrêne, Philippe Dubois, Roberto Lazzaroni and Philippe Leclère

Abstract This chapter discusses the applicability of different AFM-based techniques with force sensitivity of a few pN for mapping the nanostructure and quantifying the nanoscale mechanical properties of the surface of complex polymer coatings based on silicone oligomers in order to use them as bioadhesives. The AFM modes used are Peak Force Tapping and Contact with Si_3N_4 and chemically-modified (CH_3-terminated alkanethiols, COOH-terminated alkanethiols) probe tips both in air and aqueous media. Studying nanostructured films of

T.C. Ngo · R. Lazzaroni · P. Leclère (✉)
Laboratory for Chemistry of Novel Materials, Center of Innovation and Research in Materials and Polymers (CIRMAP), Research, Institute for Materials Science and Engineering, University of Mons (UMONS), 20 Place du Parc, 7000 Mons, Belgium
e-mail: Philippe.Leclere@umons.ac.be

R. Lazzaroni
e-mail: Roberto.Lazzaroni@umons.ac.be

Present Address:
T.C. Ngo
Institute of Research and Development, Duy Tan University, K7/25 Quang Trung, Da Nang, Vietnam

R. Kalinova · R. Mincheva · P. Dubois
Laboratory of Polymeric and Composite Materials, Center of Innovation and Research in Materials and Polymers (CIRMAP), Research, Institute for Materials Science and Engineering University of Mons (UMONS), 20 Place du Parc, 7000 Mons, Belgium
e-mail: rosica.mincheva@umons.ac.be

P. Dubois
e-mail: Philippe.Dubois@umons.ac.be

Present Address:
R. Kalinova
Institute of Polymers, Bulgarian Academy of Science, Acad. G. Bonchev Str., Block 103-a, 1113 Sofia, Bulgaria

L. Heepe et al. (eds.), *Bio-inspired Structured Adhesives*, Biologically-Inspired Systems 9, DOI 10.1007/978-3-319-59114-8_15

block copolymers containing a polydimethylsiloxane (PDMS) segment and a segment of poly(acrylic acid) (PAA) or poly[(2-dimethylamino) ethyl methacrylate] (PDMAEMA) led to a better understanding of the interaction of the polymer chains with solvent molecules or chains of another polymer in the self-assembly process. Stiffness mapping by PFT-AFM has allowed identifying the difference in mechanical properties between two polymer constituting blocks. The effect of copolymer concentration and solvents on the surface morphologies was also studied in details. In more complicated situation, the copolymers were used to modify the surface properties of elastomeric PDMS coatings and the PDMS surface properties before and after immersion in water were also evaluated. AFM-based nano-mechanical testing showed that the surface reorganization significantly affects the morphology and the adhesion properties of the silicone coatings. The observed broadening of the adhesion distribution is believed due to the different interactions between hydrophobic/hydrophilic surfaces and the silicon probe tip in aqueous solution. Finally, the nature of the tip-surface interaction forces was clarified by employing functionalized AFM tips. The adhesion force mapping with hydrophobic tips (CH_3-terminated alkanethiols) for 10 wt% PDMS-*b*-PDMAEMA-filled coatings before and after immersion in water showed larger forces for the coatings before immersion, thus confirming that the interaction forces between two hydrophobic surfaces are stronger than those between one hydrophobic and one hydrophilic surface. In addition, the interaction forces between amino groups of the PDMAEMA and COOH-terminated tips were investigated as a function of the pH and the ionic strength of aqueous media. Progressive stretching and continuous desorption of individual copolymer chains from the tip surface were recorded. The dynamic changes of polymer desorption were also reported by recording force curves at various pulling speeds and contact times. Furthermore, the adhesion force maps recorded at high ionic strength (in 0.1 M NaCl solution) showed a major decrease in the adhesion frequency and plateau forces, indicating a loss of polyelectrolyte properties. Bio-adhesion experiments with mussels were then performed

A. Beaussart · Y. Dufrêne
Biochemistry, Biophysics and Genetics of Microorganisms (BBGM), Université catholique de Louvain (UCL), Croix du Sud 1, 1348 Louvain-la-Neuve, Belgium
e-mail: audrey.beaussart@uclouvain.be

Y. Dufrêne
e-mail: yves.dufrene@uclouvain.be

Present Address:
A. Beaussart
Laboratoire Interdisciplinaire des Environments Continentaux, UMR 7360 CNRS—Université de Lorraine, 15 Avenue du Charmois, 54500 Vandoeuvre-les-Nancy, France

E. Hennebert · P. Flammang
Laboratoire de Biologie Marine, Research Institute for Biosciences, University of Mons (UMONS), 20 Place du Parc, 7000 Mons, Belgium
e-mail: Patrick.Flammang@umons.ac.be

on the different types of substrates—unfilled PDMS coatings and PDMS coatings filled with block copolymers. The results revealed that these organisms attach preferably to block copolymer-filled coatings after immersion due concomitant molecular reorganization at the top-surface of the copolymer-filled coatings. These observations provided evidence for the significant role played by the selected amphiphilic block copolymers to promote bio-adhesion on surface-treated silicone coatings.

15.1 Introduction

Polymers have found applications in different areas as molds, foils, thin films, coatings, adhesives, joints, blends, composites as well as biomaterials, they have been also used in micro- and nanotechnology. The surfaces and interfaces of polymer materials have strong influence on their properties (such as adhesion, friction, wetting, swelling, penetrability and biological compatibility). Particularly, block copolymers occupy a large area of research due to their well-controlled molecular architecture and chemical composition. Those copolymers especially exhibit the combined properties of different components that offer widespread applications in industry as well as daily-life. Among all copolymers, amphiphilic block copolymers are one of most interesting classes of synthetic systems for creating self-assembled nanostructures [1–3].

Among all polymers, polydimethylsiloxane is of great particular interest. Particularly, the cross-linked PDMS called silicone rubber or silicone elastomer has demonstrated a long history of use in multiscale structured surfaces [4]. PDMS has also been widely used for a number of biotechnological, biomedical applications due to its extraordinary features (such as excellent flexibility, nontoxicity, easy sealing with other materials, oxidative stability, and relatively low temperature of curing) [5–7].

PDMS possess extremely high flexibility and high mobility of polymer chains compared to other polymers, which in fact directly correlates with the very low value of its glass transition temperature ($T_g^{PDMS} \sim -125$ °C). The flexibility of PDMS chains can result in two different conformations by exposing either methyl groups or oxygen atoms outwards at the polymer surface. In air the apolar methyl groups point outwards and occupy the surface of PDMS, which leads to hydrophobicity (Fig. 15.3a). In contact with water, because the chains are very flexible and mobile, they can restructure to expose the polar oxygen groups to the PDMS/water interface as shown in Fig. 15.1.

PDMS is also water impermeable and has long time durability in the presence of aqueous solutions [8], which is an important requirement for its possible use as underwater adhesive. However, its application range is somehow restricted by the risk of adhesion failure due to its low surface free energy (19.9 mJ/m^2) and extremely low chemical reactivity [8, 9] This can be overcome by increasing the

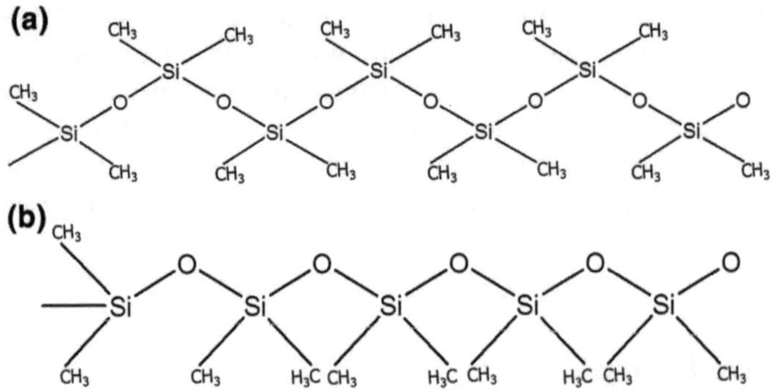

Fig. 15.1 Schematic representations showing the two possible orientations of PDMS chains

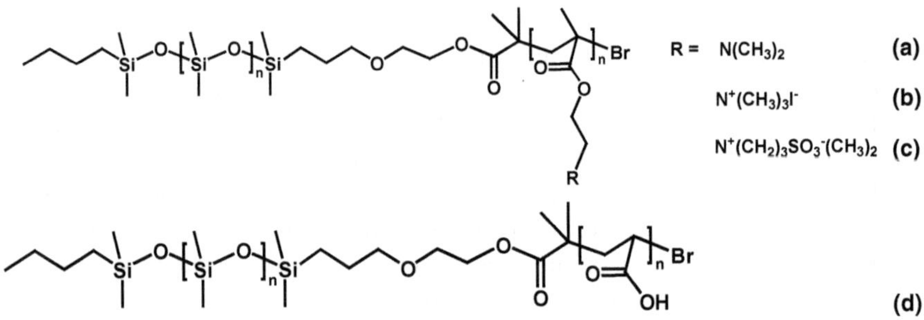

Fig. 15.2 Chemical structures of **a** polydimethylsiloxane-*b*-poly-[(2-dimethylamino) ethyl methacrylate], **b** polydimethylsiloxane-*b*-poly[(quaternized 2-dimethylamino) ethyl methacrylate], **c** polydimethylsiloxane-*b*-poly[(betainizied 2-dimethylamino) ethyl methacrylate], **d** polydimethylsiloxane-*b*-poly(acrylic acid)

PDMS surface energy and reactivity via introduction of reactive functional groups at the polymer surface. The functionalization of polymer surfaces is usually achieved by using physical or chemical processes (like plasma treatment or surface grafting) or a combination of both [10]. These techniques are able to improve the hydrophilicity of the PDMS surface without affecting its properties.

In our study, amphiphilic poly(dimethylsiloxane)-*b*-poly-[(2-dimethylamino) ethyl methacrylate] (PDMS-*b*-PDMAEMA) bearing pendent tertiary amine groups along the poly(methacrylate) block (Fig. 15.2a), and polydimethylsiloxane-*b*-poly (acrylic acid) (PDMS-*b*-PAA) bearing carboxylic groups along the poly (acrylate) sequence (Fig. 15.2d), were synthesized through atom transfer radical polymerization (ATRP). Then, in order to study the charge-type/adhesion relationship, modified PDMS-*b*-PDMAEMA was obtained by quaternization (introduction of quaternary amines, Fig. 15.2b) and betainization (introduction of zwitterions, Fig. 15.2c) reactions of the tertiary amine functions [11].

The development of scanning probe techniques plays an important role for the progress of nano-science and nanotechnology [12]. In particular, Atomic Force

Microscopy (AFM) has been used to investigate a wide range of materials (conductive, semi-conductive, insulating, ceramics, polymers, …), including soft biological systems (proteins, membranes, living cells…) [13–15] in multiple environments (air, UHV, liquid, electrochemical…). AFM is also well known as an appropriate technique to visualize the microstructure, combined with the probing of the nano- and micro-mechanical properties of material surface at the nanoscale [16].

In order to estimate the mechanical properties of a sample by AFM, a sharp probe has to establish a stable contact with the sample surface. The system usually consists of a flexible cantilever with a sharp, microfabricated probe at its end which can eventually chemically modified in order to measure local forces. In this later case, the technique is called "Chemical Force Microscopy (CFM)" [17]. Typically, these cantilevers are rectangular or V-shaped, besides other shapes of cantilevers exist for some specific applications [18]. Cantilevers are usually made of single crystal silicon of silicon nitride. The back side (opposite to the tip side) of the cantilever is generally coated with a thin metal layer (gold, platinum, or aluminum) to improve the reflectivity of the cantilever when the laser spot coming from a laser diode is back reflected into a four-quadrant photodiode. This method is the most commonly used and is named "optical lever detection". By using this kind of photodiode, it is very easy to separate the contributions of vertical and horizontal deflections of the cantilever to determine the topography and the friction properties, respectively. The sensitivity of the photodetector to the applied (low) forces depends on the shape and spring constant of the cantilever, which may vary from 0.01 to 100 N/m.

To obtain the mechanical properties, the data acquisition can occur either in the "quasi-static" or "dynamic mode". Historically, and because of the easiness of the technique, the first developed mode for the mechanical measurements involved quasi-static loading of the sample surface. In this mode, the AFM tip is brought into contact with the sample surface, pressed against it, and after withdrawn. This full approach-retract cycle is then able to provide the so-called "force-distance curve (FD curve)" by calculating the forces from the tip deflection against the piezo actuator position. Applications of various contact mechanics models to fit this FD curves allow for the calculation of the main mechanical properties at the nanometer scale such as the elastic modulus, the sample deformation, and the adhesion [16, 19]. Among the most commonly used models, depending on the applications (mainly driven by the softness and the adhesion of the sample), one can consider Derjaguin–Muller–Toporov (DMT) [20], Johnson–Kendall–Robert (JKR) [21], or other more sophisticated models [22].

AFM visualization of the mechanical properties in each pixel of the image is frequently called "Force-Volume (FV)" imaging mode. This mode if the mechanical model is appropriately chosen allows to provide the quantitative mapping of the regions of different mechanical behavior on the nanoscale and has been used to characterize many polymeric systems. Despite its usefulness of this important mode, there are however several pitfalls to this technique. The first one is related to the assumptions of chosen contact mechanics model. These contact mechanics models are usually applied for simple indenter shapes (such as paraboloid, sphere, cone, and pyramid), and the nanoscale dimensions of AFM-based measurement can strongly influence the calculated elastic properties. Thus, this is important to note that to perform quantitative calculations, the tip shape should be precisely characterized [for instance by Scanning Electron Microscopy (SEM)]. The fact that the simple contact mechanics models is ideally flat sample topography in also an important assumption. Another issue to carefully consider for the FV mapping is related to the fact that a single measurement is performed at a constant indentation rate and this rate is limited by the nonlinearities of piezo-actuator extension [23]. The FV mode is very time consuming which is a last but not least drawback. Even if "Fast Force Volume" mode has been recently introduced [24], in order to overcome these drawbacks, dynamic measurement modes were developed.

In dynamic methods, high frequency vibrations are applied to the tip–sample contact through the actuation of the cantilever, the sample, or both. Among them, Tapping Mode is the most famous [25]. Typically, Tapping Mode AFM is performed in "amplitude modulation" mode: the height of the cantilever position is constantly adjusted to keep a constant ratio of the tip vibrational amplitude in contact with the sample surface to its oscillation amplitude in air, thus imaging topographical features of the sample. The phase shift of the vibrations relative to the excitation vibrations bears information about the energy dissipation by the tip into the sample. The phase imaging technique can be used to produce excellent material contrast, especially in systems with close mechanical properties, without any additional post processing of the measurement data. Due to the fact that forces existing between the surface and the tip are not directly measured in TM-AFM, this technique does not allow for quantitative measurements of the elastic modulus of the material. However, it was shown since more than two decades that it can be used to perform contrast variation in some cases [26].

In contrast to TM-AFM, Force Modulation Microscopy measurement (FMM) is performed in contact mode. The cantilever is vertically displaced to maintain constant deflection while the tip is imaging the surface, while at the same time the cantilever is vibrated at a frequency much lower than the resonance frequency [27]. In Contact Resonance AFM (CR-AFM), similar to the FMM, scanning is also performed in the contact mode and cantilever is dynamically excited at high frequency range at relatively low amplitude. However, in CR-AFM vibrations of the tip are kept close to the resonance frequency of the cantilever, which is not the case for FMM [28]. With these two modes, the relative Young modulus can be acquired as well by using a calibration sample with well-known elastic properties. This calibration process has to be done for each new experiment which can be rapidly fastidious. Anyway, the simplest way is to perform qualitative imaging by CR-AFM, which is executed by keeping the vibration frequency constant and monitoring the change in amplitude as the tip scans the sample. The contrast of the image can be tuned by the changes of the oscillation frequencies around the resonance peak or by switching to different vibrational modes [29]. To be complete, another mode of operation of CR-AFM is a spectroscopic mode. In this mode, the vibration amplitude is recorded for the range of vibrational frequencies, establishing the full shape of the resonance curve at each point. Knowing the resonance frequency allows one to apply analysis similar to the one in the FMM method to calculate stiffness of the tip–sample contact [30]. In CR-AFM, it is also possible to track the position of the resonance peak by keeping the frequency window close to each previous resonance position for each subsequent point [31]. Once again, this method is time consuming and is better applicable to stiff samples. Therefore, several improved methods have been proposed such as 'Dual AC resonance tracking (DART) where two separate frequencies are applied to the vibrating cantilever [32]. *In fine*, the "ultrasonic AFM (UAFM)" technique is similar to FMM and CR-AFM in that the tip is constantly pressed against the sample surface and the vibrations are induced into the tip–sample contact with the applied frequency much higher than the natural resonance of the cantilever and the amplitude of this frequency is constantly changing, while the average tip deflection is monitored [33].

Pulsed Force (PF) modes have also to be consider if the quantitative mapping of the mechanical properties is needed. Firstly introduced by Marti [34] in 1997, the methodology has been declined by many manufacturers or laboratories and even the principle is basically the same, PF is nowadays known with different names such as PeakForce Tapping (PFT) [35], Intermittent Contact Resonance AFM [36], Digital Pulsed Force Microscopy [37], HybriD mode [38], Quantitative Imaging (QI) AFM [39], PinPoint Mode [40], Ringing Mode [41], … . In PF, each measurement cycle (corresponding to one pixel of the image) starts above the surface, then the tip approaches and makes a very brief contact with the surface resulting in a small indentation. The depth of the indentation of the tip is controlled by the maximum force exerted on the sample by the cantilever (set point). The cycle is finalized by withdrawing the tip from the surface up to the initial baseline static deflection. By limiting the set point from few nN to few pN (depending on the sample stiffness), indentations as small as 1–2 nm can be performed, therefore very soft, nondestructive measurement conditions could be achieved.

Recently, multifrequency techniques have been proposed and seem very promising if one can adapt the most appropriate model(s) to the recorded raw data and is able to reconstruct the force distance curves closer to the actual ones. To name a few here, we can cite: HarmoniX^{TM} [42], Band Excitation AFM [43], Intermodulation AFM [44], mode-synthesizing atomic force microscopy [45], G-mode AFM [46], …

Since it appears that the mechanical properties of a sample surface can be determined by different AFM techniques that depends on the way that the measurement is done aiming at the correct or the most accurate description of the tip-surface interactions, in this work we deliberately focused on two of them, namely Chemical Force Microscopy and Peak Force Tapping (PFT). PFT is a recent technique that allows simultaneously the recording of data on the topographic profile (i.e., the surface morphology) and the surface nano-mechanical properties with high resolution in different media including water. This is more difficult to quantitatively provide these properties with the other scanning probe techniques described above (mainly for tip and detector calibrations which are mandatory to provide quantitative measurements). This technique gives a number of advantages compared to other modes. PFT-AFM operates at a frequency of approximately 2 kHz, the time required for obtaining high-resolution images of those properties is less than that in the case with approach-retract curves recorded at 1 Hz. PFT can generate mechanical maps at higher resolution (up to 2048 pixels × 2048 pixels) compared to 64 pixels × 64 pixels at best in Force-Volume imaging. Contact-mode topographic image and FV-adhesion map are recorded separately while PFT provides simultaneously topographic and mechanical images with high resolution and relatively high-speed imaging.

In addition, Peak Force operation offers the benefits of both contact and Tapping mode: one can not only control constantly the force between the tip and the sample but also minimize the lateral forces by intermittently contacting the sample. Therefore, it is possible to minimize damages for the tip and the studied sample, which is of great interest for imaging of soft and delicate biological materials. PFT operates in a non-resonant mode and at low force; this is particularly advantageous in liquids. Using V-shaped silicon nitride probes with a low spring constant (less than 1 N m^{-1}), PFT-AFM gives higher force sensitivity compared to Tapping mode. The cantilever stiffness will be chosen based on the sample stiffness in order to cause enough deformation of the sample and still retain high force sensitivity. The probe is calibrated on a hard and clean surface prior to the experiment, in order to directly quantify the force as well as the mechanical properties.

In our study, the morphology and microstructure of PDMS-*b*-PAA and PDMS-*b*-PDMAEMA block copolymer thin films and their adhesion properties were firstly investigated using Atomic Force Microscopy. Depending on the copolymer structure, chemical composition and on the nature of the solvent, a variety of morphologies were obtained through the self-assembly of the block copolymer chains. These four copolymers were then in a second step used for modifying the surface properties of elastomeric PDMS coatings. The idea is to achieve a reversible adhesion between the ionic segments of the amphiphilic block copolymers incorporated in silicone coatings in aqueous media as a function of specific and controllable conditions (pH, ionic strength, temperature) interactions (as shown in Fig. 15.3).

In this context, the surface restructuring of PDMS in water was investigated by Chen et al. [47] using sum frequency generation (SFG) vibrational spectroscopy. The results showed that the methyl groups of PDMS tilt more toward the surface upon contact with water. Beigbeder et al. [48] also demonstrated some decrease in the surface hydrophobicity when PDMS chains get in contact with water. This is

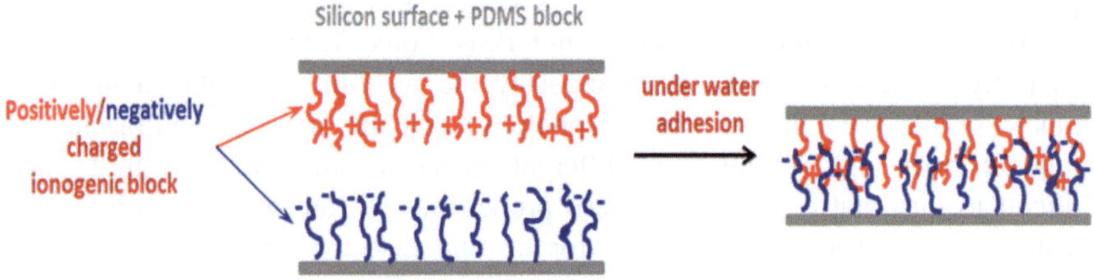

Fig. 15.3 Schematic representation of reversible adhesive in aqueous media

suggested to be a consequence of the reorganization of the siloxane backbone and the methyl groups at the coating surface. In order to functionalized PDMS coatings, many approaches have been proposed and most of them are described in the recent review by Kalinova et al. [49].

More recently, the addition of copolymers to PDMS has been proposed as a valuable pathway to tune the anti-fouling properties of silicone. Martinelli et al. [50, 51] have prepared coatings by blending amphiphilic copolymers containing poly (ethylene glycol) (PEG)-fluoroalkyl acrylate and polysiloxane methacrylate with a PDMS matrix in order to investigate the effect of the copolymer on the biological performance. The results of the immersion tests showed a better performance towards hard fouling for coatings containing the amphiphilic copolymer, compared to those without copolymer. AFM analyses showed the changes in the surface morphology of the coatings after immersion, possibly due to the swelling of the PEGylated portions at the surface [50].

Several techniques were used to evaluate the effect of the block copolymer on the PDMS surface properties before and after immersion in water. Contact angle measurements served to measure surface wettability. X-ray photoelectron spectroscopy (XPS) was then used in order to determine the surface chemical composition. Peak Force Tapping AFM and Chemical Force Microscopy were applied to investigate the surface morphology and to quantitatively measure the adhesion on the nanoscale. Finally, the bio-adhesion properties were tested by experiments with mussels, a major biological model for underwater adhesion [52–54].

15.2 Self-assembly of PDMS-Based Block Copolymers

In the case of diblock copolymers, depending on their structure, chemical composition and used solvent, a variety of morphologies can be obtained. It is known that if both block components exhibit similar solubility parameters, they are soluble in the same solvents. Repulsion between the blocks leads to different morphologies that are thermodynamically driven [55]. In contrast, a variety of self-assembled aggregates called micelles can be formed in dilute solution when the solvent preferentially solvates one of the blocks.

As already known in the literature, the strength of polymer-solvent interaction is crucially related to the solvent quality which is commonly assessed by the Hildebrand solubility parameter (δ) [56]. According to this approach, a good solvent must have a very similar solubility parameter to that of the polymers. Based on the solubility parameters presented in Table 15.1, one can conclude that THF is a selective solvent for PDMS and PDMAEMA blocks while the PAA blocks are better soluble in MeOH.

In this study, two block copolymers including PDMS-*b*-PDMAEMA (M_n^{PDMS} = 10,000, $M_n^{PDMAEMA}$ = 15,600) and PDMS-*b*-PAA (M_n^{PDMS} = 10,000, M_n^{PAA} = 12,100) were dissolved in THF and a mixture of THF and MeOH, respectively, at

Table 15.1 Solubility parameters for the solvents and the block copolymer components [89–91]

Solvent or polymer	Solubility parameter δ $(MPa)^{1/2}$
THF	18.5
MeOH	29.7
THF/MeOH	24.9
PDMS	14.6
PAA	24.6
PDMAEMA	18.8

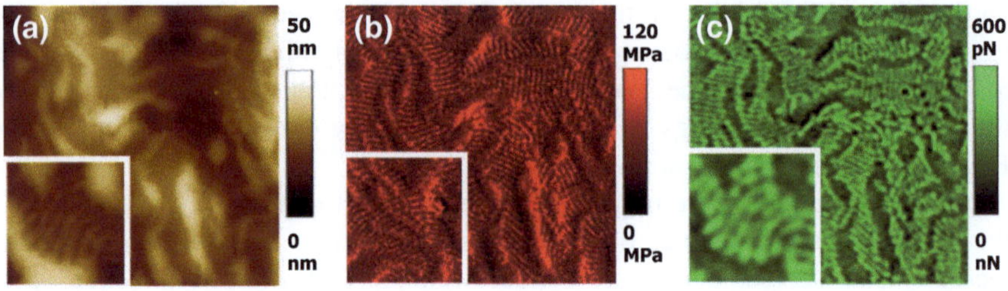

Fig. 15.4 PFT-AFM images (2 μm × 2 μm and a zoom 400 nm × 400 nm) of a PDMS-*b*-PDMAEMA block copolymer thin film obtained from a 0.1 g/mL copolymer/THF solution (M_n^{PDMS} = 10,000, $M_n^{PDMAEMA}$ = 15,600). **a** Height; **b** stiffness; **c** adhesion

concentration of 0.1 g/mL. The films were then prepared by spin coating at 4000 rpm for 120 s. The desired films are formed after the evaporation of the solvent. AFM measurements were performed in Peak Force Tapping with a Dimension Icon in air and in ambient conditions. V-shaped silicon nitride probes (k ~ 0.3–0.4 N m^{-1}) cantilever were used. The oscillation of the probe tip was operated in the vertical direction at a frequency of approximately 2 kHz. The probe was calibrated on a stiff surface prior to the experiments, in order to quantify the tip-sample forces. All images were digitally recorded with a resolution of 512 pixels/line. The scan rate was kept at 0.5 Hz. The Nanoscope image processing software was used for image analysis.

*15.2.1 Self-assembly of PDMS-*b*-PDMAEMA Copolymers in THF Non-selective Solvent*

Figure 15.4a–c, respectively displays topographic, elastic modulus and adhesion images obtained simultaneously on the same area of the surface for a PDMS-*b*-PDMAEMA copolymer film. The topographic image shows a lamellar morphology consisting of alternating stripes of the polymer components.

The Young modulus derived by using the DMT model (vide supra) mapping shows the alternation of dark and bright stripes (Fig. 15.4b) accounting for the different stiffness of the constituent polymer components. The darker domains are considered to be the softer PDMS (stripe width: 14 nm), while the brighter areas correspond to lamellae of stiffer PDMAEMA (stripe width: 26 nm). In parallel, the adhesion image (Fig. 15.4c) demonstrates that the stiffer PDMAEMA regions are less adhesive (darker stripes in the adhesion image) than the softer PDMS (brighter stripes in the adhesion image). These data correlate well with observations on the modulus map, since it is easier for the AFM tip to separate from a harder surface than from a softer one.

15.2.2 Self Assembly of PDMS-b-PAA Block Copolymers

PFT images presented in Fig. 15.5 illustrate a spherical morphology. The aggregation size is 50 ± 4 nm. The formation of spherical aggregations is the result of the micro-phase separation of block copolymers in selective solvents [57]. Depending on the solvent quality and solubility parameter (δ) [56, 58, 59], the micelle-like aggregates can be formed as a result of the association of the insoluble blocks [60–62].

In this case, the solubility parameter of the PAA ($\delta = 24.6\ \mathrm{MPa}^{1/2}$) is almost the same to one of THF/MeOH mixture ($\delta = 24.9\ \mathrm{MPa}^{1/2}$) (Table 15.1), so one may suggest that the less soluble PDMS blocks self-aggregate into the core to minimize contact with the solvent and the soluble PAA chains form the diffuse corona of the micellar-like aggregations. As observed, the spheres are relatively homogenous in terms of stiffness (no contrast among the spheres on the stiffness image in Fig. 15.5). In addition, the adhesion image shows lower adhesion for bright spheres on the height image (dark ones on the adhesion image) and stronger adhesion for dark spheres on the height image (bright ones on the adhesion image). The different adhesion may be caused by the different contact areas between the AFM tip and the surface, which is one of the major factors determining the adhesion strength [63].

After we have found out that PDMS-*b*-PAA copolymer in the THF/MeOH mixture (1:1) forms micellar-like structures, our interest was set to investigate the

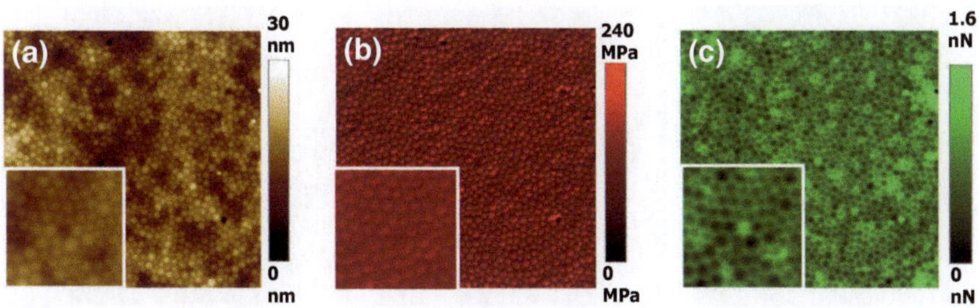

Fig. 15.5 PFT-AFM images (2 μm × 2 μm and a zoom 500 nm × 500 nm) of PDMS-*b*-PAA copolymer ($M_n^{PDMS} = 10{,}000$, $M_n^{PAA} = 12{,}100$) in an equal mixture of THF and MeOH (1:1) (0.1 g/mL). **a** Height; **b** stiffness and **c** adhesion

influence of the copolymer concentration on the size of the micelles. Copolymer solutions with different concentrations (20; 50 and 100 mg/mL) in a 1:1 ratio mixture of THF and MeOH were used. The images presented in Fig. 15.6 show the same spherical morphology for copolymer films prepared from three different concentrations. This observation leads us to the conclusion that the morphology is independent of the concentration. An increase in the aggregation size is observed (32 ± 3; 42 ± 3 and 50 ± 4 nm, at concentration 20; 50 and 100 mg/mL, respectively).

The fact that the micelle size depends on the concentration is likely related to the equilibrium between the aggregates and the isolated molecules in the solution. By increasing the concentration, the number of molecules in equilibrium to form the micelle (aggregation number) is favored in order to decrease the interface area between the solvent and the aggregating blocks [3]. This leads to an increase in the radius of the micelle core, which reflects on the size of the micelle.

As previously mentioned, the solvent has also a strong influence on micellar-like morphologies. Depending on the nature of the solvent, the self-assembly of block copolymers leads to formation of various morphologies. In this study, the influence of different solvents, THF (selective solvent for PDMS), MeOH (selective solvent for PAA) and also the mixture of both of them, on surface morphologies was investigated in detail.

The height image presented in Fig. 15.7a shows cylindrical objects (rod-like aggregates) for the PDMS-*b*-PAA copolymer in pure THF. In this case, it is expected that the PAA sequences assemble into the core while the PDMS blocks form the corona. The cylinders have an apparently uniform width of 56 ± 7 nm, but are rather polydisperse in length, lying in the 90–550 nm range. Most aggregates appear to be bent rather than straight cylinders. The results obtained with the (9:1) THF/MeOH mixture show the presence of both spherical and cylindrical micelles (Fig. 15.7b). This suggests that these two types of organizations have similar stabilities in the solvent mixture. With further addition of MeOH, moving to the (1:1) THF/MeOH mixture and then to pure MeOH (by increasing the content of MeOH to the (1:1) in THF/MeOH mixture and then moving to pure MeOH), a fully

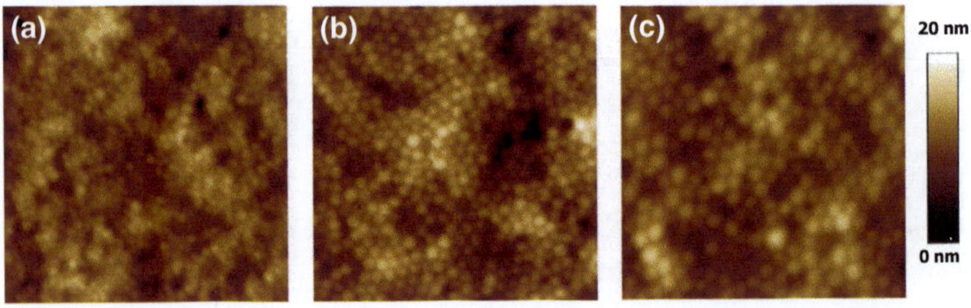

Fig. 15.6 Topography images (1 μm × 1 μm) of PDMS-*b*-PAA copolymer ($M_n^{PDMS} = 10{,}000$, $M_n^{PAA} = 12{,}100$) in THF/MeOH mixture (1:1) at different concentrations **a** 20 mg/mL; **b** 50 mg/mL and **c** 100 mg/mL

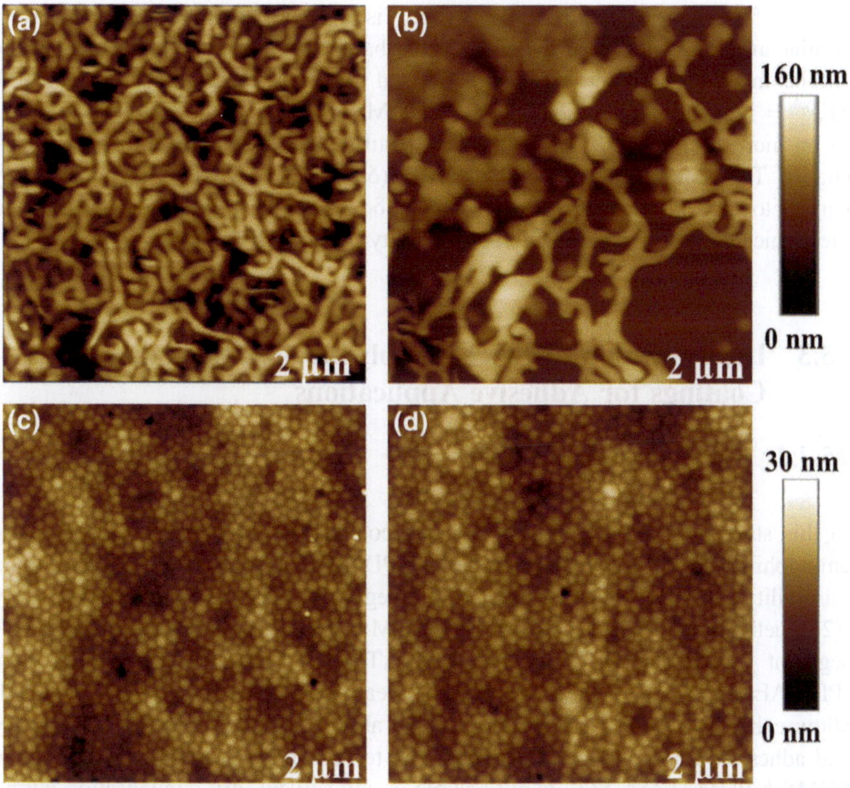

Fig. 15.7 PFT-AFM height images (2 μm × 2 μm) of PDMS-*b*-PAA copolymer in different solvents (100 mg/mL) **a** from pure THF, **b** from THF/MeOH mixture (9:1), **c** from THF/MeOH mixture (1:1), **d** from pure MeOH

spherical morphology is recovered and the size of the spheres is more homogenous (Fig. 15.7c, d, respectively).

The presence of different morphologies is probably related to polymer-solvent interactions (which are related to the difference between the solubility parameters of the polymer and the solvent). The formation of cylindrical aggregates may be caused by the particular orientation of the PAA cores in THF as its solubility parameter (δ_{PAA} = 24.6 MPa$^{1/2}$) is relatively close to the value of THF (δ_{THF} = 18.5 MPa$^{1/2}$) compared to the solubility parameter between PDMS (δ_{PDMS} = 14.6 MPa$^{1/2}$) and MeOH (δ_{MeOH} = 29.7 MPa$^{1/2}$).

With the addition of MeOH, the quality of the solvent for the PDMS blocks is expected to decrease gradually. This leads to the appearance of spherical micelles. Since MeOH is a selective solvent for the PAA blocks, it can swell the PAA cores and increases the mobility of the PAA chains. Therefore, the presence of MeOH leads to a reorganization of the system towards spherical objects. The sizes of the micelles generated in different solvents [THF, THF: MeOH (1:1) and MeOH] are

56 ± 7, 50 ± 4 and 57 ± 6, respectively. It is observed that the aggregates have a similar average size (considering the error bars). This is more likely due to the similar molecular weights of the PDMS and the PAA blocks (M_n^{PDMS} = 10,000, M_n^{PAA} = 12,100). In the case of the 1:1 THF/MeOH mixture, the micelle size seems to be more homogenous, probably as a result of a better solvation of the PAA chains. The solubility parameter of PAA (δ_{PAA} = 24.6 $MPa^{1/2}$) is indeed very similar to that of the THF/MeOH mixture ($\delta_{mixture}$ = 24.9 $MPa^{1/2}$). It means that these micelles also have considerable stability in the THF/MeOH mixture [64].

15.3 Incorporation of Block Copolymer into Silicone Coatings for Adhesive Applications

15.3.1 Coating Preparation

In this study, the modification of silicone coatings is achieved by incorporating amphiphilic block copolymers containing a PDMS segment which provides a good miscibility with the PDMS matrix and a segment of poly(acrylic acid) or poly ((2-dimethylamino) ethyl methacrylate) (PDMAEMA), or a modified-PDMAEMA segment [with quaternary amine (PDMATMA-I^-) or zwitterionic functions (PDMAEMA-SO_3^-)]. These second blocks bearing different functional groups will allow enhancing the surface wettability and also modifying the surface chemistry and adhesion after immersion in water, due to their hydrophilic character. As the PDMS-*b*-PDMAEMA-SO_3^- is not soluble in any solvent, dry condensation-curing is found to be a suitable technique to incorporate all copolymers in the silicone matrix.

In order to prepare these coatings, room temperature vulcanizing (RTV) silicone was purchased from ABCR, Germany. It consists of two parts: namely A and B, which have to be mixed together in a 10:1 weight ratio in order to obtain cross-linked materials. Part A initially contains a polymer base (silanol-terminated PDMS) and a crosslinking agent [poly (diethoxysiloxane)]. Part B contains a catalyst [DBTL (dibutyl tin dilaurate)] and trimethylsiloxy-terminated PDMS (PDMS-CH_3) was used to dilute the catalyst. A filler [silicon dioxide (silica), hexamethylsilazane surface-treated, ABCR Germany] was further added in order to improve the mechanical stability of the cured silicone.

The block copolymer-filled coatings were prepared as follows: first PDMS-OH and 2 wt% silica were mixed with a mechanical stirrer for 30 min. Then the block copolymer (5, 10, or 20 wt% for PDMS-*b*-PDMAEMA and 10 wt% for PDMS-*b*-PDMAEMA-SO_3^-, 10 wt% PDMS-*b*-PDMATMA-I^-, 10 wt% PDMS-*b*-PAA) in powder form was roll-milled within part A of the condensation-curing silicone. A three-roll milling machine (Exakt 80E, Exakt Advanced Technologies GmbH, Germany) set at room temperature, counter-rotation mode at 150 rpm, gap between the back and middle rolls 15 μm and between the middle and the front rolls 5 μm was used. The roll-milling process was repeated three times. The obtained mixture was degassed in a vacuum oven, followed by addition of the crosslinker under stirring (silanol-terminated PDMS/crosslinker = 91/9 mass ratio). Then part B (PDMS-CH_3 and the catalyst—DBTL, 1 wt%) (10:1 mass ratio) was added and the mixture stirred at 500 rpm for 1 min. In this case, DBTL is a typical catalyst for condensation curing. It plays an important role in the crosslinking reaction between silanol-terminated PDMS and poly (diethoxysiloxane) (as shown in Fig. 15.8).

Finally, each mixture was spread on glass substrates (microscopy glass slides) by means of the Doctor-Blade method, which resulted in a coating thickness of about 300 μm. The glass slides (size 25 × 75 × 1 mm, VWR) were treated with a silane coupling agent (Primer OS1200, Dow Corning, Belgium) prior to deposition of the silicone coatings.

15.3.2 Contact Angle Measurements

Sessile drop water contact angle measurements on coatings before and after immersion in water were carried out with a Kruss DSA 10 apparatus. The hydrophobic/hydrophilic character of the PDMS layer was evaluated by measuring the static contact angle between the surface of the coating and demineralized water drops (ca. 30 μL). Advancing and receding contact angles were measured by increasing by 20 μL (growing phase) and then decreasing by 30 μL (contracting phase) the drop volume. A waiting period of 20 s, called the relaxing phase, was applied between the growing phase and the contracting phase. The data presented are the average of a minimum of five measurements on more than three different samples. All measurements were performed in ambient air at room temperature.

It is well known that adhesion is strongly dependent on the surface properties, such as wettability. In order to elucidate this point, the changes in the surface

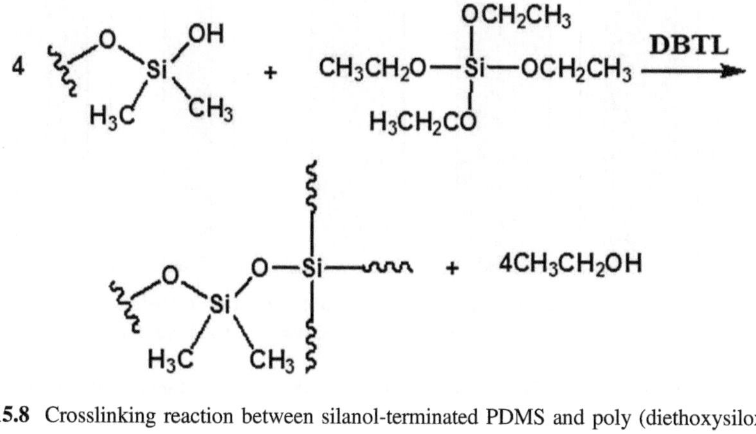

Fig. 15.8 Crosslinking reaction between silanol-terminated PDMS and poly (diethoxysiloxane)

wettability of PDMS upon addition of the copolymers were investigated in air by static contact angle (CA) and surface tension (γ_S) measurements. As underwater adhesion is aimed, measurements were performed before and after immersion in water for predetermined period(s). As is clear from the static contact angle measurements in air and before immersion in water, all the coatings exhibit a hydrophobic behavior, with contact angles exceeding 100° (Fig. 15.9). It means that all coatings exhibit hydrophobic behaviour before immersion in water; the presence of the copolymers does not seem to influence the coating wettability. Then, a gradual decrease of contact angle values was observed after immersing for 1, 3 and 4 weeks (as shown in Fig. 15.9).

According to the literature [48], after immersion in water the CA value measured for unfilled hydrosilylation-curing PDMS coatings dropped from 109° to 96°, showing some decrease in the surface hydrophobicity. This is reported to be a consequence of a reversible reorganization of the siloxane backbone and the methyl groups at the coating surface. The reversibility of the process for unfilled PDMS

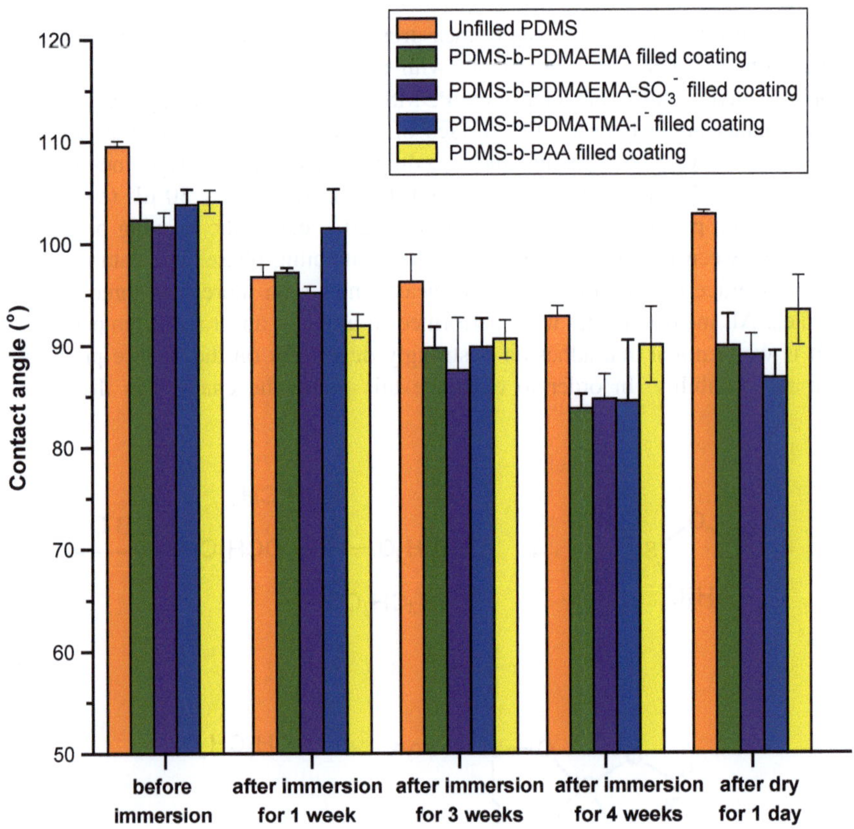

Fig. 15.9 Contact angles of PDMS and copolymer/PDMS coatings before and after immersion in water for different time

was confirmed by CA after drying at 80 °C for 24 h under reduced pressure (Fig. 15.9). Even if the exact nature of the process is not fully understood yet, it is supposed to be due to changes in the initial packing arrangement of the polymer chains, leading to surface reorganization and variations in terms of roughness and morphology [65]. In the case of the block copolymer-filled coatings, one observes a strong decrease in contact angle, from 102° to about 84° for PDMS-*b*-PDMAEMA; from 101° to about 85° for PDMS-*b*-PDMAEMA-SO_3^-, from 104° to 90° for PDMS-*b*-PAA filled coatings after 4 weeks of immersion. In fact, the average CA of copolymer-filled PDMS shifts to lower values (86°) in comparison with unfilled PDMS (96°), thus suggesting the occurrence of other processes besides the rearrangement of the polysiloxane chains. During the near-surface reorganization of the films, the amphiphilic block copolymer chains incorporated in PDMS coating could rearrange and move to locate at the outermost surface (as depicted in Fig. 15.10).

To confirm this hypothesis, coatings filled with 5, 10, 20 wt% of the PDMS-*b*-PDMAEMA block copolymer were prepared. The contact angle data for the coatings before and after immersion in water are presented in Fig. 15.11.

Before immersion, the wettability of the coating surface is not affected by the presence of the block copolymer, as the static contact angle remains in all cases close to 102°. Upon immersion in water, the contact angle of all coatings decreases with time, but it does much faster for 20 wt%-filled coatings than for the 5 wt%-filled ones. After immersion for 4 weeks, the contact angles for 10 and 20 wt% are quite similar, indicating that the maximum level of surface modification has been reached. Moreover, the reversibility of the process seems to be lost (Fig. 15.9): the CA of the copolymer/PDMS coatings insignificantly changed after drying (80 °C, 24 h, reduced pressure), thus confirming the presence of block copolymers on the coating surface.

These results are further confirmed by surface tension measurements. The surface tension of the coatings was evaluated also by measuring the contact angles with polar (water) and non-polar (diiodomethane and n-hexadecane) liquids and calculated using the Owens-Wendt-Young equation:

$$\gamma_L(1+\cos\theta)/2\gamma_L{}^{d1/2} = \gamma_S{}^{d1/2} + \gamma_S{}^{p1/2}\left(\gamma_L{}^{p}/\gamma_L{}^{d}\right)^{1/2}$$

where θ is the contact angle, γ^d and γ^p refer to surface tension arising from dispersion forces and polar interactions, respectively, and the subscripts L and S refer to the contact-angle test liquids and solid, respectively. Table 15.2 shows the surface tension (γ_s) values of all coatings at 20 °C.

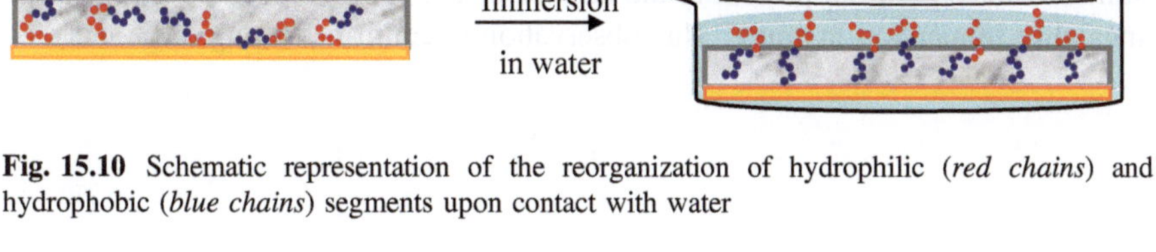

Fig. 15.10 Schematic representation of the reorganization of hydrophilic (*red chains*) and hydrophobic (*blue chains*) segments upon contact with water

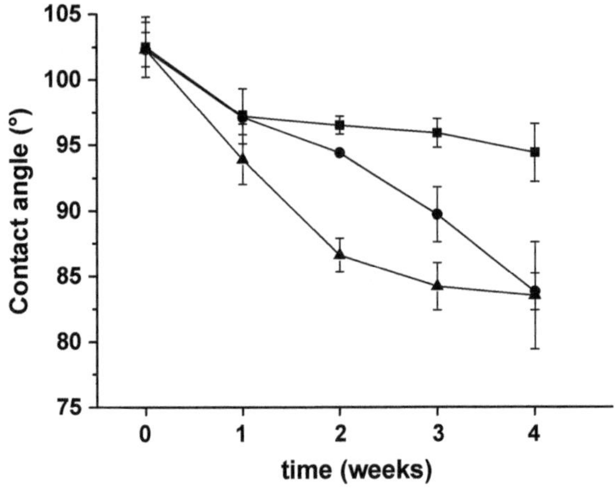

Fig. 15.11 Evolution of the contact angle (mean ± SD, n ≥ 5) of coatings filled with different amounts of PDMS-*b*-PDMAEMA copolymer (*filled square* 5 wt%; *filled circle* 10 wt%; *filled triangle* 20 wt%) with water-immersion time

Table 15.2 Surface tension (γ_s, mN m^{-1}) measurements of the condensation-cured silicone coatings

Samples	Surface tension (γ_s, mN m^{-1})		
	Before	After immersion	
		1 week	4 weeks
Unfilled PDMS	23.2	24.0	24.9
PDMS-*b*-PDMAEMA	23.2	25.8	28.1
PDMS-*b*-PDMAEMA-SO_3^-	23.8	26.5	29.8
PDMS-*b*-PDMAEMA-I^-	22.9	23.8	29.8
PDMS-*b*-PAA	23.4	23.8	26.3

Before immersion in water, the unfilled and copolymer-filled PDMS coatings exhibited relatively similar surface tension values of about 23 mN m^{-1}. Immersion leads to an increase in γ_s from 23.2 to 24.9 mN m^{-1} for unfilled PDMS. In the case of the copolymer-filled coatings, one observes a strong increase in the surface tension after immersion for 4 weeks, especially for PDMS-*b*-PDMAEMA; PDMS-*b*-PDMAEMA-SO_3^- and PDMS-*b*-PMAETMA-I^- for which γ_s rises from 23.2 to 28.1; from 23.8 to 29.8 and from 22.9 to 29.8 mN m^{-1}, respectively. Based on these results, one can imagine that the surface restructuring of PDMS chains allows penetration of water molecules in the near-surface layer. As a consequence, water molecules provoke conformational changes in the PDMS-*b*-PDMAEMA (PDMAEMA-SO_3^-) aggregates (see Surface Morphology section), bringing PDMAEMA (PDMAEMA-SO_3^-) segments to the surface. Thus, nitrogen and/or sulfur atoms should be present at the surface of the copolymer-filled PDMS coatings after immersion in water. This observation is confirmed using XPS analysis.

15.3.3 X-ray Photoelectron Spectroscopy Measurements

The surface composition of block copolymer-filled coatings was analyzed with X-ray Photoelectron Spectroscopy, using a VG ESCALAB 220iXL spectrometer. The XPS data were collected using monochromatic AlKα radiation at 1486.6 eV. Photoelectrons were collected from a 1 mm^2 sample area at take-off angle of 90°, giving an estimated probe depth of about 10 nm. For each sample, a survey spectrum was recorded with a 100 eV pass energy and 240 W of electron beam power, followed by high-resolution spectra for the C1s, O1s, Si2p and N1s peaks with a 31 eV pass energy. Atomic compositions were derived from peak areas using photoionization cross-sections calculated by Scofield, corrected for the dependence of the escape depth on the kinetic energy of the electrons and corrected for the analyzer transmission function of the spectrometer.

As immersed in water, all the coatings tend to be more hydrophilic. The molecular reorganization of PDMS and block copolymers upon contact with water gives the opportunity for the hydrophilic blocks to be exposed at the coating surface. In order to obtain information about the chemical composition of the pristine and the water-treated PDMS and copolymer-filled PDMS surfaces, XPS experiments were carried out on two coatings filled with PDMS-*b*-PDMAEMA and PDMS-*b*-PDMAEMA-SO_3^- copolymers. The results from the survey spectra of both coatings before immersion demonstrate the presence of three elements: silicon, oxygen and carbon typical of pristine PDMS, with peaks at 148, 102, 285, and 533 eV assigned to the Si_{2s}, Si_{2p}, C_{1s} and O_{1s}, respectively (Fig. 15.12a, c). After immersion in water, a new peak assigned to the N_{1s} level of PDMS-*b*-PDMAEMA appears in the spectrum (Fig. 15.12b, d).

The XPS spectra in the C_{1s} region (Fig. 15.13) were chosen to explain the copolymer-filled PDMS surface composition changes after immersion. As expected, only one peak at 284.4 eV, corresponding to the methyl groups of PDMS, is present for both coatings before immersion (Fig. 15.13a, c). After immersion in water for 4 weeks, the PDMS-*b*-PDMAEMA-filled coating shows a complex C_{1s} signal which can be curve-fitted into five components using the Voigt function (Fig. 15.13b). The component at 284.4 eV corresponds to the methyl groups of the PDMS backbone and the other four components, with binding energies between 284.9 and 288.8 eV, are typical of the PDMAEMA block. The component at 284.9 corresponds to –C–C/H groups; the contribution at 288.8 eV is due to COO groups, and the binding energy at 286.4 eV corresponds to the C–N and C–O-groups, which cannot be resolved [46]. The fourth component at 285.4 eV is characteristic of carbon atoms adjacent to the carbonyl group of the ester moiety (*C–(C=O)–O).

The C_{1s} fitting for the PDMS-*b*-PDMAEMA-SO_3^- filled systems reveals the presence of the peak at 284.4 eV corresponding to methyl groups of the PDMS block, along with five other distinct peaks for the PDMAEMA-SO_3^- block: 285.0 eV for C–C/H groups, at 285.7 eV for *C–COOH groups, at 286.6 eV for C–N groups, 287.7 eV for (C=O)–O–C* groups, at 288.8 eV for *CO (O) groups. The presence of $N^+(CH_2)_3SO_3^-$ groups is confirmed by peaks at 403.4 eV for

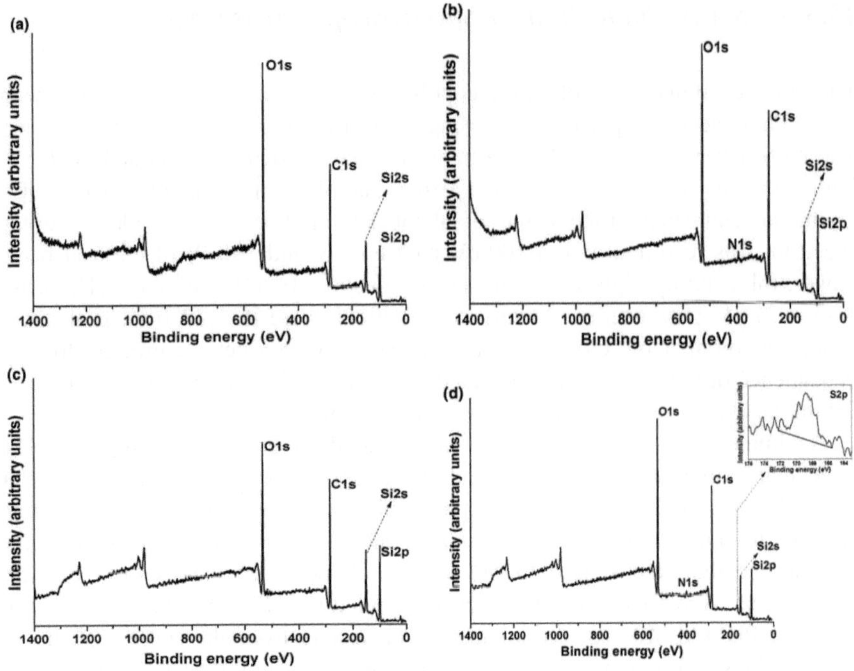

Fig. 15.12 XPS Survey spectra of 10 wt% PDMS-*b*-PDMAEMA filled silicone coatings **a** before; **b** after immersion and of 10 wt% PDMS-*b*-PDMAEMA-SO_3^- filled silicone coatings **c** before; **d** after immersion

quaternized nitrogen and at 168.0 eV for the sulfur atoms (Fig. 15.12d). The XPS results thus indicate that the PDMAEMA and PDMAEMA-SO_3^- blocks originally located underneath the surface of the coatings move to the surface upon prolonged contact with water. Moreover, considering that the repeating unit in PDMS polymer is Si $(CH_3)_2O$ and that of PDMAEMA is $C_8H_{15}O_2N$, one can assess the density of PDMAEMA after immersion in water by considering the Si/N atomic ratio after immersion: Si = 95.26% and N = 4.74%. Thus, for 10 wt% copolymer-filled PDMS coatings the PDMAEMA moieties occupy 4.74% of the surface of the sample (the thickness probed by the VG Escalab 220iXL spectrometer is estimated to be 10 nm).

15.3.4 Atomic Force Microscopy (AFM) Measurements in Air

AFM measurements were performed on the thin films using a Dimension Icon (Bruker Nano Inc., Santa Barbara, CA) and a Multimode 8 (driven by a

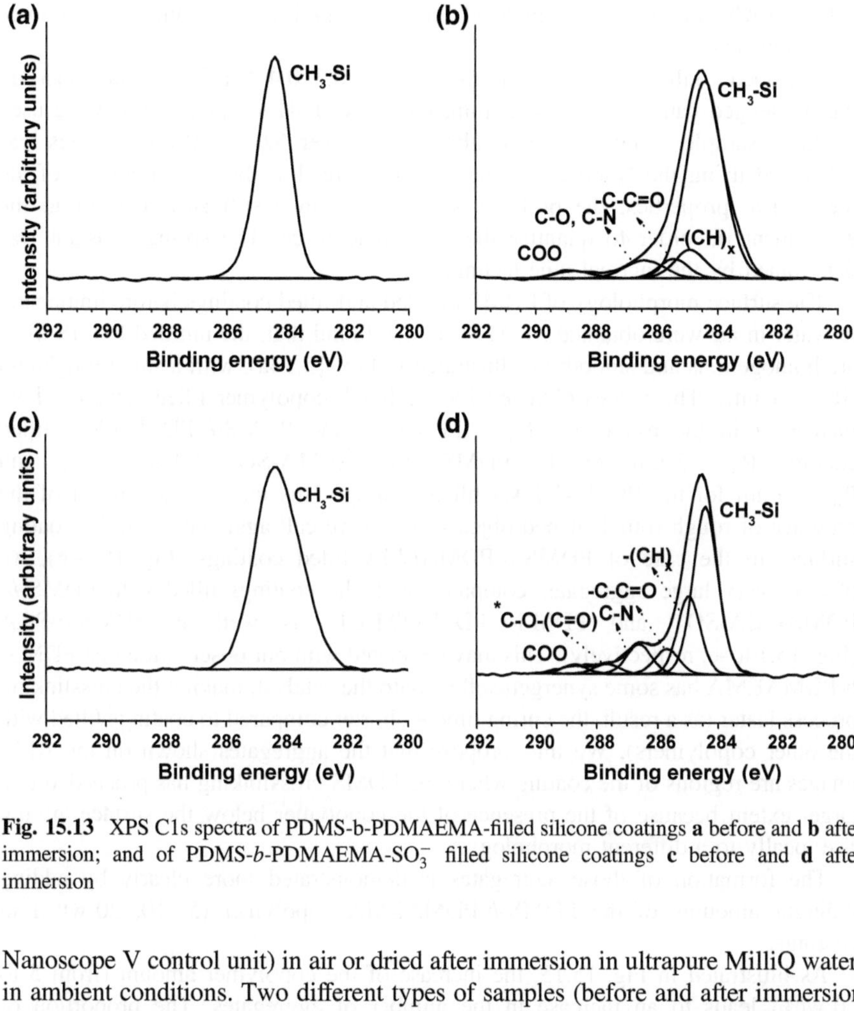

Fig. 15.13 XPS C1s spectra of PDMS-b-PDMAEMA-filled silicone coatings **a** before and **b** after immersion; and of PDMS-*b*-PDMAEMA-SO_3^- filled silicone coatings **c** before and **d** after immersion

Nanoscope V control unit) in air or dried after immersion in ultrapure MilliQ water, in ambient conditions. Two different types of samples (before and after immersion in water) were studied. The immersed coatings were dried with a nitrogen flow prior to measurement in air. For the samples analyzed in water, measurements with a specific cantilever holder were carried out in a fluid cell, in order to evaluate the variation in the surface morphology of the coatings upon immersion.

Imaging was performed in contact mode and Peak Force Tapping with a minimal applied force (250 pN for hydrophilic tips or 500 pN for hydrophobic tips). Unless otherwise specified, all force measurements were recorded with an approach and retraction speed of 1000 nm/s and a contact time lower than 50 ms. Adhesion maps (in force volume) were obtained by recording 32 × 32 force distance curves on areas of 5 μm × 5 μm, calculating the maximum adhesion peak and displaying the value as a gray pixel. PFT images were digitally recorded with a resolution of 512 pixels/line. The scan rate was kept at 0.5 Hz. For high ionic strength experiments,

0.1 M NaCl solutions were injected into the liquid cell 15 min prior to force measurements.

V-shaped silicon nitride probes (k ~ 0.3–0.4 N m^{-1}) with 115 μm narrow-legged cantilevers, CH_3-terminated and COOH-terminated tips were used on these samples. The surface roughness (R_a) over 5.0 × 5.0 μm^2 images was calculated using the Nanoscope Analysis software. For the measurements of the mechanical properties, the probe was calibrated on a stiff surface prior to the experiment, in order to quantify the tip-sample force. The spring constant was determined by the thermal tune technique.

The surface morphology of PDMS unfilled and filled coatings before immersion in water in air were obtained by AFM. It was found that, the unfilled PDMS films are homogenous and smooth (as illustrated in Fig. 15.14a), with a small roughness (R_a ~ 1 nm). The values obtained for the block copolymer filled samples show increasing in the roughness: R_a ~ 14 nm for the PDMS-*b*-PDMAEMA filled coating, R_a ~ 3 nm for the PDMS-*b*-PDMAEMA-SO_3^- filled coating and R_a ~ 6 nm for the PDMS-*b*-PAA filled coating. This increase is a result of the presence of rough round-shaped objects, which we call aggregates, on the coating surface. In the case of PDMS-*b*-PDMAEMA filled coatings (Fig. 15.14b), we observe very large aggregates compared with the coatings filled with PDMS-*b*-PDMAEMA-SO_3^- and PDMS-*b*-PDMATMA-I^- as well as PDMS-*b*-PAA (Fig. 15.14c–e, respectively). This may be related with our observation that PDMS-*b*-PDMAEMA has some synergetic effect onto the catalyst, making the crosslinking process faster (as a result, the curing time is shorter compared to coatings filled with the other copolymers). We thus propose that the aggregates shown on the AFM images are regions of the coating where the PDMS crosslinking has proceeded to a larger extent because of the presence of the copolymer below the surface, giving rise locally to a different morphology.

The formation of those aggregates is demonstrated more clearly by adding different amounts of the PDMS-*b*-PDMAEMA copolymer (5, 10, 20 wt%) in coatings.

As illustrated in Fig. 15.15, the increase of the copolymer amount (from 5 to 20 wt%) leads to an increase in the number of aggregates. The proportion of aggregates can be estimated by directly measuring the number and the size of the objects present on the images. These proportions are 11, 30 and 47% for the coating containing 5, 10, and 20 wt% copolymer, respectively. This observation suggests that the aggregates are: (i) related to the presence of copolymer; (ii) not the copolymer itself (since the XPS data show the absence of copolymer on the surface). This leads to our hypothesis that the aggregates are regions with a higher density of crosslinking because of a local increase in the amount of copolymer below the surface. The synergistic effect [66] of the PDMS-*b*-PDMAEMA copolymer on the crosslinking reaction is probably due to the amino groups of PDMAEMA, which can improve the activity of the DBTL catalyst during the crosslinking process. This would explain that the formation of aggregates is less prominent when the polymer is protonated (PDMS-*b*-PDMAEMA-SO_3^-) or

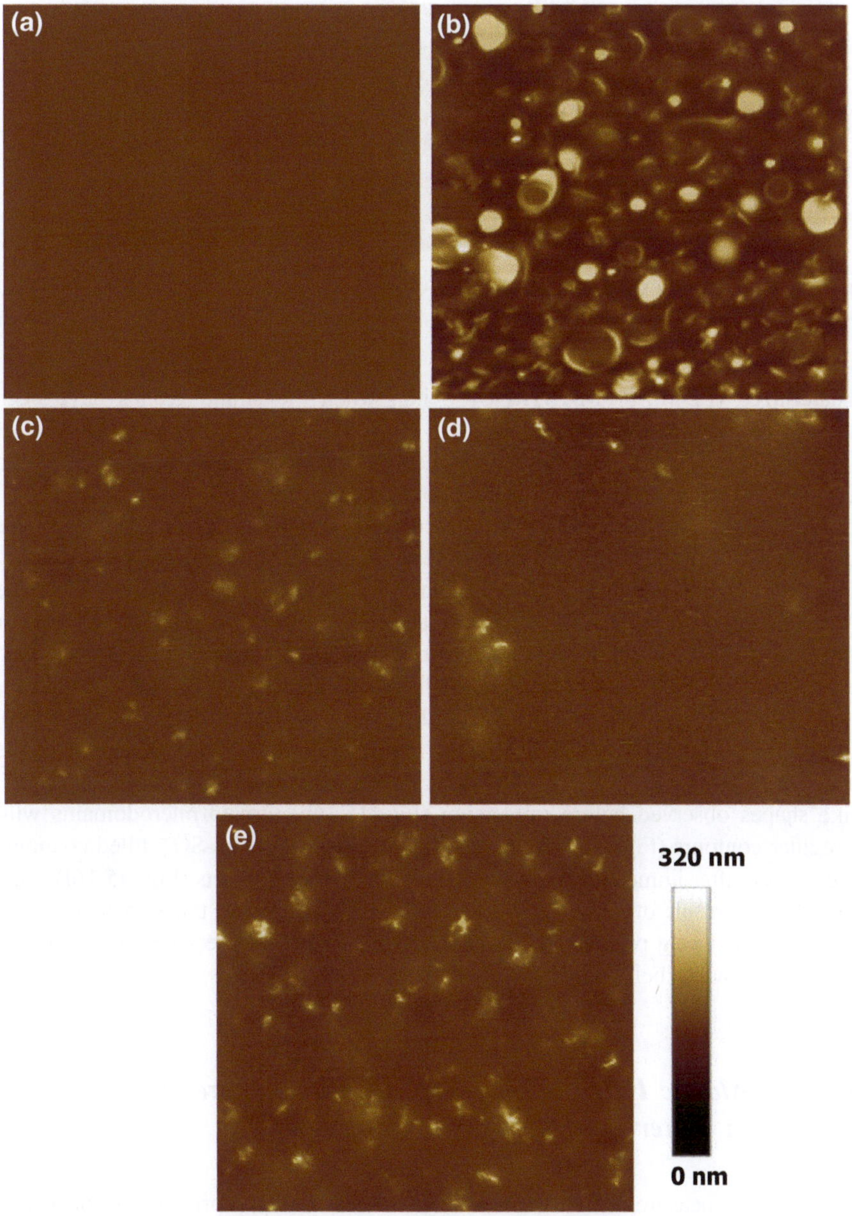

Fig. 15.14 Topography images (5 μm × 5 μm) of different coatings: **a** PDMS unfilled; **b** filled with 10 wt% PDMS-*b*-PDMAEMA; **c** filled with 10 wt% PDMS-*b*-PDMAEMA-SO_3^-; **d** filled with 10 wt% PDMS-*b*-PDMATMA-I^-; **e** filled with 10 wt% PDMS-*b*-PAA copolymers

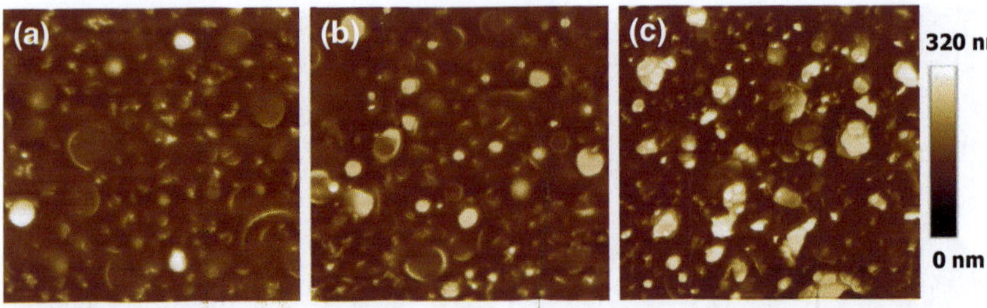

Fig. 15.15 PFT-AFM height images (5 μm × 5 μm) of PDMS filled with different amounts of PDMS-*b*-PDMAEMA copolymer before immersion in water **a** 5 wt%; **b** 10 wt% and **c** 20 wt%

quaternized (PDMS-*b*-PMAETMA-I^-), i.e., when the amino groups have been turned into ammonium groups.

Following the initial experiments, some extra measurements were carried out on dry samples immediately after their removal from water.

The unfilled PDMS coating after immersion shows a slight change in the surface topography (with the appearance of some higher spots on the surface, as shown in Fig. 15.16d) and an increase of the roughness (from R_a ~ 1 to 3 nm). It may be the result of the surface reorganization due to the flexibility of PDMS chains and/or some penetration of water in the silicone coatings. An increase in roughness is also observed for the PDMS-*b*-PDMAEMA filled coatings (from R_a ~ 14 to 20 nm) and the PDMS-*b*-PDMAEMA-SO_3^- filled coatings (from R_a ~ 3 to 6 nm). Interestingly, the morphology is clearly modified for those two materials: in the case of the PDMS-*b*-PDMAEMA filled coatings, the microdomains with "round" like shapes observed before immersion (Fig. 15.16b) turn to microdomains with irregular contours (Fig. 15.16e). In the PDMS-*b*-PDMAEMA-SO_3^- filled coatings, the surface after immersion shows elongated or star-like objects (Fig. 15.16f). This may be the result of the swelling of the aggregates during the immersion step, followed by drying prior to the AFM measurements (compare with the AFM data recorded in water below).

15.3.5 Atomic Force Microscopy (AFM) Measurements in Water

Quantitative measurements of the adhesion force are appropriately performed in liquid by using PFT technique, in order to eliminate the effect of the thin adsorbed water layer present on the surfaces in air. The results obtained for 10% PDMS-*b*-PDMAEMA filled coatings are shown in Fig. 15.17.

The aggregates are still present when the coating is immersed, but their shape appears to have changed, as seen in the height images (insets of Fig. 15.17a, b); the aggregates, which are rather dense in the dry coating, turn to a "flower-like" shape

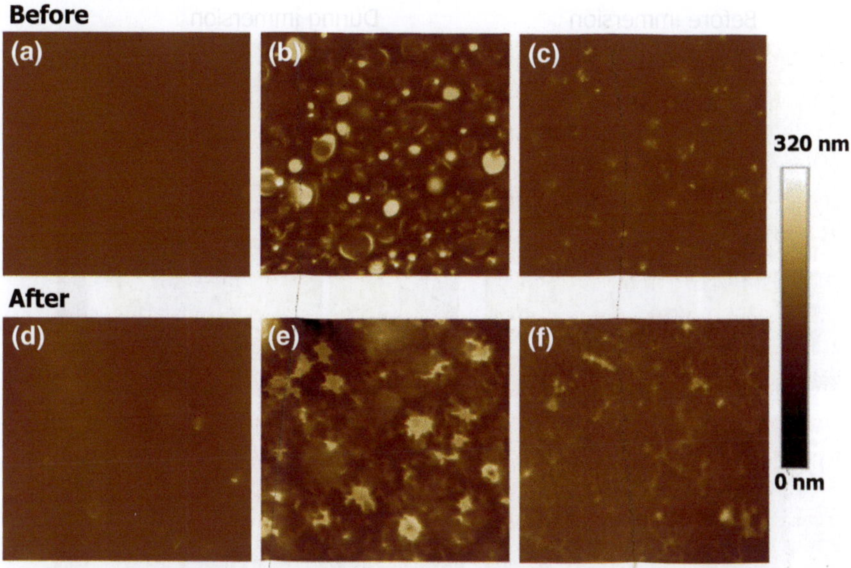

Fig. 15.16 Height images (5 µm × 5 µm) of **a**, **d** unfilled PDMS, **b**, **e** filled with 10 wt% PDMS-*b*-PDMAEMA and **c**, **f** filled with 10 wt% PDMS-*b*-PDMAEMA-SO_3^- before and after immersion in water, respectively

after immersion in water. This is believed to be the result of the reorganization of copolymer chains with the migration of hydrophilic chains on top of the surface (as indicated by the increase in wettability and the appearance of peaks typical for the hydrophilic blocks in the XPS spectra). This migration may be somewhat hampered in the areas where the PDMS is more densely cross-linked, compared to other parts of the surface, which may lead to the open aggregates on the coating surface after immersion.

Regarding the mechanical properties, it is interesting to note that, when immersed in water, these aggregates are softer than the matrix (they appear darker in the stiffness image), owing to water absorption by hydrophilic chains at the surface of the coatings. In addition, one observes that the highly cross-linked aggregates are less adhesive than the surface of the matrix (as shown in Fig. 15.17e, f). The corresponding adhesion histograms (Fig. 15.18) demonstrate the modification of adhesive distribution on the surface after immersion.

Figure 15.18a shows a narrow distribution of adhesion values (most values are found around 1800 pN) for the initial coating, while a broader distribution is obtained in the case of the immersed coating (Fig. 15.18b): the adhesion values lie in the 100–2500 pN range. The higher peak in Fig. 15.18a corresponds to the adhesion value of the PDMS matrix while the contribution around 500 pN is due to the aggregates, which are less adhesive. The adhesion shows a tendency to decrease after immersion, as can be seen in Fig. 15.18b. The adhesion values around 500 pN are still attributed to the aggregates. Adhesion values around 1800 pN become less

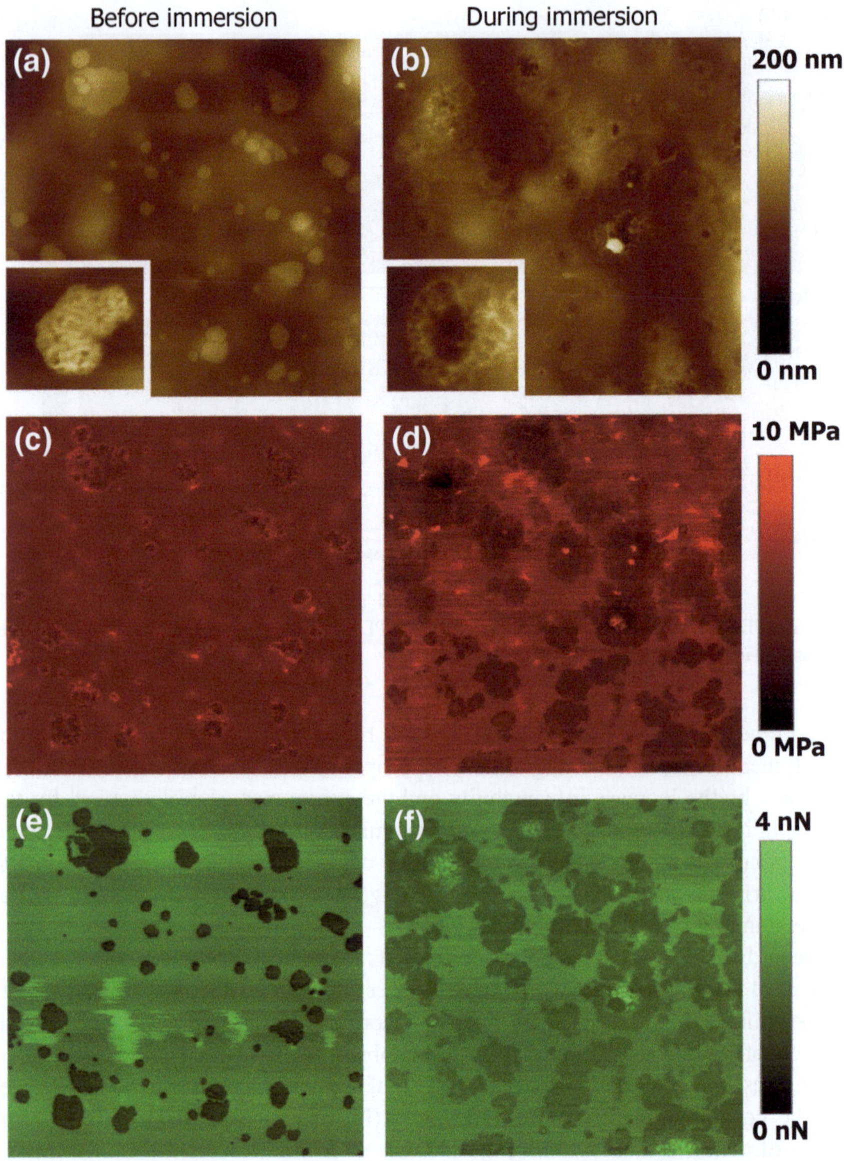

Fig. 15.17 PFT-AFM images obtained in water **a**, **b** height; **c**, **d** stiffness and **e**, **f** adhesion of 10 wt% PDMS-*b*-PDMAEMA filled coatings before and during immersion in water, respectively

frequent whereas an increase in frequency of the adhesion force ranging from 1000 to 1500 pN is observed (Fig. 15.18b). As shown above by the contact angle measurements, the coating surface after immersion becomes more hydrophilic, due to the reorganisation of copolymer chains. The modification of the adhesion can be

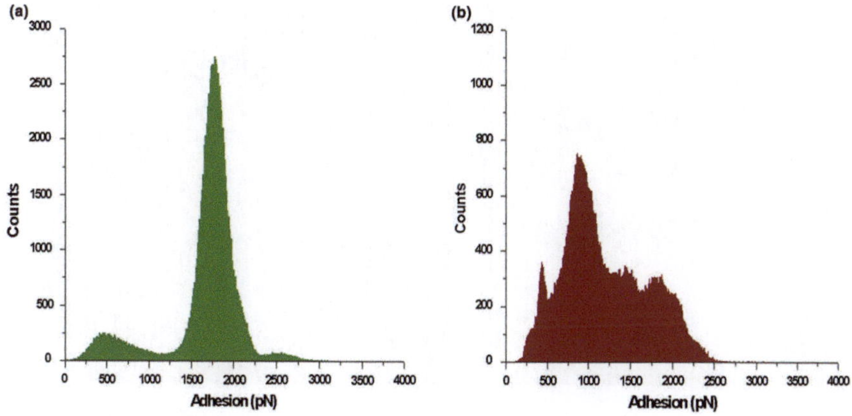

Fig. 15.18 Adhesion histograms of 10 wt% PDMS-*b*-PDMAEMA filled coatings obtained from AFM imaging in water: **a** initial coating (immersion time = 0); **b** immersed coating (immersion time = 4 weeks)

related to the different interactions between hydrophilic/hydrophobic surfaces and the silicon tip in aqueous solution. Based on the water structure rearrangement when an initially hydrophobic or hydrophilic surface comes into contact with water, Israelachvili et al. [67, 68] demonstrated that the orientation of the water molecules in the first layer next to hydrophobic or hydrophilic solid surfaces is quite different. They observed the existence of tangential alignment of the water molecules near hydrophobic surfaces, whereas normal alignment of the water molecules takes place at hydrophilic surfaces. The molecular reorganization of water on the hydrophobic substrate involves entropic effects due to the increase of hydrogen bond formation, which leads to a stronger attractive force with a hydrophilic probe surface, compared to hydrophilic substrates [69, 70]. In our case, the silicon tip is covered by a thin SiO_2 layer, which makes it hydrophilic in water. Hence, the interaction force between two hydrophilic surfaces is lower than that between one hydrophobic and one hydrophilic surface. We therefore believe that the broadening of the adhesion distribution of the immersed coatings reflects the reduction of hydrophobicity at the surface (as indicated by the decrease of contact angle after immersion), as well as the different adhesion forces between the hydrophilic/hydrophobic surfaces with the silicon tip in water. The difference of tip-surface interaction forces will then be clarified by employing AFM functionalized tips.

Figure 15.19 shows contact mode AFM topographic images recorded in milliQ water of 10 wt% PDMS-*b*-PDMAEMA copolymer-filled silicone coatings before and after immersion in water. It was found that the aggregates present on the coating surfaces before immersion (Fig. 15.18a) turn to a "flower-like" shape after immersion (Fig. 15.19b). This observation is in a good agreement with the results obtained by Peak Force Tapping (Fig. 15.16a, b). It means that contact mode AFM with minimal applied force does not damage the coating surface. Immersion in water led to a rearrangement of the coating structure, with the transfer of the

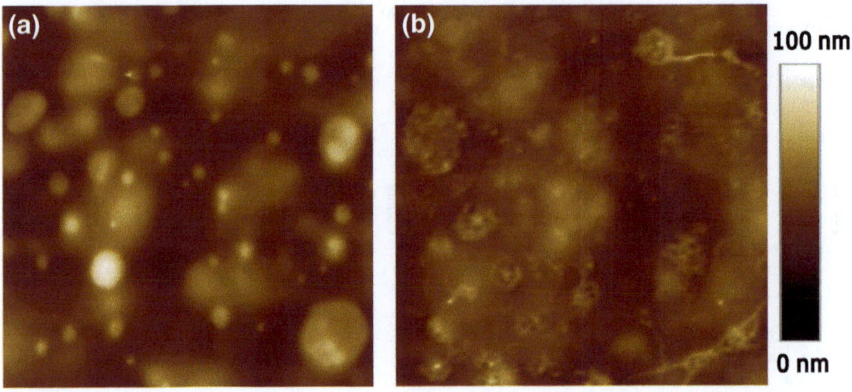

Fig. 15.19 Contact mode AFM height images (5 μm × 5 μm) of 10 wt% PDMS-*b*-PDMAEMA-filled silicone coatings **a** initial coating (immersion time: 0); **b** immersed coating (immersion time: 4 weeks)

hydrophilic blocks to the surface. This reorganization resulted in changes of the surface morphology and adhesive properties of the coatings.

As described above, Chemical Force Microscopy can also provide the adhesion properties of silicone-based coatings measured by using functionalized probe tips. Here we first used adhesion force mapping with hydrophobic tips (CH_3-terminated alkanethiols) to probe the local hydrophobic properties of the copolymer-filled coating surfaces before and after prolonged immersion in water.

Figure 15.20a, b shows the adhesion force maps, adhesion force histograms and representative force curves, recorded in milliQ water with hydrophobic tips on the 10 wt% PDMS-*b*-PDMAEMA copolymer-filled silicone coating before prolonged immersion in water. Most force curves recorded across the surfaces showed large adhesion forces, with a mean magnitude of 5.5 ± 0.4 nN (n = 2048 force curves from two maps obtained using different tips and samples). Comparison with the data obtained on reference surfaces [69] revealed that the polymer surface had a marked hydrophobic character, equivalent to that of a self-assembled monolayer composed of 80% CH_3-terminated and 20% OH-terminated alkanethiols. In addition, adhesion maps showed homogeneous contrast, meaning the surface is entirely hydrophobic. Taken together, these observations indicate that the surface of the initial copolymer-filled coating is homogeneously covered with hydrophobic PDMS.

The observations for immersed coatings show lower adhesion forces (as seen in Fig. 15.20c, d), indicating that those coatings are more hydrophilic after prolonged immersion in water. Interestingly, the adhesion maps were heterogeneous, hydrophilic nanoscale patches (0.9 ± 0.4 nN) being observed in a more hydrophobic matrix (3.1 ± 0.3 nN). As these patches correlate with the heterogeneous surface morphology (Fig. 15.19c), we believe this heterogeneous hydrophobic contrast reflects the coexistence of the hydrophilic PDMAEMA and the hydrophobic PDMS

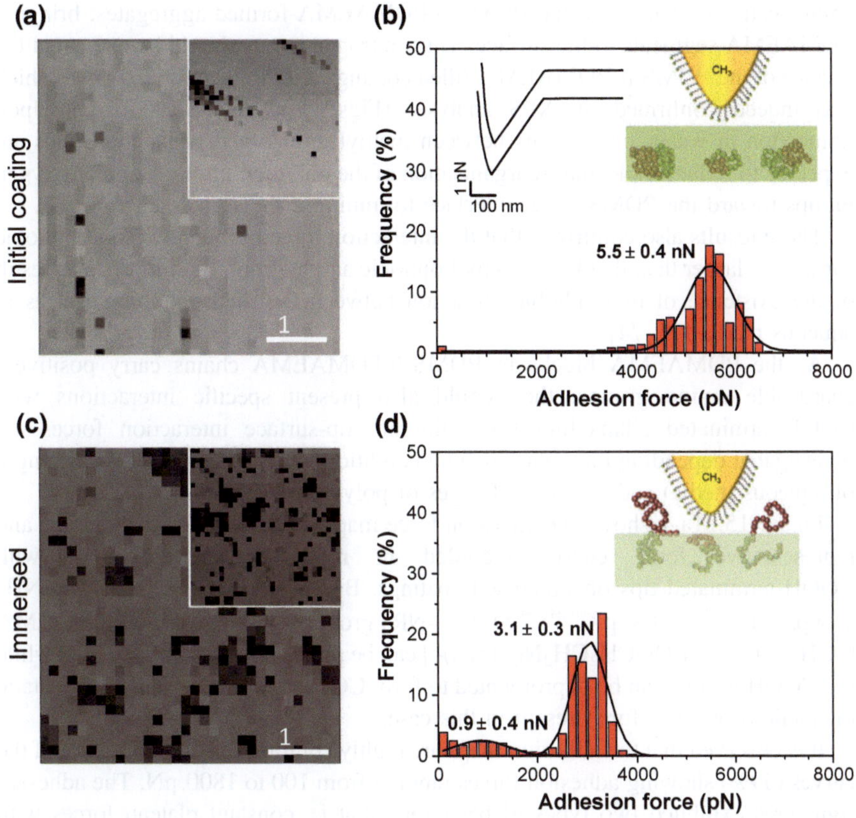

Fig. 15.20 Adhesion force maps (**a**, **c**) (5 μm × 5 μm) and **b**, **d** corresponding adhesion force histograms recorded in milliQ water with CH_3-terminated tips on **a**, **b** initial samples and **c**, **d** immersed samples

chains on top of the immersed-coating surface. That the surface hydrophobicity of silicone coatings decreased upon contact with water is consistent with earlier contact angle measurements (as shown in Fig. 15.9) [70–72]. Static contact angle measurements showed that both unfilled PDMS and PDMS-*b*-PDMAEMA-filled PDMS coatings exhibit a hydrophobic behaviour, with contact angles exceeding 100° before immersion in water. After 4 weeks of immersion, the contact angle value measured for unfilled PDMS coatings decreased to 96°, an effect believed to be a consequence of a reorganization of the siloxane backbone and the methyl groups at the coating surface. Interestingly, for the block copolymer-filled coatings, the average value of the contact angle shifted to even lower values (down to 84°), thus suggesting the occurrence of other processes, besides the rearrangement of the polysiloxane chains. Based on these results, one can imagine that the surface restructuring of flexible PDMS chains allows penetration of water molecules into the near-surface layer. As a consequence, water molecules would provoke conformational changes in the PDMS-*b*-PDMAEMA-formed aggregates, bringing PDMAEMA segments to the surface. Thus, nitrogen atoms should be present at the surface of the PDMS-*b*-PDMAEMA-filled coatings after immersion in water, which was indeed confirmed by XPS analyses (Figs. 15.12 and 15.13) [72]. Upon immersion in water, interactions between methyl groups and water molecules are expected to induce molecular reorganisation of the polymer, tilting the polar oxygen groups toward the PDMS-water interface to minimise the surface energy [47].

These results also confirmed that the interaction force between two hydrophobic surfaces is larger than that between hydrophobic and hydrophilic surfaces as a result of the existence of hydrophobic attraction between two hydrophobic surfaces in aqueous media [73, 74].

As the PDMAEMA blocks in PDMS-*b*-PDMAEMA chains carry positively chargeable amino groups, they could also present specific interactions with COOH-terminated alkanethiol tips. Different tip-surface interaction forces are investigated depending on environmental conditions (the pH and the ionic strength of aqueous media) and dynamic changes of polyelectrolyte desorption.

Figure 15.21a–c shows the adhesion force maps, adhesion force histograms and representative force curves recorded in milliQ water (pH $\sim$ 6) with COOH-terminated tips on immersed coatings. Based on pK_a of COOH and NH_2 groups (pK_a^{COOH}: 4.8; pK_a^{NH}: 7.5), the $-NR_2$ groups ($R{=}CH_3$) of the PDMAEMA [$-CH_2-C(CH_3)(COOCH_2CH_2N(CH_3)_2)-$] can be protonated to form $-NHR_2^+$ while the COOH groups can be deprotonated to form COO^- in milliQ water. Electrostatic interactions are therefore present in this case.

It is observed that the adhesion maps are highly contrasted, a large fraction of the curves (59%) showing adhesion forces ranging from 100 to 1800 pN. The adhesive signatures exhibited two types of behaviors, that is, constant plateau forces with extended rupture lengths (14%) or single linear force peaks with short rupture length (45%). The initial force peaks are attributed to the rupture of cohesive electrostatic bonds between the charged tip and the sample surface, and the plateau forces to the stretching of individual copolymer chains [75, 76].

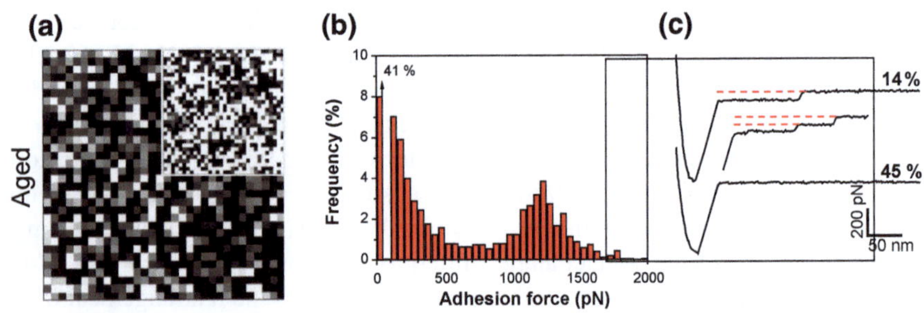

Fig. 15.21 **a** Adhesion force map (5 μm × 5 μm), **b** corresponding adhesion force histogram, and **c** representative force curves recorded using COO^--terminated tips on immersed sample in milliQ water

Plateau forces exhibit uniform rupture forces of 82 ± 7 pN magnitude, but variable rupture lengths (from 70 to 700 nm) due to the location of molecules at different points on the tip and/or on the sample surfaces [76]. These forces correspond to either continuous desorption (Fig. 15.22a, b) of individual polymer chains which detach, together with the retraction of cantilever to the rest position, or the progressive stretching (Fig. 15.22c, d) of single PDMAEMA chains from the tip surface [75–78]. In the latter case, during approach-retraction cycles, the polymer chains tend to bridge the tip and the substrate, and then desorb from the substrate. The first peak is characteristic of an attractive force that decreases rapidly and linearly with distance. After the jump-off-contact, the cantilever returns to the original position (zero deflection). The tip and sample seem to be detached, but some molecules are still bound by bonds which are stretched along the subsequent attractive surge until the force reaches a rupture value [76]. The interruption of 200 pN at a distance of 150 nm (green circle in Fig. 15.21d) indicates the detachment due to the rupture of the specific bonds between these molecules.

In addition, multiple plateaus are sometimes observed, reflecting detachment of multiples chain (Fig. 15.22b, e) [79, 80]. Note that adhesion maps did not show nanoscale domains and that the adhesion frequency (59%) was much larger than the fraction (10 wt%) of block copolymer introduced in the cross-linked PDMS matrix. This observation confirms that the copolymer preferentially locates at the coating surface.

To assess whether these polyelectrolyte interactions can be controlled by environmental conditions, we probed immersed copolymer-coatings at high ionic strength (in 0.1 M NaCl solution). Figure 15.23 shows the force data obtained using the same tips and samples as in Fig. 15.20 following injection of 0.1 M NaCl solutions.

As can be seen, the presence of salts led to a major decrease in the adhesion frequency from 59 to 16%, and the disappearance of most plateau forces, from 14 to 1%. The adhesion forces are also diminished down to less than 1200 pN with the most frequent values around 100 pN (84%). Since NaCl is an electrolyte, it is ionized into Na^+ and Cl^- ions in water. The $-NHR_2^+/COO^-$ interactions are therefore affected due to the formation of electrostatic interactions between COO^-/Na^+ and $-NHR_2^+/Cl^-$ ions. This leads to a strong decrease of the attractive interactions between the tip and the surface, explaining the decrease of the measured adhesion values.

In this systematic study, we also investigated the dynamics of polymer desorption by recording force curves at various pulling speeds and contact times.

As can be seen in Fig. 15.24a, b, plateau forces did not depend on the retraction speed, a behaviour clearly different from receptor-ligand unbinding forces [81] and protein unfolding forces [82]. This phenomenon, similar to that observed for amyloids [83] and polymer chains [79, 84], indicates that the measurements were made near thermodynamic equilibrium, thus the bonds involved in the desorption process dissociate and re-associate on a much faster timescale than the retraction speed of the tip. Figure 15.23c shows that increasing the pulling speed from 100 to 10,000 nm/s drastically decreased the plateau frequency, indicating that the occurrence of desorption events depends on the separation rate. To check whether

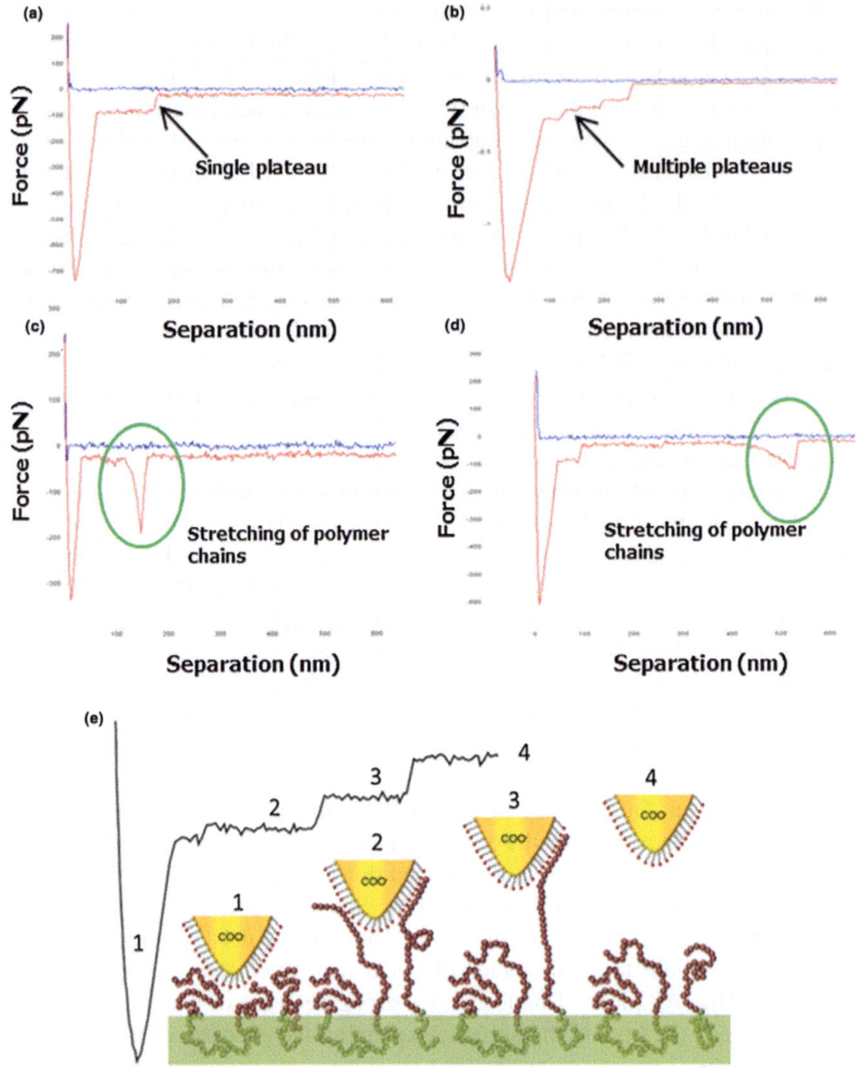

Fig. 15.22 **a–d**Representative force curves recorded for 10 wt% PDMS-*b*-PDMAEMA-filled silicone coatings after immersion in water: **a**, **b** plateau force curves represent continuous desorption of polymer chains from the tip surface **a** single plateau, **b** multiple plateaus; **c**, **d** force curves represent the progressive stretching of single polymer chain; **e** scheme representing detachment of multiples chains

this could be due to a time dependence in the desorption interaction, the contact time was varied while keeping the pulling rate constant (1000 nm/s). Under these conditions, no variation in plateau force frequency was observed (Fig. 15.24d), suggesting that the electrostatic interaction bridging the polymer to the tip is a fast process.

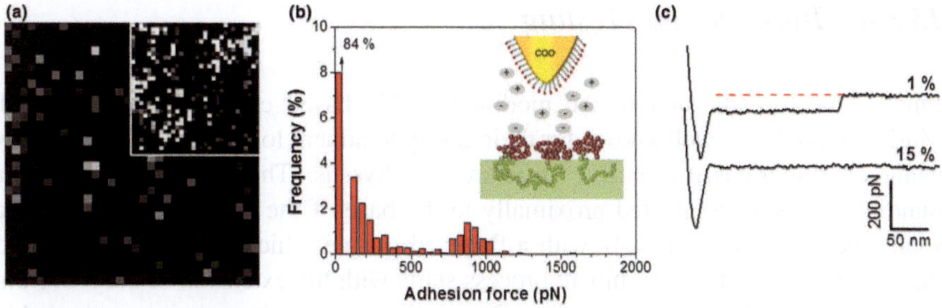

Fig. 15.23 **a** Adhesion force map (5 µm × 5 µm, *gray scale* 1000 pN), **b** corresponding adhesion force histogram, and **c** representative force curves recorded using COO^{-}-terminated tips on immersed sample in 0.1 M NaCl solution

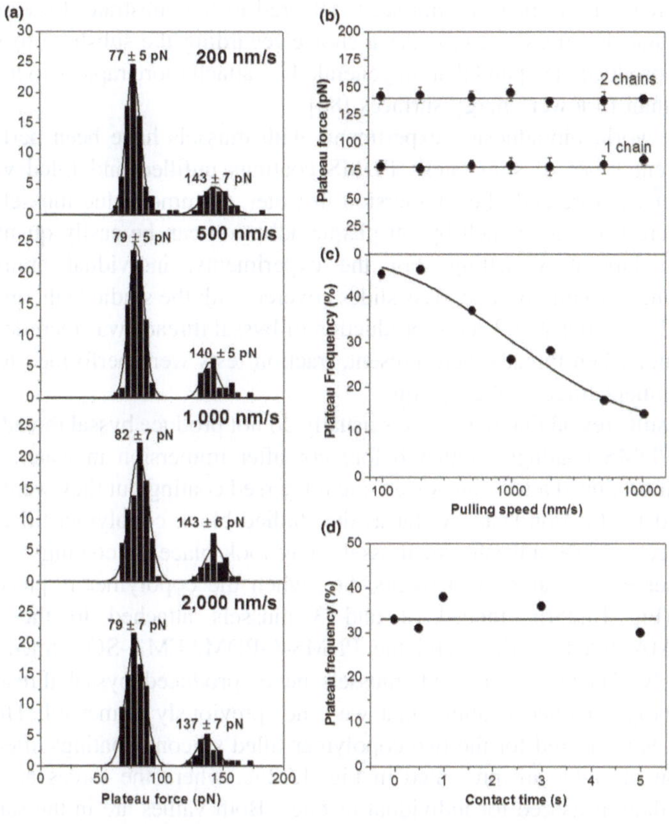

Fig. 15.24 **a** Histograms showing the distribution of plateau forces at different pulling speeds; **b**, **c** dependence of the mean plateau force (**b**) and force plateau frequency (**c**) on the pulling speed. **d** Dependence of the plateau force frequency on contact time measured at a constant pulling speed of 1000 nm/s

15.3.6 Bio-adhesion Testing

Finally, the bio-adhesion of the modified PDMS-based coatings were analyzed. Marine mussels are well known for their ability to adhere to a variety of surfaces by using a protein-based adhesive structure, the byssus. The byssus consists of a bundle of threads connected proximally to the base of the animal foot, within the shell, and terminating distally with a flattened plaque which mediates adhesion to the substrate [80]. The attachment process starts with the extension of the foot from the shell and contact with the surface. During this contact, the proteins stockpiled in the foot are secreted in a groove running along the length of the foot where they self-assemble. After a few seconds, the foot retracts, leaving a plaque adhering to the substratum with a complete thread which emerges from the groove as the foot is withdrawn [85]. This process is repeated several times and when the complete set of byssus threads is formed, the mussel is tethered to the substrate. Previous studies reported that the mussels can make a choice regarding the substrate to which the byssus is produced [85] and that, in general, they attach more rapidly to high energy surfaces than to lower energy surfaces [86].

In this work, bio-adhesion experiments with mussels have been performed on the different types of substrates—PDMS coatings unfilled and filled with block copolymers, before and after immersion in water. Common blue mussels, *Mytilus edulis*, were used, for which byssal plaque adhesion can be easily quantified in a controlled laboratory setting. For the experiments, individual animals were rubber-banded to microscope glass slides covered with the studied silicone coatings (Fig. 15.25a). After 2 or 3 days, production of byssal threads was recorded for each mussel and, when threads were present, traction tests were performed to measure the detachment force of their plaque.

The results reveal that the mussels usually do not produce byssal threads onto the unfilled PDMS coatings, neither before nor after immersion in water. Only one individual produced a few threads on one immersed coating but they were too small to be used for traction tests. As far as the studied block copolymer-filled coatings were concerned, the adhesion of mussels only took place on coatings which were first immersed in water for 4 weeks, i.e., when the copolymer is present at the surface (Fig. 15.25b). Indeed, 2 and 3 mussels attached to the PDMS-*b*-PDMAEMA filled coatings and the PDMS-*b*-PDMAEMA-SO_3^- filled coatings, respectively. On the other hand, mussels never produced byssal threads on the block copolymer-filled coatings that were not previously immersed. The detachment forces measured for the two copolymer-filled silicone coatings after 4 weeks immersion in water are presented in Fig. 15.25c, where the forces are averaged from the data measured for individual plaques. Both values are in the same range: 0.068 ± 0.018 N for the PDMS-*b*-PDMAEMA filled coatings and 0.079 ± 0.029 N for the PDMS-*b*-PDMAEMA-SO_3^- filled coatings.

This clearly highlights that both the immersion treatment and the presence of block copolymers play a significant role in the attachment of mussels to the coatings. When looking at the surface energy of the coatings, the mussels attach to the

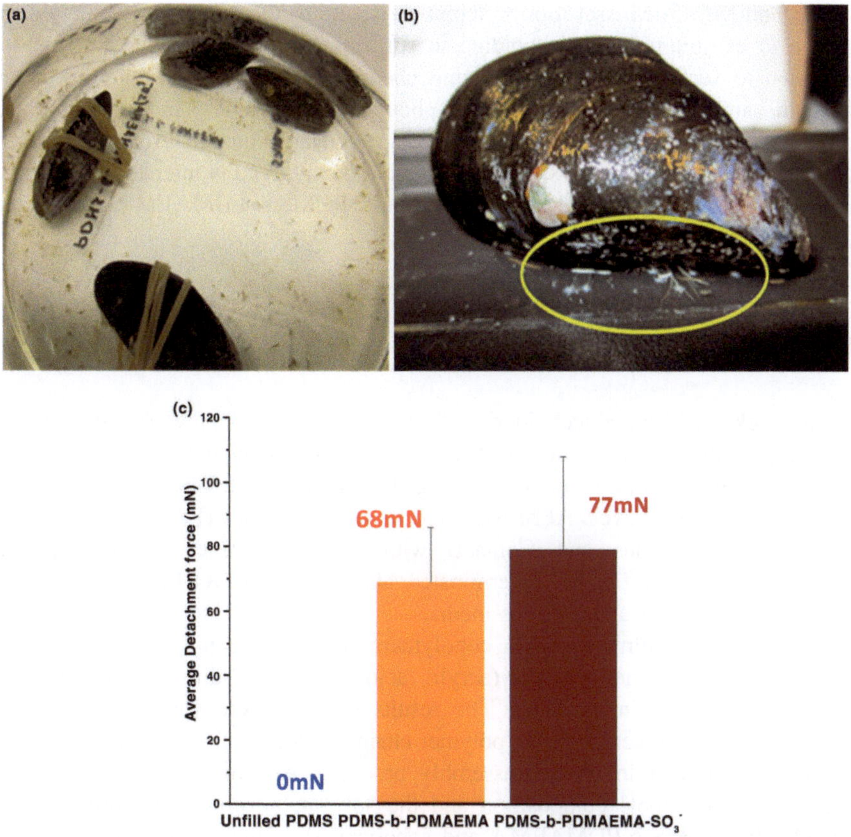

Fig. 15.25 **a** Adhesion experiments with mussels; **b** byssal threads produced onto a block copolymer-filled silicone coating; **c** average detachment forces measured for the two copolymer-filled silicone coatings after 4 weeks immersion in water

surfaces with higher energy (i.e., block copolymer-filled coatings after immersion in water). Although there are controversial opinions in the literature, our results are in good agreement with the studies of Crisp et al. [85] and Waite [54]. Indeed, in these studies, mussels prefer surfaces with higher critical surface energy and their plaques are more strongly attached to slate and glass surfaces than to plastic acetal (acetate), paraffin wax and polytetrafluoroethylene (PTFE) surfaces.

In addition, the chemical aspects related to the attachment of the mussel byssus to the coatings can be explained in terms of biological adhesion. Aquatic surfaces in nature are usually charged and in equilibrium with their environment, populated by an electrical double layer of ions [87]. Surface adsorption of underwater bioadhesives likely occurs by exchange of surface-bound ligands by amino acid sidechains, driven primarily by the relative affinities and effective concentrations of polymeric functional groups. In mussel byssal plaques, two individual adhesive proteins

Mfp-3 and Mfp-5 (Mussel foot proteins) are found at the interface and have been shown to exhibit remarkable binding to surfaces [88]. The versatility of mussel adhesion to surfaces with wide-ranging chemical and physical properties stems from the amino acid 3, 4-dihydroxy-L-phenylalanine (DOPA) but also from the other amino acids which can be hydrophobic, polar or charged. Among the later, phosphorylated serine and lysine residues could be involved in interactions with the PDMS-*b*-PDMAEMA filled coatings and the PDMS-*b*-PDMAEMA-SO_3^- filled coatings, respectively.

15.4 Conclusions

In this work, AFM-based techniques with force sensitivity of a few pN were utilized for mapping the nanostructure and quantifying the nanoscale mechanical properties of the surface of complex polymer coatings based on silicone oligomers in order to use them as bioadhesives. AFM was used in various modes (Peak Force Tapping Mechanical properties and Contact) with different probe tips [Si_3N_4 and chemically-modified tips (CH_3-terminated alkanethiols, COOH-terminated alkanethiols)] both in air and aqueous media.

Nanostructured films of block copolymers containing a polydimethylsiloxane segment and a segment of poly(acrylic acid) or poly((2-dimethylamino) ethyl methacrylate) were investigated. The results have allowed for a better understanding of the interaction of the polymer chains with solvent molecules or chains of another polymer in the self-assembly process. Depending on the copolymer structure, its chemical composition and the solvent, a variety of morphologies, lamellar for PDMS-*b*-PDMAEMA and spherical or cylindrical morphologies for PDMS-*b*-PAA, were obtained through the self-assembly of the copolymer chains. Stiffness mapping by PFT-AFM has allowed identifying the difference in mechanical properties between two polymer constituting blocks The effect of copolymer concentration and solvents on the surface morphologies was also studied in details. The results showed that the micellar morphology is independent of the concentration but dependent on the solvent while the micelle size can be controlled by the concentration. Spherical micelles were always obtained in MeOH and THF/MeOH (1:1) mixture when changing the concentration from 20 to 100 mg/mL). Cylindrical micelles were observed in when THF was used as a solvent as a result of a particular orientation of the PAA cores.

Then, these copolymers were used to modify the surface properties of elastomeric PDMS coatings. The effect of the copolymer on the PDMS surface properties before and after immersion in water was also evaluated by various techniques. Contact angle measurements showed that the block copolymer is not presented on the silicone coatings before immersion in water, but an increase in wettability of the coatings was observed after prolonged immersion in water, as a result of the reorganization of the copolymer chains with the migration of hydrophilic chains on top of the coating surface. Consistently, XPS results indicated the appearance of peaks typical for the hydrophilic blocks (PAA, PDMAEMA, and PDMAEMA-SO_3^-) at the surface. In addition, AFM-based nano-mechanical testing showed that the surface reorganization significantly affects the morphology and the adhesion properties of the silicone coatings. The observed broadening of the adhesion distribution is believed due to the different interactions between hydrophobic/hydrophilic surfaces and the silicon probe tip in aqueous solution.

The nature of the tip-surface interaction forces was clarified by employing functionalized AFM tips. The adhesion force mapping with hydrophobic tips (CH_3-terminated alkanethiols) for 10 wt% PDMS-*b*-PDMAEMA-filled coatings before and after immersion in water was recorded by using chemical force microscopy. The adhesion force maps showed larger forces for the coatings before immersion, thus confirming that the interaction forces between two hydrophobic surfaces are stronger than those between one hydrophobic and one hydrophilic surface. In addition, the interaction forces between amino groups of the PDMAEMA and COOH-terminated tips were investigated as a function of the pH and the ionic strength of aqueous media. The observation of a plateau in the force curves represents the progressive stretching and continuous desorption of individual copolymer chains from the tip surface. These plateau curves exhibit uniform rupture forces, but variable rupture lengths. Multiple plateaus were observed, reflecting detachment of multiple chains. The dynamic changes of polymer desorption were also reported by recording force curves at various pulling speeds and contact times. The results showed that the plateau frequency decreased with increasing the pulling speed from 100 to 10,000 nm/s, but no variation in plateau force frequency was observed when varying contact time, indicating that the electrostatic interaction bridging the polymer to the tip is a fast process. Furthermore, the adhesion force maps recorded at high ionic strength (in 0.1 M NaCl solution) showed a major decrease in the adhesion frequency and plateau forces, indicating a loss of polyelectrolyte properties.

Bio-adhesion experiments with mussels were then performed on the different types of substrates—unfilled PDMS coatings and PDMS coatings filled with block copolymers. The results revealed that these organisms attach preferably to block copolymer-filled coatings after immersion. In this case, after immersion in water and concomitant molecular reorganization at the top-surface of the copolymer-filled coatings, adhesion of mussels occurs and can be quantified. These observations provided evidence for the significant role played by the selected amphiphilic block copolymers to promote bio-adhesion on surface-treated silicone coatings.

Acknowledgements The work is supported by the ARC BIOMIME project (ARC AUWB-2008-08/12-UMH15), the Science Policy Office of the Belgian Federal Government (Belspo IAP-PAI 7/05 *Functional Supramolecular Systems*—FS2), and Fund for Scientific Research of Belgium (FRS-FNRS). The authors also express their thanks to the European Cooperation in Science and Technology (COST) Action TD0906 (2009–2013): «*Biological Adhesives: from Biology to Biomimetics*». Ph.L., Y.D., and P.F. are Senior Research Associate and Research Directors of FRS-FNRS (Belgium), respectively.

References

1. S.J. Holder, N.A.J.M. Sommerdijk, New micellar morphologies from amphiphilic block copolymers: disks, toroids and bicontinuous micelles. Polym. Chem. **2**, 1018–1028 (2011)
2. L. Zhang, A. Eisenberg, Multiple morphologies of "crew-cut" aggregates of polystyrene-b-poly(acrylic acid) block copolymers. Science **268**, 1728–1731 (1995)
3. L. Zhang, A. Eisenberg, Multiple morphologies and characteristics of "crew-cut" micelle-like aggregates of polystyrene-b-poly(acrylic acid) diblock copolymers in aqueous solutions. J. Am. Chem. Soc. **118**, 3168–3181 (1996)
4. J.-C. Huang, Fabrication of a gecko seta-like structure using polydimethylsiloxane. Int. J. Adhes. Adhes. **36**, 25–31 (2012)
5. D. Xiao, H. Zhang, M. Wirth, Chemical modification of the surface of poly(dimethylsiloxane) by atom-transfer radical polymerization of acrylamide. Langmuir **18**, 9971–9976 (2002)
6. J.C. McDonald, D.C. Duffy, J.R. Anderson, D.T. Chiu, H. Wu, O.J.A. Schueller, G.M. Whitesides, Fabrication of microfluidic systems in poly(dimethylsiloxane). Electrophoresis **21**, 27–40 (2000)
7. X. Ren, M. Bachman, C. Sims, G.P. Li, N. Allbritton, Electroosmotic properties of microfluidic channels composed of poly(dimethylsiloxane). J. Chromatogr. B Biomed. Sci. Appl. **762**, 117–125 (2001)
8. J. Roth, V. Albrecht, M. Nitschke, C. Bellmann, F. Simon, S. Zschoche, S. Michel, C. Luhmann, K. Grundke, B. Voit, Surface functionalization of silicone rubber for permanent adhesion improvement. Langmuir **24**, 12603–12611 (2008)
9. D.T. Eddington, J.P. Puccinelli, D.J. Beebe, Thermal aging and reduced hydrophobic recovery of polydimethylsiloxane. Sens. Actuator B Chem. **114**, 170–172 (2006)
10. D. Bodas, C. Khan-Malek, Formation of more stable hydrophilic surfaces of PDMS by plasma and chemical treatments. Microelectron. Eng. **83**, 1277–1279 (2006)
11. R Kalinova, *Imparting (bio)adhesive properties to silicone coatings: the effect of functionalized diblock copolymers*. Ph.D. Thesis, University of Mons (2013)
12. C. Gerber, H.P. Lang, How the doors to the nanoworld were opened. Nat. Nanotechnol. **1**, 3–5 (2006)
13. A. Engel, D.J. Muller, Observing single biomolecules at work with the atomic force microscope. Nat. Struct. Mol. Biol. **7**, 715–718 (2000)
14. J.K.H. Hörber, M.J. Miles, Scanning probe evolution in biology. Science **302**, 1002–1005 (2003)
15. G. Kada, F. Kienberger, P. Hinterdorfer, Atomic force microscopy in bionanotechnology. Nano Today **3**, 12–19 (2008)
16. M. Chyasnavichyus, S.L. Young, V. Tsukruk, Recent advances in micromechanical characterization of polymer, biomaterial, and cell surfaces with atomic force microscopy. Jpn. J. Appl. Phys. **54**, 08LA02-1-13 (2015)
17. A. Noy, D.V. Vezenov, C.M. Lieber, Chemical force microscopy. Annu. Rev. Mater. Res. **27**, 381–421 (1997)
18. B. Bhushan, O. Marti, in *Handbook of Nanotechnology*, ed. by B. Bhushan (Springer, Heidelberg, 2010), pp. 573–617
19. J Melcher, D Kiracofe, S Hu, A Raman, VEDA 2.0 (Virtual Environment for Dynamic AFM) (2008), http://nanohub.org/resources/adac. doi:10.4231/D38W3821D. Accessed 09 May 2016
20. B.V. Derjaguin, V.M. Muller, Y.P. Toporov, Effect of contact deformations on the adhesion of particles. J. Colloid Interface Sci. **53**, 314–325 (1975)
21. K.L. Johnson, K. Kendall, A.D. Roberts, Surface energy and the contact of elastic solids. Proc. R. Soc. Lond. A **324**, 301–313 (1971)
22. S.A. Chizhik, Z. Huang, V.V. Gorbunov, N.K. Myshkin, V.V. Tsukruk, Micromechanical properties of elastic polymeric materials as probed by scanning force microscopy. Langmuir **14**, 2606–2609 (1998)

23. K.O. van der Werf, C.A.J. Putman, B.G. de Grooth, J. Greve, Adhesion force imaging in air and liquid by adhesion mode atomic force microscopy. Appl. Phys. Lett. **65**, 1195–1197 (1994)
24. S. Seghezza, S. Dante, A. Diaspro, C. Canale, High resolution nanomechanical characterization of multi-domain model membranes by fast force volume. J. Mol. Recognit. **28**, 742–750 (2015)
25. Q. Zhong, D. Inniss, K. Kjoller, V.B. Elings, Fractured polymer/silica fiber surface studied by tapping mode atomic force microscopy. Surf. Sci. Lett. **290**, L688–L692 (1993)
26. P. Leclère, R. Lazzaroni, J.L. Brédas, J.M. Yu, P. Dubois, R. Jérôme, Microdomain morphology analysis of block copolymers by atomic force microscopy with phase detection imaging. Langmuir **12**, 4317–4320 (1996)
27. P. Maivald, H. Butt, S. Gould, C. Prater, B. Drake, J. Gurley, V. Elings, P. Hansma, Using force modulation to image surface elasticities with the atomic force microscopy. Nanotechnology **2**, 103–106 (1991)
28. G. Stan, S.W. King, R.F. Cook, Nanoscale mapping of contact stiffness and damping by contact resonance atomic force microscopy. Nanotechnology **23**, 215703 (2012)
29. D.C. Hurley, M. Kopycinska-Müller, D. Julthongpiput, M.J. Fasolka, D.C. Hurley, M. Kopycinska-Müller, D. Julthongpiput, M.J. Fasolka, Appl. Surf. Sci. **253**, 1274–1281 (2006)
30. U. Rabe, S. Amelio, M. Kopycinska, S. Hirsekorn, M. Kempf, M. Göken, W. Arnold, Imaging and measurement of local mechanical material properties by atomic force acoustic microscopy. Surf. Interface Anal. **33**, 65–67 (2002)
31. R. Arinéro, G. Lévêque, P. Girard, J.Y. Ferrandis, Image processing for resonance frequency mapping in atomic force modulation microscopy. Rev. Sci. Instrum. **78**, 023703-1–023703-6 (2007)
32. B.J. Rodriguez, C. Callahan, S.V. Kalinin, R. Proksch, Dual-frequency resonance-tracking atomic force microscopy. Nanotechnology **18**, 475504 (2007)
33. F. Dinelli, H. Assender, N. Takeda, G. Briggs, O. Kolosov, Elastic mapping of heterogeneous nanostructures with ultrasonic force microscopy (UFM). Surf. Interface Anal. **27**, 562–567 (1999)
34. A. Rosa-Zeiser, E. Weilandt, S. Hild, O. Marti, The simultaneous measurement of elastic, electrostatic and adhesive properties by scanning force microscopy: pulsed-force mode operation. Meas. Sci. Technol. **8**, 1333–1338 (1997)
35. T.J. Young, M.A. Monclus, T.L. Burnett, W.R. Broughton, S.L. Ogin, P.A. Smith, The use of the PeakForceTM quantitative nanomechanical mapping AFM-based method for high-resolution Young's modulus measurement of polymers. Meas. Sci. Technol. **22**, 125703 (2011)
36. G. Stan, R.S. Gates, Intermittent contact resonance atomic force microscopy. Nanotechnology **25**(24), 245702 (2014)
37. M.J. Holzwarth, A.M. Gigler, O. Marti, Digital pulsed force mode—determining local mechanical properties of HeLa cells. Imaging Microsc. **8**, 37–38 (2006)
38. S.N. Magonov, *Expanding Atomic Force Microscopy with HybriD Mode Imaging*. NT-MDT Application Note 087 (2014)
39. C. Braunsmann, J. Seifert, J. Rheinlaender, T.E. Schäffer, High-speed force mapping on living cells with a small cantilever atomic force microscope. Rev. Sci. Instrum. **85**, 073703 (2014)
40. B. Choi, *Park PinPoint™ Mode for Cell Biology* (Private Communication, Boston, MA, 2015)
41. I. Sokolov, *Ringing Mode™* (Private Communication, Boston, MA, 2015)
42. O. Sahin, S.N. Magonov, C. Su, C.F. Quate, O. Solgaard, An atomic force microscopy tip designed to measure time-varying nanomechanical forces. Nat. Nanotechnol. **2**, 507–514 (2007)
43. S. Jesse, S.V. Kalinin, R. Proksch, A.P. Baddorf, B.J. Rodriguez, The band excitation method in scanning probe microscopy for rapid mapping of energy dissipation on the nanoscale. Nanotechnology **18**, 435503 (2007)

44. D. Platz, E.A. Tholen, D. Pessen, D.B. Haviland, Intermodulation atomic force microscopy. Appl. Phys. Lett. **92**, 153106 (2008)
45. P. Vitry, E. Bourillot, C. Plassard, Y. Lacroute, L. Tetard, E. Lesniewska, Advances in quantitative nanoscale subsurface imaging by mode-synthesizing atomic force microscopy. Appl. Phys. Lett. **105**, 053110 (2014)
46. A. Belianinov, S.V. Kalinin, S. Jesse, Complete information acquisition in dynamic force microscopy. Nat. Commun. **6**, 6550 (2015)
47. C. Chen, J. Wang, Z. Chen, Surface restructuring behavior of various types of poly (dimethylsiloxane) in water detected by SFG. Langmuir **20**(23), 10186–10193 (2004)
48. A. Beigbeder, R. Mincheva, M. Claes, P. Brocorens, R. Lazzaroni, P. Dubois, On the effect of carbon nanotubes on the wettability and surface morphology of hydrosilylation-curing silicone coatings. J. Nanostruct. Polym. Nanocompos. **5**, 37–43 (2009)
49. R. Kalinova, R. Mincheva, P. Dubois, Imparting adhesion property to silicone materials: challenges and solutions. Rev. Adhes. Adhes. **2**, 30–55 (2014)
50. I. Marabotti, A. Morelli, L.M. Orsini, E. Martinelli, G. Galli, E. Chiellini, E.M. Lien, M.E. Pettitt, M.E. Callow, J.A. Callow, S.L. Conlan, R.J. Mutton, A.S. Clare, A. Kocijan, C. Donik, M. Jenko, Fluorinated/siloxane copolymer blends for fouling release: chemical characterisation and biological evaluation with algae and barnacles. Biofouling **25**, 481–493 (2009)
51. E. Martinelli, M.K. Sarvothaman, G. Galli, M.E. Pettitt, M.E. Callow, J.A. Callow, S.L. Conlan, A.S. Clare, A.B. Sugiharto, C. Davies, D. Williams, Poly(dimethyl siloxane) (PDMS) network blends of amphiphilic acrylic copolymers with poly(ethylene glycol)-fluoroalkyl side chains for fouling-release coatings. II. Laboratory assays and field immersion trials. Biofouling **28**, 571–582 (2012)
52. H. Lee, B.P. Lee, P.B. Messersmith, A reversible wet/dry adhesive inspired by mussels and geckos. Nature **448**, 338–341 (2007)
53. J.H. Waite, Nature's underwater adhesive specialist. Int. J. Adhes. Adhes. **7**, 9–14 (1987)
54. J.H. Waite, Adhesion a la moule. Integr. Comp. Biol. **42**, 1172–1180 (2002)
55. L. Leibler, Theory of microphase separation in block copolymers. Macromolecules **13**, 1602–1617 (1980)
56. J.H. Hildebrand, R.L. Scott, *The Solubility of Non-electrolytes*, 3rd edn. (Reinhold, New York, 1975)
57. Z. Tuzar, P. Kratochvíl, Block and graft copolymer micelles in solution. Adv. Colloid Interface Sci. **6**, 201–232 (1976)
58. J.H. Hildebrand, R.L. Scott, *ReguZar Solutions* (Prentice-Hall, Englewood Cliffs, 1962)
59. J.H. Hildebrand, J.M. Prausnitz, R.L. Scott, *Regular and Related Solutions* (van Nostrand-Reinhold, Princeton, NJ, 1970)
60. C. Price, in *Developments in Block Copolymers*, ed. by I. Goodman (Applied Science Publishers, Barking, 1982), pp. 39–79
61. J.G. Selb, Y. Gallot, in *Developments in Block Copolymers*, vol. 2, ed. by I. Goodman (Applied Science Publishers, London, 1985), p. 327
62. Z. Tuzar, P. Kratochvil, in *Surface and Colloid Science*, vol. 15, ed. by E. Matijevic (Plenum Press, New York, 1993), pp. 1–83
63. O. Dos Santos Ferreira, E. Gelinck, D. de Graaf, H. Fischer, Adhesion experiments using an AFM—parameters of influence. Appl. Surf. Sci. **257**, 48–55 (2010)
64. R. Kalinova, C. Ngo, R. Mincheva, R. Lazzaroni, P. Leclère, P. Dubois, From cylindrical to spherical nanosized micelles by self-assembly of poly(dimethylsiloxane)-b-poly(acrylicacid) diblock copolymers. Polym. Bull. (2015). doi:10.1007/s00289-016-1598-2
65. M.-E. Vlachopoulou, A. Tserepi, K. Beltsios, G. Boulousis, E. Gogolides, Nanostructuring of PDMS surfaces: dependence on casting solvents. Microelectron. Eng. **84**, 1476–1479 (2007)
66. L.B. Thiele, R. Frommelt, H. Faserforschung, Z. Textiltechnik, Polymerforschung **28**, 405 (1977)
67. J.N. Israelachvili, P.M. McGuiggan, Forces between surfaces in liquids. Science **241**, 795–800 (1988)

68. E.E. Meyer, K.J. Rosenberg, J. Israelachvili, Recent progress in understanding hydrophobic interactions. Proc. Natl. Acad. Sci. USA **103**, 15739–15746 (2006)
69. D. Alsteens, E. Dague, P.G. Rouxhet, A.R. Baulard, Y.F. Dufrêne, Direct measurement of hydrophobic forces on cell surfaces using AFM. Langmuir **23**, 11977–11979 (2007)
70. A. Beigbeder, M. Jeusette, R. Mincheva, M. Claes, P. Brocorens, R. Lazzaroni, P. Dubois, On the effect of carbon nanotubes on the wettability and surface morphology of hydrosilylation-curing silicone coatings. JNPN **5**, 37–43 (2009)
71. E. Duquesne, J. Habimana, P. Degée, P. Dubois, Synthesis of silicone-methacrylate copolymers by ATRP using a nickel-based supported catalyst. Macromol. Chem. Phys. **207**, 1116–1125 (2006)
72. T.C. Ngo, R. Kalinova, D. Cossement, E. Hennebert, R. Mincheva, R. Snyders, P. Flammang, P. Dubois, R. Lazzaroni, Ph Leclère, Modification of the adhesive properties of silicone-based coatings by block copolymers. Langmuir **30**, 358–368 (2013)
73. E.W. van der Vegte, G. Hadziioannou, Acid–base properties and the chemical imaging of surface-bound functional groups studied with scanning force microscopy. J. Phys. Chem. B **101**, 9563–9569 (1997)
74. J.N. Israelachvilli, *Intermolecular and Surface Forces* (Academic Press, New York, 1992)
75. T. Hugel, M. Seitz, The study of molecular interactions by AFM force spectroscopy. Macromol. Rapid Commun. **22**, 989–1016 (2001)
76. B. Cappella, G. Dietler, Force-distance curves by atomic force microscopy. Surf. Sci. Rep. **34**, 1–104 (1999)
77. R. Barattin, N. Voyer, Chemical modifications of AFM tips for the study of molecular recognition events. Chem. Commun. **13**, 1513–1532 (2008)
78. C. Jérôme, N. Willet, R. Jérôme, A.S. Duwez, Electrografting of polymers onto AFM tips: a novel approach for chemical force microscopy and force spectroscopy. Chem. Phys. Chem. **5**, 147–149 (2004)
79. C. Friedsam, M. Seitz, H.E. Gaub, Investigation of polyelectrolyte desorption by single molecule force spectroscopy. J. Phys. Condens. Matter **16**, S2369–S2382 (2004)
80. J.H. Waite, Adhesion in byssally attached bivalves. Biol. Rev. **58**, 209–231 (1983)
81. R. Merkel, P. Nassoy, A. Leung, K. Ritchie, E. Evans, Energy landscapes of receptor–ligand bonds explored with dynamic force spectroscopy. Nature **397**, 50–53 (1999)
82. M. Rief, M. Gautel, F. Oesterhelt, J.M. Fernandez, H.E. Gaub, Reversible unfolding of individual titin immunoglobulin domains by AFM. Science **276**, 1109–1112 (1997)
83. D. Alsteens, C.B. Ramsook, P.N. Lipke, Y.F. Dufrêne, Unzipping a functional microbial amyloid. ACS Nano **6**, 7703–7711 (2012)
84. M.A. Nash, H.E. Gaub, Single-molecule adhesion of a stimuli-responsive oligo(ethylene glycol) copolymer to gold. ACS Nano **6**, 10735–10742 (2012)
85. D.J. Crisp, G. Walker, G.A. Young, A.B. Yule, Adhesion and substrate choice in mussels and barnacles. J. Colloid Interface Sci. **104**, 40–50 (1985)
86. N. Aldred, L.K. Ista, M.E. Callow, J.A. Callow, G.P. Lopez, A.S. Clare, Mussel (*Mytilus edulis*) byssus deposition in response to variations in surface wettability. J. R. Soc. Interface **3**, 37–43 (2006)
87. R.J. Stewart, T.C. Ransom, V. Hlady, Natural underwater adhesives. J. Polym. Sci. Part B Polym. Phys. **49**, 757–771 (2011)
88. J. Yu, Y. Kan, M. Rapp, E. Danner, W. Wei, S. Das, D.R. Miller, Y. Chen, J.H. Waite, J.N. Israelachvili, Adaptive hydrophobic and hydrophilic interactions of mussel foot proteins with organic thin films. Proc. Natl. Acad. Sci. USA **110**, 15680–15685 (2013)
89. A.F.M. Barton, Solubility parameters. Chem. Rev. **75**(6), 731–753 (1975)
90. S. O'Driscoll, G. Demirel, R.A. Farrell, T.G. Fitzgerald, C. O'Mahony, J.D. Holmes, M.A. Morris, The morphology and structure of PS-b-P4VP block copolymer films by solvent annealing: effect of the solvent parameter. Polym. Adv. Technol. **22**, 915–923 (2011)
91. J. Brandrup, E.H. Immergut, E.A. Grulke (eds.), *Polymer Handbook*, 4th edn. (Wiley, New York, 1999)

Index

A
Adhesion, 22, 81, 108, 113, 124, 133, 136, 141, 143–148, 158, 160, 165, 168, 170, 173, 175, 179, 247, 249, 261, 269, 277, 281, 284, 291–295, 299, 300, 304–307, 310–313, 316–318, 323, 326–333, 335–337, 339
Adhesion experiments, 284
Adhesion force mapping, 304, 330, 339
Adhesion frequency, 304, 333, 339
Adhesion stress, 281
Adhesive, 113
Adhesive layer, 203
Adhesive pad, 28
Adhesive plaque, 64, 73
Adhesive tape, 98
Affinity, 282
Aggregate, 313
Air bubbles, 263
Ambient humidity, 28
Amontons-kotelnikov, 182
Amplitude modulation, 308
Anchorages, 64, 66
Anchors, 64, 74, 76
Animal locomotion, 96
Anisotropic, 66
Anisotropic adhesive properties, 208, 214
Anisotropy, 102
Anisotropy of adhesion, 207
Antagonistic, 102
Anti-attachment, 34
Approach-retract, 307, 310
Aquaplaning, 30
Archard model, 195
Aspect ratio, 107, 111, 129, 216
Atomic Force Microscopy (AFM), 203, 307–309, 311–313, 322, 324, 326, 329, 338, 339
Attachment, 33–36, 38, 42, 44
Attachment ability, 22, 25, 26, 28
Attachment devices, 141, 159, 168, 170
Attachment disc, 64, 67, 70, 73
Attachment pad, 29
Autumn, 5

B
Beetle, 23, 211
Bending behavior, 101
Bibliometric, 2
Bioadhesive, 303, 337, 338
Bioinspired, 3, 83, 207, 223, 240
Biomimetic, 2, 3, 18
Block copolymer, 310, 311, 312, 316, 319, 324, 333, 338
Bottom-up, 119, 122
Bristle Model, 192
Burridge-knopoff, 199
Burstness, 3, 5, 7, 18
Byssus, 73, 74

C
Cantilever, 307–310, 312, 323, 333
Cantor-Borodich, 196
Capillarity, 28, 29
Capillary, 29
Capillary force, 28, 29
Carbon black, 233, 234
Carbon nanotube, 119, 122, 210, 222
Ceiling, 102
Centrality, 5, 7, 18
Chemical, 33, 36, 38
Chemical bonds, 183
Chemical force microscopy, 307, 309, 311, 330, 339
Chemical/physical, 35
Chemistry, 42, 45
CiteSpace, 2
Coefficient of friction, 210

L. Heepe et al. (eds.), *Bio-inspired Structured Adhesives*,
Biologically-Inspired Systems 9, DOI 10.1007/978-3-319-59114-8

Cohesive zone, 278
Cold-pressing, 128
Complementary, 291, 292, 294, 298–300
Complementary surfaces, 292, 299
Composite, 109, 233
Contact angle, 317–320, 328, 329, 331, 338
Contact mechanics, 141, 168, 307, 308
Contact resonance AFM, 308, 309
Contact splitting, 82, 95
Contact tapering, 81
Contamination, 238, 240
Contra-lateral, 96
Coulomb, 182
Crack, 278
Crack propagation, 247, 249
Crack-trap, 291, 292, 299
Crack velocity, 101
Creep, 232, 240
Critical angle, 100
Critical pre-strain, 101
Crosslinking, 316, 317, 324
Cylindrical micelles, 314, 338

D
Derjaguin-Muller-Toporov, 307
Diblock copolymer, 134
Dipping, 281
Discoidal, 23, 30
Double peeling, 101, 105
Dragline, 64–67, 69, 70, 72, 76
Durability, 287

E
Edge angle, 277, 280, 283, 288
Effective elastic modulus, 108, 109
Elastic deformation, 101
Elastic modulus, 108, 118, 124, 130, 136
Elastomers, 230, 236
Electrohydrodynamic, 226, 229
Electrospinning, 67, 74
Electrospun, 71
Energy dissipation, 29, 101

F
Fabrication, 19
FEM, 88
Fibrillar, 3, 5, 7, 14, 141, 152, 161, 167, 168, 175, 208, 216
Fibrillar adhesives, 289
Fibrillar structures, 82
Flat-punch, 250, 267, 268, 274
Flaw tolerance, 67
Flexible tissue, 141, 147–150
Force-distance, 284, 288
Force-distance curve, 307
Force transducer, 24
Force-Volume imaging, 310
Fractals, 196
Fractal surface, 141, 142, 144, 158, 160, 171
Fracture toughness, 67
Frequency, 308–310, 312, 328, 334, 335, 339
Friction, 22
Frictional sliding, 101

G
Gait pattern, 103
Gecko, 28, 89, 114, 277
Gecko adhesive, 221
Gecko pad, 213, 208
Gelatin, 293
Grading, 81
Greenwood and Williamson, 187, 189
Griffith energy balance, 83
Gumfoot, 69

H
Hairy, 112, 113, 115, 136
Hierarchical, 74, 181, 195, 207, 213, 217
Hierarchical branching, 88
Hierarchical gecko hair models, 207
Hierarchical spring model, 215
Hierarchical structure, 81, 118, 136
Hierarchy, 180
Hildebrand solubility parameter, 311
Hot-pressing, 128
Humidity, 22, 23, 25, 26, 29, 117, 135
Hydrophilic, 304, 316, 317, 319, 321, 323, 327–330, 332, 338, 339
Hydrophobic, 36, 39, 235, 304, 317–319, 323, 329–332, 338, 339
Hysteresis, 101

I
Inhibition, 228, 231
Insect, 22, 95, 102
Intrinsic adhesion stress, 279, 286
Ionic strength, 304, 310, 323, 332, 333, 339
Iterative, 208

J
Janus pillars, 223
Johnson-Kendall-Robert, 287, 307

K
Kendall, 83, 91, 95
Kragelsky Rod-Assembly Model, 190

L
Level, 180

M
Magnetic, 233, 234
Material gradient, 141, 152, 153, 155, 159, 176, 215
Material properties, 28, 29
Maugis, 206
Maximum, 100
Mechanical properties, 303, 304, 307, 309, 338
Micelle, 314, 316, 338
Microemulsion, 22
Microfibers, 283
Modelling, 83
Mold, 225–227
Morphology, 304, 310–312, 314, 315, 319, 320, 323, 324, 326, 330, 338, 339
MSAMS, 270, 274
Multi-asperity, 202
Multi-level, 208
Multiple peeling, 67, 84, 96, 103
Multi-scale, 201, 207, 217
Multi-step adhesion function, 206
Mushroom, 69, 72, 251, 255, 257
Mushroom-like, 125–127, 129, 130, 136, 211, 216
Mushroom-like fiber, 278, 280, 282
Mushroom-shaped, 98, 101, 229, 235
Mussel, 336–338
Mussel adhesion, 338

N
Nano-composite, 75
Nanofibers, 76
Nano-hairs, 213
Nanopillar array, 134
Nanopillars, 117, 133
Nano-topography, 194
Nucleation process, 266
Numerical simulations, 147, 173

O
Optimal elasticity, 141, 144, 147
Optimum, 67, 104
Optimum angle, 100
Organogel, 292, 300

P
PAA, 304
PDMAEMA, 304
PDMS, 225, 228, 291–295, 298–300
Peak force tapping, 303, 309, 311, 312, 323, 329, 338
Peeling, 66, 67, 69
Peeling angle, 85, 98, 101, 104
Peeling force, 87, 99
Peeling line, 83, 91
Peeling model, 95
Peeling tester, 99
Photolithography, 224
Photoresist, 224, 228
Physical mechanisms, 182
Physical properties, 38
Pillar array, 119, 122, 124, 126, 128, 129
Piriform, 64, 67
PMMA, 23
Poly(acrylic acid), 304, 306, 316, 338
Poly((2-dimethylamino) ethyl methacrylate), 304, 316, 338
Polydimethylsiloxane, 304–306, 338
Polyelectrolyte, 304, 332, 333, 339
Polyimide, 222, 233
Polymeric nanorods, 214
Polymethylmethacrylate, 23
Polyurethane, 232, 281, 283
Porous, 110, 134, 136
Prandtl-tomlinson, 185
Pre-strain, 99, 104
Pre-tension, 97
Probe tip, 303, 304, 312, 330, 338, 339
Profilometre, 203
Properties, 33, 35, 36
Pull-off experiments, 287
Pull-off force, 284, 285
Pull-off stress, 277, 280, 285, 288

R
Repeatability, 288
Resilin, 29
Roll-to-roll, 235
Rough, 37–39, 42, 44
Roughness, 33, 44, 45, 111, 131, 180
Rubber-like, 29

S
Safety factor, 103
Scale, 181
Scanning Electron Microscopy, 95
Schmittbuhl hierarchical model, 199
Secretion, 29, 104
Selective solvent, 311, 314, 315
Self-assembly, 304, 310, 311, 312, 314, 338
Self-cleaning, 19, 212
Self-stabilising, 67
Self-stabilization, 102, 105
Sergienko-Bukharov Model, 191
Setae, 22, 23

Sexual dimorphism, 30
Shape memory, 225
Shear, 217
Shear-pressing, 128
Silicon, 224, 227
Single fiber, 284
Single peeling, 105
Smart glue, 76
Smooth, 36–39, 42
Smooth pad, 115
Solid defect, 263
Solvent, 230, 232
Spatula, 23, 95, 223, 224
Spatular, 112, 114, 131, 133
Spiders, 28
Spider's Leg, 215
Spider silk, 64
Spider web anchorages, 87
Spinnerets, 64, 66
Spinning, 71
Spring constant, 307
Stability, 113, 118
Stance phase, 103
Statistical interpretation, 202
Steady-state, 100
Stick-slip, 197
Stiffness, 304, 309, 310, 312, 313, 327, 328, 338
Stress concentrations, 88, 90
Stress intensity factors, 281
Structural threads, 67
Structured material, 110
Structure/geometry, 18
SU-8, 229
Superhydrophilic, 39
Superhydrophilicity, 37
Superhydrophobic, 39
Superhydrophobicity, 37
Superomnipobicity, 37
Surface chemistry, 30
Surface energy, 230, 238
Surface structures, 35
Surface tension, 318–320
Swing phase, 103
Synthetic, 1, 7, 9, 17

T
Tapered geometries, 91
Tarsal setae of beetles, 215
Teflon, 65
Temperature, 22
Tensile strength, 66
Texture, 45
Textured, 39
Thin film peeling, 83
Tilt angle, 102
Tilted Configuration, 132
Tilt-Tolerancy, 269
Tip geometry, 125, 128
Tip shape, 278
Tip size, 277
Tip-to-base ratio, 285
Top-down, 119, 122
Topographica, 36
Topographical properties, 33
Traction force, 23, 25, 28, 30
Tune, 87
Two-photon lithography, 224

U
USPTO, 17

V
Van der Waals, 118, 143, 144, 147, 148, 154, 164, 171–173, 183, 203
Van der Waals forces, 28, 111, 112, 130, 142
Viscoelastic, 29, 73
Viscoelastic effects, 287
Viscosity, 30
V-shape, 96

W
Wet adhesion, 22
Work of adhesion, 101, 287
WOS, 17

X
X-ray Photoelectron Spectroscopy (XPS), 311, 321

Y
Young modulus, 29, 308, 312

Z
Zhuravlev, 187

MIX
Papier aus verantwortungsvollen Quellen
Paper from responsible sources
FSC® C105338

If you have any concerns about our products,
you can contact us on
ProductSafety@springernature.com

In case Publisher is established outside the EU,
the EU authorized representative is:
Springer Nature Customer Service Center GmbH
Europaplatz 3, 69115 Heidelberg, Germany

Printed by Libri Plureos GmbH
in Hamburg, Germany